Birkhäuser

Claude Brezinski • Dominique Tournès

André-Louis Cholesky

Mathematician, Topographer and Army Officer

Birkhäuser

Claude Brezinski
Laboratoire Paul Painlevé
Université des Sciences
et Technologies de Lille
Villeneuve d'Ascq
France

Dominique Tournès
Laboratoire d'Informatique
et de Mathématiques
Université de La Réunion
Sainte Clotilde
France

ISBN 978-3-319-08134-2 ISBN 978-3-319-08135-9 (eBook)
DOI 10.1007/978-3-319-08135-9
Springer Cham Heidelberg New York Dordrecht London

Library of Congress Control Number: 2014945358

Photo on the cover: André-Louis Cholesky © Cholesky's family

Printed on acid-free paper

Springer Basel is part of Springer Science+Business Media (www.birkhauser-science.com)

... [l]a misera vita non trapassi sanza ... lasciare
di noi alcuna memoria nelle menti de mortali ...

*... the miserable life should not pass without ... leaving
some memory of ourselves in the minds of mortals ...*

Leonardo da Vinci
Codex Urbinus, Vatican

Contents

Preface

The name of Cholesky is well known in two communities of scientists: applied mathematicians and topographers. Any applied mathematician knows Cholesky's method for solving certain systems of linear equations (see Chapter 4). Ask an applied mathematician who Cholesky was. Maybe, she/he will know some details about his life since, for some years now, information has been provided on various forums and biographical papers have been published (see the bibliography at the end of the book). Cholesky's method is also used by topographers who have to solve such linear systems when applying the least squares method for minimizing the sum of the squares of the errors. But topographers also know Cholesky for his method of double–run leveling (see Section 3.3). If you see one of them in the street, ask him if he knows this name. At least in France, the answer is always yes, but his biography is usually unknown. This method is still taught nowadays, and it can be found in modern textbooks on topography; see, for example, [117]. Putting his name (also with the spelling Choleski) into a search engine on the internet returns several hundreds of thousands of pages.

The genesis of this book was quite a long one. Some ten years ago, not many details were known on the life of Cholesky although his biography, in fact an eulogy, was published in French in 1922 [41], and was translated into English in 1975 (see Section 8.3). But this biography did not circulate much and, from time to time, questions were asked on internet forums. It was mainly known that he was a French officer, that he was born in 1875, and was killed at the end of World War I. The only detailed information we had came from *École Polytechnique* in Paris where he had been a student. But they were restricted to the marks he obtained at the exams there. In France, a law stipulates that personal archives are open to the public 120 years after the birth of that person. Thus, 120 years and one day after the anniversary of Cholesky's birth, I went to the archives of the Army at the Fort de Vincennes near Paris where I was able to access all the documents about him [2]. Then, I wrote a first biography on him [52]. After the publication of this paper, Yves Dumont, at that time a professor at the *Université de la Réunion*, built a web site on Cholesky.

Then, in 2003, Claudine Billoux, the archivist of *École Polytechnique*, received a phone call from a grandson of Cholesky. He was calling to say that the family was willing to donate the archives in their possession to the *École Polytechnique*. However, he asked that the materials be picked up as quickly as possible, otherwise he would be obliged to throw them away. The problem for Mrs. Billoux was the volume of these documents: one cubic meter. But they were in Bordeaux, in the southwest of France, and the budget cuts made it difficult to rapidly obtain a car and a driver. Fortunately, another grandson of Cholesky, Michel Gross, was able to bring the papers to his house near Paris. He also offered to help in classifying these archives. However, although an engineer, he was not a mathematician and he contacted Yves Dumont who directed him to me. Thus, I was quite surprised, in December 2003, to receive a letter from this grandson informing me that the family had given all the documents to the *École Polytechnique* and asking if I could help him and Mrs. Billoux to classify them. Of course, I accepted and we almost immediately found the unpublished manuscript where Cholesky describes his famous method for solving systems of linear equations [46].

From what can be known, the documents constituting now the *Fonds André Cholesky* which is at *École Polytechnique* were given to his widow after the death of Cholesky on the front in 1918. After the death of the latter, in 1944, these documents came into the hands of one of his daughters, Hélène, who lived in Morocco and had her mother with her for several years. In 1945, Hélène and her family moved to Bordeaux where the documents remained until 2003. According to the recollections of the grandchildren of Cholesky, none of these documents were exploited by the family during the time they were in her possession. At most, they aroused the curiosity of the young children of the family who found it fun to fill in some military blank forms.

With Michel Gross, we classified all these documents and published papers relating our discoveries in the *Bulletin de la Société des Amis de la Bibliothèque de l'École Polytechnique* (see the bibliography at the end of the volume, in particular [59, 86]). In October 2005, I was invited by Dominique Tournès to give talks, one of them on Cholesky, at the *Université de la Réunion* where he was (and still is) a professor. Then, in 2010, I was contacted by Frédéric De Ligt, a professor of mathematics and the president of the *Association des Professeurs de Mathématiques de l'Enseignement Public* of the Région Poitou–Charentes who invited me to give a talk on Cholesky at the Lycée Bellevue in Saintes. Frédéric was living in Montguyon, the native town of Cholesky. He brought me a book on the history of the city of Montguyon written by Raymond Nuvet, the vice–mayor of this city where some details about the family of Cholesky were given. I contacted him and this is how the search for a more complete genealogy of the family started.

After that, I returned many times to the archives of *École Polytechnique*, alone or with Dominique Tournès, for digging into the documents left by Cholesky. Since Dominique is a well–known historian of mathematics and a specialist of graphical computations, I asked him if he was willing to collaborate with me on the project of this book, in particular on an unpublished manuscript of Cholesky on graphical calculation.

Cholesky's method for solving certain types of systems of linear equations is known worldwide. It is taught to all students in applied mathematics and numerical analysis, and can be found in any book on these topics. In Section 4.7, we will see how it was disseminated. Let us only say now that Cholesky never published his method himself but that it was solely known from a paper, written by a Commandant Benoît, six years after Cholesky's death [42]. In the documents his family gave to *École Polytechnique*, as said above, Michel Gross and I almost immediately found the autographed manuscript where Cholesky described his method. This was the beginning of a great adventure.

On 1 September 2012, the name of André–Louis Cholesky was bestowed on the new cultural center of the city of Montguyon, were he was born, following a proposition of Raymond Nuvet (see the picture on the last page).

Cholesky's biography is the topic of Chapter 1. The information on his family, gathered in great part by Raymond Nuvet, is given in Chapter 2. Chapter 3 explains some generalities on topography and describes the work of a topographer. Chapter 4 is dedicated to Cholesky's method for systems of linear equations. Other works by Cholesky are reported in Chapter 5. Chapter 6 is devoted to Léon Eyrolles and the high school that he created where Cholesky was a professor. As I said above, we also found in the archives an unpublished book by Cholesky on graphical computations. In Chapter 7, Dominique Tournès comments this course, analyzes it and places it in its historical context. He also edited the book that is reproduced for the first time in Appendix C. Chapter 8 is devoted to Benoît. Various documents are in Chapter 9. The paper by Cholesky, reproduced in Appendix B, and his book transcribed in Appendix C by Dominique Tournès, belong to the *Fonds Cholesky - École Polytechnique* which holds their copyright. We were given the authorization to use all the documents it contains. A chronology of the life of Cholesky, a bibliography, and an index (where the family name *Cholesky* is not included) end the volume.

Sometimes, my translation from French may be not as accurate as I would like. However, I did my best to keep the general meaning of the text and the style of the authors. Quotations are in italics. I would like to apologize in advance for all errors and omissions the book could contain. They are all mine.

Claude Brezinski

Claude Brezinski is emeritus professor of numerical analysis at the University of Sciences and Technologies of Lille (Laboratory Paul Painlevé, UMR CNRS 8524), France. His work mainly concerns extrapolation methods, orthogonal polynomials, Padé and rational approximation, and numerical linear algebra. He also published several papers and books on the history of sciences.

Dominique Tournès is professor of mathematics and history of mathematics at the University of La Réunion (Laboratory of Mathematics and Computer Science, EA 2525), France, and associate researcher at the SPHERE laboratory (UMR 7219, CNRS and University Paris–Diderot, France). His work focuses on the history of calculation methods and instruments since the seventeenth century, particularly regarding differential equations.

Chapter 1

Biography of Cholesky

André–Louis Cholesky was born on Friday 15 October 1875, at 1:00 pm, in Montguyon, a village in the district of Jonzac (department of Charente–Inférieure, named Charente–Maritime since 1942), about 60 km northeast of Bordeaux and 65 km southwest of Angoulême, which counted around 1,500 inhabitants at that time.

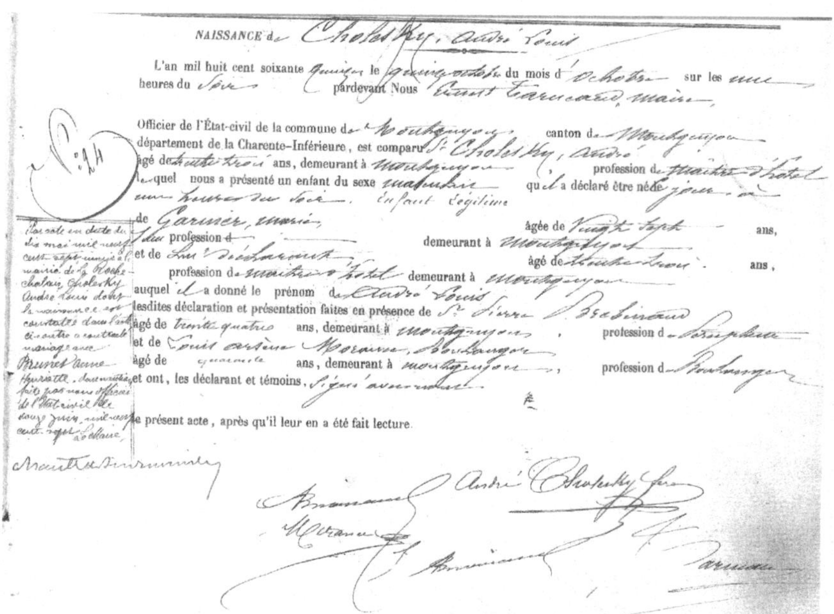
NAISSANCE de Cholesky, André Louis

L'an mil huit cent soixante … le … du mois d … sur les … heures du … pardevant Nous …

Officier de l'État-civil de la commune d… canton d…

département de la Charente-Inférieure, est comparu …

âgé de … ans, demeurant à … profession d…

lequel nous a présenté un enfant du sexe … qu'il a déclaré être né …

de … âgée de … ans,

profession d… demeurant à …

et de … âgé de … ans,

profession d… demeurant à …

auquel il a donné le prénom d…

lesdites déclaration et présentation faites en présence de …

âgé de … ans, demeurant à … profession d…

et de …

âgé de … ans, demeurant à … profession d…

et ont, les déclarant et témoins, …

le présent acte, après qu'il leur en a été fait lecture.

Birth certificate of André–Louis Cholesky

The occupation of the site is very old as evidenced by a megalithic construction, *La Pierre Folle*, around 4500 years old. Its heaviest stone weighs 30 tons, and is backed by a dolmen erected 4000 years ago. The latin name of the city, *Monte Guidonis*, is attested in 1121 and 1281. The village has a castle which goes back to 1082 but the current one is from the 13th century and was recast in the 16th century. It belonged successively to the families De Montguyon, De la Rochefoucauld and De Rohan (see http://vieuxblaye.free.fr/spip.php?article59). In 1451, the Army of King Charles VII, commanded by Jean d'Orléans, comte de Dunois, drove the English occupiers out. The Kings Henri IV and Louis XIII stayed in the castle. The fortress was set afire by lightning in 1793. The church Saint–Vincent, located in Vassiac, a suburb of Montguyon, dates from the 12th and 15th centuries. The local industries dealt with white clay, and the manufacturing of flooring and paneling. There were also a cheese–dairy, a tannery, and a piggery in the *Domaine de Fontbouillant.* A distillery of Cognac was founded in 1893 by the Richard brothers. See [185, 186] for an extensive history of this village.

André–Louis was the son of André Cholesky, a hotel keeper born in Montguyon on 9 October 1842, and of Marie Garnier, 27 years old [1]. In his officer's booklet of the Army, he is mentioned under the nickname of René, which is also the first name written on his grave.

1.1 Infancy and Studies

Almost nothing is known of the first years of André–Louis Cholesky which he spent in his native village and where he attended the primary school. Then he went to the high school (*lycée* in French) in Saint–Jean–d'Angély, a nearby town. His name appears on the commemorative plaque of the former students and professors *Morts pour la France* inside the *Lycée Audouin–Dubreuil.*

He obtained the first part of his *Baccalauréat* (the French degree ending high school) in Bordeaux, on 14 November 1892 and its second part, with honors, on 24 July 1893, again in Bordeaux. On 15 October 1895, he was admitted to *École Polytechnique*, one of the most prestigious military superior schools, after a competitive examination in Poitiers. He was admitted 87th out of 223, and he signed a 3 years engagement in the Army at the town hall of the 5th district in Paris (district where the *École Polytechnique* was located at that time). He was 1 m 75 tall (that is 5.74 ft), had light chestnut hair and chestnut eyes.

After the French Revolution of 1789, France was lacking of engineers and officers since many of them deserted or emigrated. All the universities were closed according to a decree of the *Convention Nationale.* The river and road system needed important renovations, and new infrastructures had to be built. A com-

[1] More information on Cholesky's family is given in Chapter 2.

Elementary school of Montguyon where Cholesky went
© Claude Brezinski

mission for civil engineering was created on 21 Ventôse an II [2] (11 March 1794). Jacques–Élie Lamblardie (Loches, 2 November 1747 - Paris, 26 December 1797), Gaspard Monge (Beaune, 9 May 1746 - Paris, 28 July 1818) and Lazare Carnot (Nolay, 13 May 1753 - Magdebourg, now Germany, 2 August 1823) were given the task of organizing a new *École Centrale des Travaux Publics.* It was created on 7 Vendémiaire an III (28 September 1794). It was the future *École Polytechnique.* The school, installed in the *Palais Bourbon* (where now sits the National Assembly), was inaugurated on 21 December 1794 with 272 students. After a second session of exams, the school expanded to 400 students. They were taught mathematics, physics, and chemistry during three years in order to train them to enter into one of the special schools of the State, such as the *École d'Application de l'Artillerie et du Génie*, the *École des Mines*, or the *École des Ponts et Chaussées.* But the laboratories were not ready, some professors were not very good, and some lectures were attended by only 20 students. Claude–Antoine Prieur, called Prieur de la Côte d'Or (Auxonne, 22 September 1763 - Dijon, 11 August 1832), decided to reform the school, to name it *École Polytechnique* by a law of 15 Fructidor an III (1 September 1795). In 1805, Napoléon gave a military status to the school and installed it in the Latin Quarter. Considering that *"it is dangerous to give an advanced education to people who are not from rich families"*, he ended free education by imposing very high annual fees and changed the entrance tests to make

[2] Revolutionary calendar used from 1792 to 1806.

mandatory the transition by a high school, where studies were not free, and were in fact reserved to the children of the bourgeoisie. Many well–known scientists issued from it: the mathematicians Siméon Denis Poisson (Pithiviers, 21 June 1781 - Sceaux, 25 April 1840) and Louis Poinsot (Clermont–en–Beauvaisis, 3 January 1777 - Paris, 5 December 1859), the physicists Jean–Baptiste Biot (Paris, 21 April 1774 - Paris, 3 February 1862) and Augustin Jean Fresnel (Broglie, 10 May 1788 - Ville–d'Avray, 14 July 1827), the chemist Louis Joseph Gay–Lussac (Saint–Léonard–de–Noblat, 6 December 1778 - Paris 9 May 1850), and the astronomer François Arago (Estagel, 26 February 1786 - Paris, 2 October 1853), for example. On 13 avril 1816, King Louis XVIII closed the school as retaliation for indiscipline, the students having not accepted the *Restauration*. Reopened on 17 January 1817 under the name of *École Royale Polytechnique*, the students mainly remained opposed to the monarchy. During the Second Empire, it became the *École Impériale Polytechnique*, and the students returned gradually to their studies. After the defeat of 1870 by Prussia, the school actively participated in the national recovery effort. Most of the students turned to a military career but many others went to one of the civil activities needed by the country: development of railways, creation of new industries, modernisation of the cities, conquest and organization of the French colonial empire, and sciences culminating with Henri Poincaré (Nancy, 29 April 1854 - Paris, 17 July 1912) in 1873. Many prominent scientists were (and are still) professors at *École Polytechnique*.

When Cholesky was admitted, the head of *École Polytechnique* was the Général [3] Louis Joseph Nicolas André (Nuits, Côte d'Or, 29 March 1838 - Dijon, 18 March 1913), and the director of studies was Ernest Jules Pierre Mercadier (Montauban, 4 January 1838 - Paris, 27 July 1911). His teachers were Camille Marie Ennemond Jordan (Lyon, 5 January 1838 - Paris, 22 January 1922) and Marie Georges Humbert (Paris, 7 January 1859 - Paris, 22 January 1921) for mathematical analysis, Jacques Rose Ferdinand Émile Sarrau (Perpignan, 24 June 1837 - Paris, 10 May 1904) and Henri Charles Victor Jacob Léauté (Bazile, 26 April 1847 - Paris, 5 November 1916) for mechanics, Paul Émile Haag (Paris, 10 January 1843 - Paris, 26 April 1911) for geometry, Marie Alfred Cornu (Orléans, 6 March 1841 - La Chausonnerie, near Romorantin, 12 April 1902) and Henri Antoine Becquerel (Paris, 15 December 1852 - Le Croisic, 25 August 1908) for physics [4], Henri François Gal (Marseille, 15 June 1839 - 1897) for chemistry, Fernand de Dartein (Strasbourg, 9 February 1838 - Paris, 19 February 1912) for architecture, Auguste George Gabriel Duruy (Paris, 11 March 1853 - 23 March 1918), the son of the

[3] The military ranks are kept in their original language and not written in italics. The French military ranks in the ground forces are sous–lieutenant, lieutenant, capitaine, commandant (or chef d'escadron), lieutenant–colonel, colonel, général (de brigade, de division, de corps d'armée, d'armée). Maréchal is not a rank, but a dignity.

[4] He will be awarded half of the Nobel Prize for Physics in 1903 for his accidental discovery of spontaneous radioactivity in 1896, the other half going to Pierre and Marie Curie for their study of this phenomena.

Cholesky at *École Polytechnique*

historian and politician Victor Duruy (1811 - 1894), for history and literature, and the sculptor Jean–Baptiste Claude Eugène Guillaume (Montbard, Côte d'Or, 4 July 1822 - Rome, 1 March 1905) for drawing. Octave Pierre Jean Callandreau (Angoulême, 18 September 1852 - Paris, 13 February 1904) was teaching astronomy and geodesy, lectures which certainly had a great influence on the future inclination of Cholesky [64]. In 1896, Cholesky entered into the second year, 56th out of 222, and finished his studies at *École Polytechnique* in 1897, 38th out of 222.

His conduct and keeping were qualified as very good, and his military training as quite good.

After *École Polytechnique*, students had to enter into an application school of the Army. Cholesky was admitted, the 4th out of 92, as a sous–lieutenant at the *École d'Application de l'Artillerie et du Génie* in Fontainebleau, 60 km from Paris, on 1 October 1897.

This school was a military school where students had to enter after *École Polytechnique*, and whose aim was to train future officers according to their professional choice. It was created in 1794 by Lazare Carnot (Nolay, 13 May 1753 - Magdebourg, now Germany, 2 August 1823) by transferring to Metz the *École Royale d'Artillerie de Mézières* founded in 1748. The beginning of the school was difficult; the students lacked means of instruction since almost all the books and useful objects of interest had been assigned to *École Polytechnique* and to *École des Ponts et Chaussées*. The teachers were paid irregularly. It was not until 1797 that the second commander reported that all students finally had a uniform. The programs of the courses were also difficult to adjust, in particular those on fortifications whose program generated endless discussions with the *École Polytechnique*. In 1807, the school was merged with the *École d'Artillerie de Châlons*. During the Restauration, as was the case at *École Polytechnique*, the students were mostly against the regime. The year 1848 saw the victory of the moderate republicans, with among them the mathematician Jean Victor Poncelet (Metz, 1 July 1788 - Paris, 22 December 1867) who was a professor there. After years of difficulties, the school entered into a stable period. Classes continued normally until the war of 1870. After the defeat, Metz was no longer a French town, and the school was transferred to Fontainebleau one year later. At the beginning, the classrooms and the administration offices were installed in buildings of the *Héronnières* in the castle. The students were accommodated in the old rooms of the stablemen, converted for them, above the stables of the *Carrousel*. Later, the school expanded by encompassing other dependencies of the castle: the pavilion *Henri IV* for the rooms of the sous–lieutenants and the pavilion of the *Princes* for the aspirants housed in dormitories. A large shooting polygon was cut in the forest. The instruction, initially modeled on that given in Metz, evolved into a more practical education.

By the time of Cholesky, students had to spend one year at the school and then to serve for one year in a regiment. They had courses on artillery, fortification, construction, military art, mechanics, applied sciences, veterinary medicine for horses, military history, and geography of France and its Colonies. The lectures on artillery were completed by 55 days of application, with 5 drawings and 9 memoirs on ballistics, firing problems, the study of firing tables, surveys of buildings and factories, projects for machines, and geological sections of the grounds. There was also a course on topography which was given, before the arrival of Cholesky, by Lieutenant–colonel Charles–Moyse Goulier (Richelieu, 31 January 1818 - Paris, 14 August 1891), the inventor, among other devices, of various alidades and a clinometer (an instrument for measuring the angles of a slope), devices that will be later described by Cholesky in his books. The topographical exercises required of the students quite serious qualifications in drawing, but Goulier mentioned their extreme disabilities in that regard. Also, from 1897, students were given optional lessons in the skill of bicycling.

It is possible to find on the internet the textbooks on topography that Cholesky might have used in his courses. They are:

Cours de topographie by Auguste–Virgile–Jean Lehagre,
Instruments et procédés de lever by A. Lehagre,
Étude du terrain by Albert–M.–René Romieux,
Méthodes de levers by A. Romieux,
Instruction sur l'étude géologique et topographique d'un terrain by A. Romieux,
Instruction sur le dessin topographique by H. Bosson,
Géologie by R. Tétart,
Conventions du dessin topographique by L. Jardinet,
Instruments et procédés topographiques by L. Jardinet,
Instruction pratique sur la reconnaissance par les perspectives by L. Jardinet,
Instruction pratique sur le lever expédié by L. Jardinet,
1re et 2e études de figuré de terrain by L. Guillemard and L. Jardinet.

All the other courses he followed can also be found similarly. Among them:

Cours de mécanique appliquée aux machines by Sautereau Du Part,
Cours d'artillerie. 3e partie. Du matériel. 3e section. Projectiles et fusées by A. Lachaussée.

Let us mention that Capitaine Ferdinand Ferber (Lyon, 8 February 1862 - Beuvrequen, 21 September 1909), a pioneer of aviation in France, was appointed as a professor in this school in 1897.

On 8 September 1898, Cholesky was assigned to the 10th Artillery Regiment. He graduated from the school in 1899, 5th out of 86.

1.2 The topographer

On 1 October 1899, Cholesky is appointed lieutenant en second at the 22th Artillery Regiment. From 17 January 1902 to 27 June, he is sent to Tunisia for a mission, and again from 21 November to the 1st of May 1903. From 31 December 1903 to 6 June 1904, he is in Algeria. Pictures and the corresponding maps of this region at that time can be found in [12]. On 24 June 1905, he is assigned to the *Service Géographique de l'État Major de l'Armée* (Geographical Service of the Staff of the Army). He is noticed for a keen intelligence, a great facility for math-

Cholesky as a student at *École Polytechnique*
© Cholesky's family

ematics, original and even paradoxical ideas, always marked by a high elevation of feelings, and that he sustained with great warmth.

At the end of the 17th century, a special service for surveying and drawing maps was created. These officers took the title of *Ingénieurs des Camps et des Armées*. François Michel Le Tellier, Marquis de Louvois (Paris, 18 January 1641 - Versailles, 16 July 1691) instituted the *Dépôt de la Guerre* in 1688. At its origin, its aim was only to gather and classify the archives in his mansion. By a decree of 8 June 1871, the *Dépôt de la Guerre* was suppressed and merged into the General Staff of the Army. François Perrier (Valleraugue, 18 April 1833 - Montpellier, 20 February 1888), who spent some time surveying in North Africa in 1879 and will become a général, led a strong campaign towards the organization of a geographical service in the Army. By a decree dated 25 October 1881, the *Dépôt de la Guerre* was reconstituted under the form of a Section of Geography with four offices: one of geodesy, one of topography, engraving and workshops, one of archives and library, one of accounting and map sales. Perrier was appointed as head of the section of geodesy. On 10 January 1882, he became assistant director of the geographical section of the Staff of the Army. But the new arrangement did not meet Perrier's ideas who wanted to gather under him all the state cartographic organisms, and to exclude some others. He obtained partial satisfaction in 1885. On 24 May 1887, the *Dépôt de la Guerre* was suppressed and replaced by the *Service Géographique de l'Armée* (Geographical Service of the Army) with Perrier, général since January 1887, as its director. He died on 20 February 1888. After a short interim by Colonel Foucher, Colonel Victor Bernard Derrécagaix (Bayonne, 14 September 1833 - Anglet, 11 April 1915) succeeded Perrier. Promoted général, he was replaced by the Général Gaston Ovide de la Noë (Limoux, 17 May 1836 - Paris, 1 September 1902) which advocated the establishment of a map of France at the scale 1:50,000. It will be his successor, the Général Antonin Léon Bassot (Renève, 6 April 1841 - Paris, 17 January 1917), who will be responsible for this task. In 1903, the Général Henri Marie Auguste Berthaut (Épinal, 1 January 1848 - Paris, 18 December 1937) took the responsibility of the Service. The mapping of France, and the works in Algeria and Tunisia went on. In 1908, Berthaut created a topographical office in Casablanca. As we will see later, he was succeeded by Joseph Émile Robert Bourgeois (Sainte–Marie–aux–Mines, 21 February 1857 - Paris, 13 November 1945, X 1876, who will later become général and will be elected to the Academy of Sciences, Section of Geography and Navigation, on 18 June 1916, and Senator from 1920 to 1936) in 1912.

1.2.1 France

Around 1900, following the revision of the meridian of Paris, a new cadastral triangulation of France had been decided as well as the extent of the meridian of Lyon. These missions were assigned to the Section of Geodesy together with establishment of the map of Algeria, and a precise geometric leveling of this country. The problem of the adjustment (or compensation) of networks (corrections to be brought to the angles measured) concerned many officers of the Geographical Service, eager to find a simple, fast and accurate method. According to the obituary written by Commandant Benoît [41] (see Section 8.3), it was at this occasion that Cholesky imagined his method for solving the equations of condition by the method of least squares. This ingenious method of computation was immediately very helpful.

Let us quote the report [11]:

> *The part of the Meridian of Lyon between the parallels of Paris and the Mean Parallel was recognized in 1904: north and center, by Capitaine Durand; in the south, by Commandant de Fonlongue. The locations of all signals were fully defined, and the height to give at each of them exactly determined with the help of the scale of reconnaissance.*

In 1905, these two geodesists had made the observations of the east of the Mean Parallel and of the inclusion of that parallel into the Meridian of Lyon. Both having left, at the end of the same year, for joining the French Geodesic Mission to Ecuador, the observations of the Meridian of Lyon were assigned, in May 1906, to Capitaine Lamotte, assisted by Capitaine Benoît.

The program of 1906 consisted in completing the connection of the Meridian to the parallel of Paris, and to continue the observations by an extension from this parallel to the South.

In 1904, Capitaine Durand had completed the observations of the stations of Vaudémont, Hautmont, and Haudompré with the weight 24; but the signals of Croix–Marguerite and Champfleury had not been built.

In 1906, it was thus needed to build the last two signals, as Chailluz, La Serre, Mont–Poupet and Pierre, to observe in Moncel and Essey for completing the connection of the parallel of Paris to the Meridian of Lyon, to resume the observations in Hautmont and Haudompré; and to continue south through the stations of Croix–Marguerite, Champfleury, Chailluz and La Serre.

This program was run from point to point.

Cholesky took part in these campaigns of measurement in the valley of the Rhône river, in the region of Dauphiné, in the department of Isère and in the Cevennes massif, near Pierre–en–Besse (10 June - 7 July), at Mont–Poupet (8 - 25 July), in Montellier near Mount Aigoual (July and September 1905) and La Charpenne. These operations are described in [13]. It is important to notice the presence of Capitaine Benoît, who will write the obituary of Cholesky [41], and publish his method for solving systems of linear equations in 1924 [42].

On 26 September 1905, Cholesky was promoted to premier lieutenant. On Friday 10 May 1907 at the town hall of La Roche–Chalais (department of Dordogne, 18 km southeast of Montguyon), he married, by a ministerial approval of 22 April 1907, his first cousin Anne Henriette Brunet, born on 27 June 1882. She will die from anthrax in 1944 in the house of Françoise Popoff, Cholesky's sister, in Rabat (Morocco) where she was first buried. Her body was then repatriated to La Roche–Chalais. She was the daughter of François Brunet, aged 52, owner of a farm, and Anne Garnier, Cholesky's aunt. At that time, he lived at 33bis rue Rosa Bonheur, in Paris 15th. As we will see, they will have two sons, one posthumously, and two daughters. Cholesky's father did not attend the wedding but gave a written consent. A Calvinist minister was present at the ceremony.

The measurements of the meridian of Lyon continued. They were first delayed by rain and fog, but were then favored by exceptionally nice weather. It was the same at Mount–Poupet due to solar mists. Signals were constructed at La

The wedding of Cholesky

Mouillère (in two days), at La Chalentinne and at Nivigne (in three weeks, until late September, following a long period of very intense haze). Other signals were built at Crêt de la Neige, at Grand Colombier (in the massif of Bauges, 2045 m), and at Granier (in the massif of Chartreuse, 1933 m) in less than a month. Measurements at La Tournette (in the massif of Bornes, 2351 m) and at Charvin (in the massif of Aravis, 2409 m) were also conducted. There is a controversy about Crêt de la Neige which is the highest summit (1720 m) of the Jura massif, in the department of Ain. On 6 September 1907, Cholesky built his lightning rod, not at the exact summit, but at the lower summit (1718 m), called by the ancients Molard des Paulmes, 410 meters southwest, because the visibility to a distant signal was easier (see his notebook No. 3). This led to the wrong belief that the summit was indicated by the signal built by Cholesky. In 1907, the portion of the chain of the mean parallel from La Serre–Chailluz to Montellier–Grand Colombier was completed [9, 10]. Some of these names are mentioned in his notebooks. The name of Cholesky also appears in the list of officiers having participated in the operations in 1908 but, as we will see, he was in Crete at that time and he himself scratched his own name on his copy of [14]. A detailed history of cartography in the north of the Alps can be found in [126]. The conception and realization of the new map of France are described in [6].

1.2.2 Crete

Since Antiquity, the history of Greece has always been quite complicated. And this is also the case for Crete.

Greece became independent only in 1830 after a long period of Turkish domination. But Crete was still under Ottoman domination since the mid–17th century. Around 1890, in the face of the intensification of the acts of violence all over the island, the Great European Powers (France, United Kingdom, Russia, Germany, Italy, and Austria–Hungary) obliged the Sultan to grant a new constitution to the island in August 1896. But the Turks were determined to undermine this new constitution, and they undertook massive violence against the population's Christian majority. The Greek government, pressured by its public opinion and not seeing any intervention coming from the Great Powers, decided to send troops to help the Cretan population. The commander of the Greek Army force landed at Kolymbari on 1 February 1897, declared that he was taking over the island in the name of the King of the Hellenes, and announced the union of Crete with Greece. But the Great European Powers were against this decision. This immediately led to a revolt that spread throughout the island. The Great Powers finally decided to land their troops, and the Greek Army was stopped when approaching Chania. On 16 October, a compromise led to the creation of an autonomous state under Ottoman suzerainty, guaranteed by the presence of military contingents of the Powers. The British occupied Iraklion, the Russians Rethymnon, the Italians Kissamos and Ierapetra, the French Agios Nikolaos and Sitia. All four Powers shared Chania, the capital.

On 18 November 1898, Prince George of Greece (Corfou, 24 June 1869 - Saint–Cloud, France, 25 November 1957), second son of King George I of the Hellenes, was named High Commissioner of the Cretan State for a three–year term. A sixteen–member committee composed of twelve Christians and four Muslims was appointed on 25 December to design a constitution. The first Cretan assembly was elected, and the constitution was voted on and approved. On 27 April 1899, Prince George created the first government of Crete composed of the Cretan leaders. Eleftherios Venizelos (Mournies, Crete, 23 August 1864 - Paris, 18 March 1936), a long time opponent to the Turks, was minister of Justice. But disagreements soon arose between the two men. The main source of conflict concerned their vision for the island's government. Prince George alone was allowed to negotiate with the Great Powers, and he took responsibility for discussing the union of the island with Greece, without asking his counselors. On the contrary, Venizelos recommended the creation of a Cretan Army, followed by the withdrawal of the European troops. His opinion was that the island was not autonomous, since the military forces of the Great Powers were still present and were governing through the Prince.

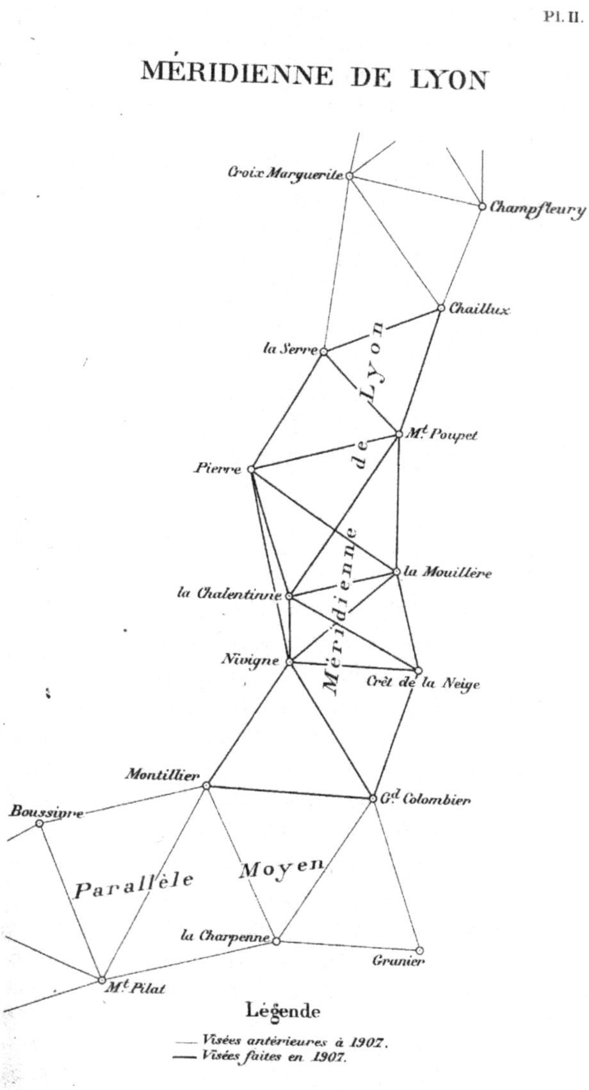

Triangulation of the Alps
© Public domain

In February 1901, the Great Powers rejected any change to the island's status. Venizelos submitted his own proposals which were not accepted by the Prince and the other councillors. Thus, he presented his resignation twice but George refused to accept it, preferring to dismiss him for insubordination on 19 March. In the two following years, Prince George was still trying to promote the union with Greece, but it was again rejected.

In 1904, insurgency against the Prince and demonstrations began throughout Crete. An armed insurrection against the Cretan government broke out in the spring of 1905 in the village Therisso. Venizelos was its leader. He wanted to persuade the Great Powers to abolish the High Commission, and to accept the idea of Crete's union with Greece. In March, the Revolutionary Assembly declared the union, and sent a memorandum to the Great Powers. Prince George rejected it. Riots and demonstrations followed. Among the European nations, Russia reacted most promptly to the events. At the beginning of June, Russian troops disembarked in Crete, took several villages and the Russian fleet bombarded several others. The British, despite their support of the existing regime, only took a few symbolic actions. The French and the Italians avoided taking part in any anti–revolutionary activities. Nevertheless, an international force was assembled at Alikianos, on the road to Therisso, ready to receive orders. Colonel Jules Clément Ladislas Lubanski (Nice, 18 December 1854 - Crete, 1906, from a cardiac syncope), commander of the French troops in Crete, organized a meeting with the rebels at Mournies, between Alikianos and Therisso, in a spot from where the European armies gathered in the valley could be seen. The talks took place on 2 July and were attended by Venizelos, Konstantinos Foumis (1860 - 1942), and Konstantinos Mános (Athens, 1869 - plane crash in the region of Lagkadas, 1913), with Colonel Lubanski receiving the insurgents' grievances. The Powers were unanimous regarding the impossibility of the island's annexation by Greece. On 17 July, martial law was declared. On 15 August, the regular assembly in Chania voted in favor of most of the reforms that Venizelos proposed. At the beginning of November, he expressed his willingness to lay down arms. The Great Powers, realizing that Prince George had lost popular support, informed him of their dissatisfaction, and accepted Venizelos' proposals concerning the system of government and the limitation of the powers of the High Commission on 2 November. On 11 November, amnesty was granted to the rebels. This led to the end of the Therisso revolt.

In February 1906, the Great Powers sent a mission to evaluate Crete's administration and finances. The mission recognized the necessity of a union with Greece. The former rebels of Therisso and the Cretan Assembly agreed to settle a Constitutive Assembly. Legislative elections took place in May 1906. The Prince's party gained the majority in the Second Constitutive Assembly which issued a resolution for the union on 30 June. When Prince George opened the parliamentary session in early July, he announced that the Great Powers, following the Commission's report, were considering to take into account the Cretans' national aspirations. Strengthened in its intentions, the assembly once again voted for

Crete's annexation by Greece and suspended its activities, waiting for the Powers' decision to be known. On 25 July 1906, at Chania, the Great Powers announced a series of reforms for the island. Venizelos was ready to accept them, but Prince George was not, and he resigned on 12 September. His father, King George I, tried to obtain a modification of the reforms, but he only gained the right to choose his son's successor. On 25 September 1906, Prince George departed secretly from his palace. He was replaced by Alexandros Zaimis (Athens, 9 November 1855 - Vienna, 15 September 1936), a former Greek prime minister, for a five–year term beginning on 18 September 1906. On 2 December, a new constitution was submitted to Zaimis, who accepted it. Order gradually returned to the island. A public administration was set up and laws passed for improving the health and educational systems. In July 1907, Antonios Michelidakis (1844 - 1923) was elected president of the Cretan Assembly. Large parts of the foreign troops began to depart from the island, and the rest were to leave over the following twelve months after receiving guarantees for the safety of the Muslim population.

André–Louis Cholesky with his wife in Crete in 1908
© Cholesky's family

From 7 November 1907 to 25 June 1908, Cholesky performed a mission to Crete. Following a proposal of Colonel Lubanski, himself a former surveyor, and a rapid recognition performed in March–April 1906 by Lieutenant–colonel Robert

Joseph Émile Bourgeois who was then head of the section of geodesy, it was decided to undertake the triangulation of the French and British sectors of the island (the departments of San Nicolo and Candia), and the topographic survey of the French sector.

The plain of Kavousi

Three officers, including Cholesky and Commandant Nicolas Arthur Lallemand (Saint–Aubin–sur–Aire, 4 November 1859 - 1946), performed the preliminary work during three months: the measurement of a base of 8 km long in the plain around Kavousi and the determination of a latitude and an astronomical azimuth in the south. Then Cholesky remained alone three more months to execute the triangulation of the French and English sectors. The surveying and the construction of the rods continued during winter. Crete, at most 57 km wide and 250 km long, rises to over 2,400 meters. In late May, on the heights of Lassithi, it was still necessary to melt snow to obtain water for the detachment. The difficulty of the work, which ended mid–June 1908, can be easily conceived. His wife, pregnant, was following him on a donkey. Unfortunately the political circumstances did not allow then to make all the topographical measurements.

Impressed by the efforts made by the Cretan government, and having received assurances regarding the safety of the Muslim population, the European powers decided to send their troops back home. On 26 August 1908, the first French contingents embarked at Chania, thus beginning the complete evacuation of the island. The Therisso revolt established Venizelos not only as the leading politician in Crete, but his reputation led him, in 1909, to become Prime Minister of Greece.

Lassithi
© Public domain

On 25 March 1909, Cholesky was promoted to capitaine en second in the 27th Artillery Regiment, and maintained in the Geographical Service. On 28 August 1909, he was removed from the list of this Service and, on 14 September, he had to join the 13th Artillery Regiment, by a ministerial decision, in order to perform his statutory period of two years as commander of the 13th battery that had just been created. It was during this period that Cholesky wrote the manuscript on his method for solving systems of linear equations.

1.2.3 Tunisia and Algeria

On April 29, 1827, the Dey of Algiers, in the midst of an argument with French diplomats over the settlement of debts, struck the French consul in the face with a fan. Two months later, the French troops landed in Algeria and blocked the

An artillery battery
© Public domain

port of Algiers for three years. In 1829, France sent an ambassador to the Dey with a proposal for negotiations, but he responded with cannon fire toward one of the French ships. Algeria was invaded in 1830, and the Dey defeated. The country became an administrative part of France. This was the beginning of the second French colonial empire. But it was only after the Franco–Prussian war of 1870–1871, and the founding of the Third Republic in 1871, that most of France's later colonial possessions were acquired. A protectorate on Tunisia was established in 1881, and another one on Morocco in 1912, where a semblance of limited sovereignty was maintained with the local ruling families remaining on the throne. Nevertheless, France controlled finances, public works, education, armed forces, security, and agriculture. The colonial project was, in principle, a civilizing mission.

On 24 September 1911, Cholesky was assigned to the particular general staff of artillery, and on 13 October of that year, to the Geographical Service of the Army led by Général Bourgeois with Capitaine Jean–Raoul Marie Chicoyneau de Lavalette du Coetlosquet (Metz, 13 February 1869 - Réméréville, 24 August 1914) as his assistant. Capitaine Noirel, who established the triangulation of the region of Chaouïa in Morocco in 1910, on which the whole topography of this country is based, and will later revise the editions of Cholesky's book on topography (see below), and Lieutenant De Fontanges (see below) also belonged to this mission. The direction of leveling in Algeria and Tunisia was entrusted to Cholesky. The head of the Section of Geodesy was the Lieutenant–colonel Lallemand, himself a brilliant geodesist. In [8, p. 73], it is mentioned:

> *The precise leveling, started in Tunisia, continued in Algeria, from 1889... Interrupted for lack of funds in 1896, the work was resumed in 1903; actively pushed since 1910 by Capitaine Cholesky, which significantly modified the methods to save time, they were able to continue steadily until 1914.*

From 27 October 1911 to 24 April 1912, Cholesky performed geodetic work in Algeria and Tunisia for the construction of tracks, roads, and railway lines. These operations are described in [15].

On 8 October 1912, Cholesky was again appointed to take part in the leveling campaign of 1912–1913 in Algeria and Tunisia. He left Paris on 22 October. He had to be in Marseille on 25 October 1912 at 1 pm and to embark for Algiers. He was entitled to an allowance of 10 French Francs (FF). a day for geodetic work and to a 5 FF mountain allowance. He was allowed to bring his military servant but not his horse. He had to be back to Paris at the end of his mission. Cholesky continued, with his usual eagerness, the work of triangulation for the mapping and the precise leveling in Algeria and Tunisia between 23 October 1912 and 17 April 1913.

André–Louis Cholesky
© Cholesky's family

At the end of these campaigns, Cholesky wrote a detailed report [16]. It begins by:

> *During the summer of 1910, it had been decided that the leveling operations in Algeria and Tunisia, previously supervised by Capitaine Archambeaud, called to other duties, will be entrusted to Capitaine Ponce, of the Section of Geodesy... Unfortunately, he died as soon as November 1910, in a railway accident.*
>
> *Commandant Lallemand, of the Section of Geodesy, was sent to Algiers after the death of Capitaine Ponce, to liquidate the situation created by the accident, and to control the leveling operations, while inspecting the geodetic brigades operating then in Algeria. He retained the lead of leveling until the end of the campaign 1910–1911.*
>
> *It was already decided that the lead of leveling would be entrusted to Capitaine Cholesky who, having left the Geographical Service in October 1909 for a two years commanding period of instruction at the 13th Artillery Regiment, would return to the Section of Geodesy in October 1911. This officer actually conducted the operations from the month of October 1911 until the beginning of May 1913.*

Then, Cholesky gave the names of the staff who worked with him, and an exact description of the work performed. A longitudinal railway line was connecting Oran, Algiers, and Constantine, and joined the Moroccan and Tunisian lines. There was a parallel one between Sidi–Bel–Abbès, Tiaret, and Trumelet, and some transversal lines. Finally, three lines were going to the south. The group of Cholesky leveled the route of the railway between Orléansville, Vialar and Trumelet in order to link the agricultural plateau of Sersou to the valley of the Cheliff river. Another brigade completed the leveling of the section of M'raïer Chegga. Considerable difficulties were encountered because of the rugged terrain and the harsh climate of the Ouarsenis massif. A section of road was also leveled between Biskra and Touggourt. Cholesky also stayed four days in the Sahara (26 - 30 March 1912) in the Oasis of Rhir. In Tunisia, the leveling of the railway from Tunis to Kalaa Djerda and from Sousse to Henchir Souatir, and the bifurcation to Slata and Kalaat es Senam were performed, as well as the continuation of the trail to Kesserine by Thala. The leveling of the road from Souk el Arba to Tebessa began in February 1912. The main Tunisian network was completed on the ground during the winter 1913–1914. The calculations were immediately reviewed, put in order, and the network was finalized and adjusted. Cholesky wrote [16]:

> [In 1910–1911]. *In total of 550 kilometers, including 100 kilometers of road and 68 of tracks, were leveled. The leveling conditions were particularly bad from Orléansville to Vialar where the track has slopes almost everywhere incompatible with the direct leveling.*

CHOLESKI, A.-L. (✠),
capitaine d'artillerie.
Capitaine hors cadres, groupe de canevas du tir du détachement d'une armée. Excellent officier ayant de beaux services, s'est distingué depuis le début de la campagne dans l'exécution des canevas de tir.

André–Louis Cholesky
© Public domain

[In 1911–1912]. *In total 622 kilometers including 350 kilometers on roads or tracks, very hilly, in particular from Ramka to Vialat, from Souk el Arba to Tabarka, and from Thala to Kasserine, that is over a length of 166 kilometers.*

[In 1912–1913]. *In total 624 kilometers including 370 kilometers on roads or tracks, with very hilly parts from Béja to Djebel Abiod, from*

Tamera to Djebel Abiod and Tabarka, from Testour to Teboursouk and around Maktar.

One has also to add around 33 kilometers leveled in one direction, namely the new railroad under construction from Béja to Pont de Trajan, that would have overlapped with the old one, and the railroad under construction to Nebeur, from Pont de l'Oued Mellègue, that would have overlapped with the leveling already made on the road from Souk el Arba to Kef. The verification of such sections are performed using the work already done between their ends.

Finally, the brigade used for the leveling of the road from Sfax to Sbeitla had previously made 152 kilometers of rapid leveling in the oasis of Oued Rhir in order to determine the conditions for the establishment of a network of channels for the sanitation of this region.

Then, the conditions under which Cholesky performed his double–run leveling procedure are given (see Section 3.3) and he concludes the introduction to this report by writing that, during these three campaigns, 400 kilometers of leveling were accomplished from Marrakeck (sic) to Casablanca, Rabat and Lalla Ito.

Médéa in Algeria
© Public domain

On 25 May 1913, Cholesky was relieved from the roster, at the assignment of the Minister of Foreign Affairs, and he was appointed head of the Topographical

Service of the Regency of Tunis [5]. On 19 June 1913, he asked the general commanding the occupation division to put a soldier from the garrison at his disposal. His request was rejected because he was serving in a non–military administration, but he was allowed to take a native soldier at his service. He lived in Villa Marie–Louise, rue du Safsaf (or Avenue Ismail Dubos, according to another document) in the Belvédère in Tunis. He stayed there until Sunday 2 August 1914, date of the mobilization against Germany, and then joined the 7th Artillery Group in Bizerte. On 15 September, he sailed from Bizerte to join the deposit of the 16th Artillery Regiment based in Issoire. He arrived in Marseille on 17 September.

1.3 The teacher

From December 1909 (and perhaps before) until January 1914 at least, Cholesky participated in the correspondence courses of the *École Spéciale des Travaux Publics, du Bâtiment et de l'Industrie* (commonly designated as ESTP) founded by Léon Eyrolles (Tulle, 14 December 1861 - Cachan, 1st December 1945) in 1891 (see Chapter 6).

This school consisted of 110 teachers and 7,000 students, internal, external or correspondents from all countries. Cholesky had to correct assignments submitted by students and to write notes for his courses. On 30 December 1909, the Deputy Director wrote him (at the address 5 rue Clodion, in the 15th district of Paris):

> *I send you in communication a part of the work of our student No. 18852. The claim of this student is justified. When a student makes such glaring blunders in one of his assignments, it is necessary to point them out to him very clearly. Too succinct corrections are of no use to the student.*

Between 30 December 1909 and January 1914, the Deputy Director sent him several reminders asking for his lecture notes. Cholesky was responsible for a course of *Topographie Générale* (General Topography), and for a course of *Calcul Graphique des Contenances* (Graphical Computation of Capacities), see Chapter 7.

The Deputy Director of ESTP asked him to prepare a course on *Notions de Cosmographie et d'Astronomie de Position* (Notions on Cosmography and Position Astronomy), and to send its synopsis, since *"the commission that established the program was of the opinion that you were best qualified to determine it"*.

In the papers left by Cholesky there are various manuscripts corresponding to all these courses, and the exercises he prepared. The following documents related to his activity as a professor are also preserved in the archives of *École Polytechnique* (for the entire description, see http://sabix.revues.org/531):

[5] *Régence de Tunis* was the former name of Tunisia especially during the Ottoman domination ending 1881.

André–Louis Cholesky
© Cholesky's family

• a handwritten course of 239 pages entitled *Complément de Topographie*, written on sheets of 15.5×20 cm. There is also a printed version with corrections from the hand of Cholesky. This is a preliminary version of his book *Topographie Générale*,

• a handwritten *Cours de Calcul Graphique* given in Appendix C. It has 83 pages of size 15.5×20 cm. It is analyzed in Chapter 7 and reproduced in Appendix C,

École Spéciale des Travaux Publics in Cachan
© Public domain

- 15 handwritten pages with the title *Instructions pour l'exécution des nivellements de précision.* It has 11 sections,

- two sheets with the title *Compléments de Topographie. Levés d'Études à la Planchette. 5 séries et 2 exercices pratiques. Tâches à remplir.* This document presents the ideas of Cholesky on the program for a course he gave at École Spéciale des Travaux Publics. The text is quite similar to the one in his book *Levés d'Études à la Planchette*, already mentioned. It ends with the following paragraph which is not contained in it:

> *Very important recommendation. The student would be mistaken a lot if he believed finding in the courses he has in hands the solutions of exercises that are offered. These exercises are mainly designed to force him to think, to prevent him to learn his lessons too strictly, by indicating that in a work as complex as a topographic survey, everything depends on the value of the operator who must consequently be used to reason about all his operations. As a student he will often benefit, when stopped by one exercise, to abandon the study of the course and seek only if the common sense did not tell him the solution. The corrected exercises will be an essential complement to the course, and not a repetition; so the student shall not be discouraged if he encounters serious difficulties in the exercises that are offered. He has to show that he knows thinking, we do not ask him more.*

No comments !

André–Louis Cholesky at his office
© Cholesky's family

1.4 The soldier

On 3 August 1914, Germany declares war on France, and the United Kingdom enters into it one day later. The German troops rapidly invade France whose forces retreat. Paris is threatened. It is necessary to send fresh troops to stop the advance of the enemy. On 6 and 7 September, Général Joseph Simon Gallieni (Saint–Beat, 24 April 1849 - Versailles, 27 May 1916), military governor of Paris, requisitions around 600 Parisians taxis to transport the soldiers on the Marne river. The invasion is stalled. Then, the *Race to the Sea* begins in Champagne and the line extends until the coast of the North Sea in November, the belligerents continually attempting to outflank each other. The race involves a number of battles, from the First Battle of the Aisne (13 to 28 September), the First Battle of Picardy (22 to 26 September), the Battle of Albert (25 to 29 September), the First Battle of Artois (27 September to 10 October), the Battle of La Bassée (10 October to 2 November), the Battle of Messines (12 October to 2 November), the Battle of Armentières (13 October to 2 November), the Battle of the Yser (18 October to 30 November), and the failure of the Germans at Ypres. The front extends from the North Sea to Switzerland.

On 24 September 1914, Cholesky was appointed as the commander of the 9th battery of the 23rd Artillery Regiment which took part in these battles. His regiment was engaged in the offensives on Cote 200 and Perthes–lès–Hurlus (where the German impressionist painter August Macke, born on 3 January 1887 in Meschede, was killed on 26 September) between Reims and Verdun. On 27 September, Cholesky had to replace the commander of the 3rd group that had been evacuated. There, he acted as chef d'escadron [6] until 18 October. After the arrival of Commandant Girard, he returned to the command of his own battery.

1.4.1 The Groups of firing grids

At the beginning of the war, the French ordnance maps at the scale 1:80,000 used the projection due to the French hydrologic engineer Rigobert Bonne (Raucourt, 1727 - Paris, 2 December 1794). After the victory of the Marne (12 September 1914), the positions of the armies froze on a line of over 800 km from the North Sea to the Vosges. The artillery, which was previously firing on sight, had now to point to invisible targets defined by their positions on a map. It was necessary to identify some points with the greatest possible precision, as happened in sieges.

The expression *plan directeur* (directing plan or fire plan) comes from the fact that it was used to point the firing batteries. It was necessary to improve the preparation of firing by using more accurate maps where the relief was represented. It was also necessary to include the positions of the enemy. Cholesky took part in this work.

The *Groupes de Canevas de Tir* (Groups of firing grids) were created in late 1914 by Général Bourgeois [31]. They had to provide grid charts, essential for the computations performed by gunners who had to shoot at invisible targets (indirect fire), and had to deal with the proliferation of local systems of coordinates. Indirect fire consists in aiming and firing a projectile in a high trajectory without relying on a direct vision of the target, as in the case of direct fire. Aiming needs to compute the azimuth and the elevation angles. It requires a precise surveying of the gun and of the target positions, and accurate maps. The task is not easy. The relief has to be rendered as accurately as possible. Such maps cannot be surveyed directly on the ground. Aerial photography then began. But the planes flew low, were unstable, and took oblique photographs that had to be brought to the horizontal. Another solution was to track the sound of the enemy's batteries by an acoustic analysis system invented by the French physicists Pierre Weiss

[6] In artillery, chef d'escadron corresponds to commandant in infantry.

André–Louis Cholesky
© Cholesky's family

(Mulhouse, 25 March 1865 - Lyon, 24 October 1940) and Aimé Auguste Cotton (Bourg–en–Bresse, 9 October 1869 - Sèvres, 16 April 1951). Cholesky wrote a report on the glow tracking of batteries, and he had a copy of *Documents on sound phenomenon accompanying the firing of cannons and the motion of projectiles* by Ernest Esclangon (Mison, 17 March 1876 - Eyrenville, 28 January 1954), a professor at the Faculty of Sciences in Toulouse. Accurate firing needs complicated calculations including such factors as the elevation of the target and of the firing position, the wind speed and its direction, the barometric pressure, the wear of the gun barrel and even of the propellant batch, and the temperature. It may include corrections of the fall of shot onto the target by observing it and calculating new angles since the geospatial relationship between the gun and the target may not be accurately known, or good quality data for non–standard conditions may not be available, or the target may move or be expected to move. Predicted fire was developed during World War I, and it became the main method of firing for field artillery. The first battle in which the fire plan consisted entirely of predicted fire was the battle of Cambrai in 1917, where the British guns were moved into pre–surveyed positions at the last moment, achieving a complete tactical surprise when they began firing. See http://en.wikipedia.org/wiki/Indirect_fire and http://en.wikipedia.org/wiki/Predicted_fire. For a detailed history of firing during World War I, see [123, pp. 126–158].

Ideally, all firing batteries have been surveyed onto a common survey grid chart. Therefore, in a note of 10 April 1915, the general who was the head of Geographical Service of the Army asked these groups their views on the choice of a single projection for the whole front. The proposal of Commandant Lavalette du Coetlosquet, who commanded the group of firing grids of the 1st Army, was chosen. It was to adopt the conformal projection proposed by the mathematician Johann Heinrich Lambert (Mulhouse, then Switzerland, 26 August 1728 - Berlin, 25 September 1777) in 1772 [158]. This projection method allowed representation of the spherical surface of the Earth by preserving angles without distorting the ground, to respect relative sizes, to keep the distance scale and the alignments in order to calculate coordinates. The decision of 18 June 1915 endorsed this choice.

On 3 January 1915, Cholesky was attached to the general commanding the artillery of 17th Army Corps for organizing the firing. On 11 February, he was assigned to the Geographical Service of the Army and attached to the group of firing grids of the Army of the Vosges. He rejoined his station on 15 February. As a member of the study commission led by Général Pierre Jean Charles Antoine Nourrisson (24 October 1862 - 15 February 1930) in Saint–Nicolas–de–Port (department of Meurthe–et–Moselle in the region of Lorraine), he arranged a number of conferences on the organization of firing to artillery officers (in May 1915 with Capitaine de Fontanges, then in the city of Saint–Dié–des–Vosges, in January 1916). They wrote a small leaflet (37 pages) summarizing these talks. According to the assessments of him, he was one of the officers who best understood and developed the role of geodesy and topography in the organization of artillery fire. Many documents on this work can be found in the archives at *École Polytechnique.*

But his duties were not confined to this. He was interested also in tracking and monitoring aircrafts, in the apparent point of emission of the sounds of gunfire, in aerial photography, in appliances for pointing the machine guns positioned on the Nieuport aircrafts. He wrote many documents on these questions, and also reported on the purpose of the groups of firing grids and the distribution of work within them, on the use of battery–cons, on the artillery firing against masked batteries, on the correction of firing in aerial combat, on the work of the cartographer officer, and more. He even began to learn English. In January 1916, he took part in the offensive in the Vosges. In July 1916, he became chief of the Group of Firing Grids of the Seventh Army, commanded by Général Étienne

Godefroy Timoléon de Villaret (Saint–Laurent–Lolmie, 17 February 1854 - Angers, 18 January 1931).

André–Louis Cholesky
© Cholesky's family

1.4.2 Romania

After two years of neutrality, though a member of a cadet branch of the Hohenzollern imperial family, King Ferdinand of Romania (Sigmaringen, Germany, 24 August 1865 - Sinaia, Romania, 20 July 1927), influenced by his wife Marie, granddaughter of Queen Victoria, entered his country into World War I on the side of the Triple Entente Powers (Great Britain, France and Russia) against the Central Powers (Germany, Austria, Hungary, Bulgaria, Turkey) on 27 August 1916. In order to enter the war on the Allied side, the Kingdom of Romania was demanding recognition of its rights over the territory of Transylvania. The Romanian Army was quite large but suffered from poor training and equipment, especially compared to that of the Germans. On the night of August 27, the Romanians launched attacks through the Carpathians and into Transylvania. The Entente incorrectly assumed that Germany would be unable to respond to the invasion, since the First Battle of the Somme and the Brusilov Offensive were at their height around this time. But, by mid–September, the Romanian offensive was stopped. Bucharest was abandoned on 6 December, and the King and the Government established themselves in Iasi, the capital of Moldavia.

Romania was then the leading producer of oil in Europe. In November of 1916, British military were sent ahead of the invading troops with the secret mission to destroy the oil stocks and to sabotage the oil wells in Ploesti. This was successfully accomplished, much to the dismay of the Germans. Pushed by the invading armies moving across the Carpathians into Moldavia, a front was established with the town of Marasesti as its center. The Romanian Army, reorganized and better equipped under the guidance of the French Général Henri Mathias Berthelot (Feurs, 7 December 1861 - Paris, 29 January 1931), won the battles of Marasesti in July of 1917 and nearby Oituz in August. These battles prevented access to the Moldavian plains by the Austro–Hungarian troops which were attempting to advance from Transylvania.

But the Russian troops quickly and chaotically abandoned their positions at the beginning of the Revolution in October 1917. An armistice was signed between Romania and Germany, and troops were demilitarized though King Ferdinand refused to accept the terms of the treaty that Germany tried to impose on Romania. The French armies crossed the Danube and entered Bucharest in November of 1918, some hours after the German troops left the city. The King returned on 30 November, and the armistice with the Central Powers was signed on 9 December.

For Romania, World War I led to its enlargement and to the fulfillment of its long–held dream of uniting Wallachia, Moldavia and Transylvania into a single Romanian entity.

From 25 September 1916 to February 1918, Cholesky was assigned to the military mission in Romania by decision of Général Berthelot, commander–in–chief. As stated: *The officers shall refrain carefully to give orders, but give advice.* There, Cholesky rendered eminent services. He served as the Technical Director of the Geographic Service and, on several official documents, he is referred to with the rank of lieutenant–colonel. In April 1916, he moved to the headquarters of the Second Army in Bacau, in the region of Moldavia. Then, from 26 November to 2 December of the same year, the Geographical Service moved to Iasi. Cholesky completely reorganized the Geographical Service of the Romanian Army. Many documents, often quite thick, relative to this organization belong to the *Fonds A. Cholesky* of *École Polytechnique.* They show his talent as an organizer, and his care to details.

Jules Émile Legras (Passy, Yonne, 25 May 1866 - Dijon, 12 May 1930) was a professor of literature at the Universities of Bordeaux and Dijon. He soon developed a passion for Russia, its literature, and its language. He made frequent and long stays (nine years for one of them) in this country, and met Chekhov and Tolstoy. Then, he taught Russian literature at the Sorbonne and wrote several books about Russia. In 1914, despite his age (50), he enrolled in the Army as a member of the French military mission on the Russian front where he organized the information offices. He met Cholesky in Romania and, in one of his books [167, pp. 302–304], he tells (translation):

> *23 October - Commandant Cholesky, that I went to visit in Iasi* [spelled Yassy] *on the employment of sight instruments for gunners, arrived the day before yesterday; we visited together a few positions of the 11th Corps, where General Chimanski gave us a princely welcome, and this morning we went back to the observatories. The commandant is accompanied by a Romanian officer: I could see how, thanks to him, we could supply. Speaking in Russian, we could fast, unless to come across a military unit; with the knowledge of Romanian, one finds in the first coming Moldavian house enough for a hearty meal. Tonight we're going to Lipkany to address the issue of maps.*
>
> *Commandant Cholesky, in charge of the Romanian geographical service, has done wonders in Romania to create a mapping. He was evacuated from Bucharest, under the noses of the German, two trains of equipment; in Iasi, he created a paper mill to supply his service, and also a zincography factory; he sought how to produce colors for printing maps; he calculated the coordinates allowing him to correct the maps of the Romanian service, and, finally, he produced a fire plan at the 1:20.000 of the whole Romanian front...*
>
> *In addition, Commandant Cholesky provides 1500 maps of each of its prints to the Russian staff in Iasi, which immediately sends them to Berditchef to mark them down, and distribute them to the Russian units of the Romanian front, without prejudice to subsequent distributions.*

Before World War I, the territory of Romania was represented on various types of topographic maps. But, during the war, the necessity for a conformal grid system became evident for controlling artillery firing. Due to Cholesky, the projection was unified by using a Lambert conformal conic grid based on the ellipsoid of Clarke of 1880. He imposed two conditions on the design. Firstly, the projection distance between the 45th and the 50th parallels shall be equal to the distance between the 50th and the 55th parallels, and both distances must be 500 km. Secondly, the scale factor at the central parallel shall be equal to the reciprocal of the scale factors at the projection limits. This scale was 1:20,000. The importance of the Lambert–Cholesky projection consisted in the fact that it allowed representation of the entire Romania under a unique projection system, with a unitary legend. The basic map, called in Romanian *Plan Director de Tragere*, consisted in 2,118 drawings. Furthermore, it was the only map achieved at the highest scale up to that moment [39,72,203,204]. The maps continued to be created after World War I, for military purposes (especially artillery). More than 1,800 map sheets from this collection were identified in archives belonging to the Faculty of Geography of the University of Bucharest, the Faculty of Geography of the University Babes Bolyai, the Institute for Cultural Memory and the National Museum of Romanian History.

On 6 July 1917, Cholesky was promoted to chef d'escadron, that is to say commandant. On 17 December 1917, Colonel Ion Pavelescu, head of the Geographical Service of the Romanian Army, named him, on behalf of the king, *Officer of the Order Steaua României with swords* during a farewell dinner (probably in Iasi) offered for the departure of the French troops from Romania. Speeches were given (both are in the *Fonds Cholesky*). After his return to France, Cholesky sent him his photo and the Colonel did the same.

1.4.3 The second battle of Picardy

After the Race to the Sea, the physiognomy of the war completely changes. A war of position begins which culminates with the battle of Verdun (21 February to 19 December 1916). On 2 April 1917, the United States enters the war. However, during the first half of 1918, the military situation of the Allies, whose commands are separated and do not follow a common strategy, becomes catastrophic.

On 21 March 1918, at 4 am, Kaiser Wilhelm II (Berlin, 27 January 1859 - Doorn, Netherlands, 4 June 1941) launches *Operation Michael* for the final victory in the vicinity of Saint–Quentin with 3,755 pieces of field artillery and 2,508 pieces of heavy artillery on a front of 70 km. Conducted by General Erich Friedrich Wilhelm Ludendorff (Kruszewnia near Posen, 9 April 1865 - Tutzing, 20 December 1937), the attack was organized at the juncture of the Franco–British armies from the sector of Montdidier up to the region of Arras on the Scarpe river. Its aim is to separate the British and the French forces, and to drive the former into the sea before the French can intervene. The enemy advances his lines on the road to Paris and, after five days, threatens the capital which is bombed by long–range artillery (the famous *Grosse Bertha* which fires from a distance of 120 km). The Allies react and, on 26 March, in the village of Doullens in Picardy, a meeting gathers the French Prime Minister Georges Clémenceau (Mouilleron–en–Pareds, 28 September 1841 - Paris, 24 November 1929), the French President Raymond Poincaré (Bar–le–Duc, 20 August 1860 - Paris, 15 October 1934), first cousin of

Cholesky in a Romanian uniform

the mathematician Henri Poincaré, and the French and the British generals. It is decided to unify the commandment under the leadership of Général Ferdinand Jean Marie Foch (Tarbes, 2 October 1851 - Paris, 20 March 1929). The American Army joins the Allies two days later. The Germans are exhausted, they are cut off from their bases. On 3 April, the German assault is definitely broken. The failure of the offensive marks the beginning of the end of the war. On this period of the war, see, for example, [83].

On 5 June 1918, Cholesky was assigned to the 2nd group of the 202th Field Artillery Regiment which was part of the 10th Army of Général Charles Marie Emmanuel Mangin (Sarrebourg, 6 July 1866 - Paris, 12 May 1925). On 11 June, he took command of the 2nd group of artillery, 24th, 25th and 26th batteries. Let us specify that a regiment consisted of three or four groups of artillery, each of them disposing of three batteries of guns commanded by a captain. Each battery was equipped with four guns of 75 mm, model 1897.

Gun of 75 mm
© Public domain

Until June, the Allies managed to halt the German offensives in extremis, especially during the Battle of the Matz (a river, 25 km long, an affluent of the Oise river), where the German breakthrough was stopped on 13 June. In July, Foch, from his headquarters in the village of Sarcus, agreed that the troops were ready to go on the offensive and set up with Field Marshal Douglas Haig (Edinburgh, 19 June 1861 - London, 29 January 1928), commander–in–chief of the British troops in France, Général Philippe Pétain (Cauchy–à–la–Tour, 24 April 1856 - Port–Joinville, 23 July 1951), commander–in–chief of the French troops, and General John Joseph Pershing (Laclede, Missouri, 13 September 1860 - Washington, D.C., 15 July 1948) commander–in–chief of the American Expeditionary Force in France, the plans for partial attacks. However, a general victorious offensive was not yet considered. The first objective was to reduce the pocket of resistance formed by the enemy during the *Michael offensive* of 31 March.

In Champagne, on the salient from Villers–Cotterêts to Château–Thierry, the armies of Generals Mangin (10th Army) and Degoutte (6th Army) counter–attack from 18 July on the Marne river, in the department of Oise, between Montdidier and Noyon. The front was held by the 3rd Army of Général Humbert and the 1st Army of Général Debeney. In the heart of the Somme's region, the offensive was led by the 3rd and 4th British armies (General Rawlinson), composed of Canadians, Australians and New Zealanders. On the Scarpe river, the 1st British Army of General Horne attempted to relieve the region of Arras. Surprised between the Marne and the Aisne rivers, the Germans fold to ten kilometers, losing more than 10,000 prisoners. On 20 July, the 5th Army of Général Berthelot entered into the battle, attacking the eastern flank of the forces and threatening the Crown Prince with overflow or encirclement. Ludendorff resigned himself to a retreat on Soissons, Fère–en–Tardenois (the native village of the sculptor Camille Claudel), and Ville–en–Tardenois, and resisted there from 28 to 31 July, but, unable to keep his positions, he brought back the front on the Aisne and the Vesle rivers where Général Fayolle arrived on 3 August 1918. The pocket of Château–Thierry was resorbed. The Germans lost 30,000 prisoners, 600 guns of all calibers, and 3,000 machine guns. Due to the site and the purpose of the attack, that battle of the *"last chance"* launched by Ludendorff, was named the *second victory of the Marne.* For this victory, Général Foch , who was commander–in–chief of the Allied Armies since 17 April 1918, was raised to the exalted position of Maréchal de France on 7 August.

Battle of the Marne - French troops with prisoners
© Public domain

The offensive of the Allies started on Thursday 8 August, *"day of mourning for the German Army"* as Ludendorff wrote. The French troops broke through the German front north of the Oise river and south of the Somme river, in the area Noyon–Montdidier–Nesle, through Lassigny and Roye. The allied troops penetrated about 16 km into the enemy lines on the first day. The deterioration of the situation was so fast that, as early as 13 August, the top German staff, worried, asked the Chancellor to initiate peace talks. On 18 August, Foch launched his general offensive simultaneously on six fronts. However, one month of fighting was needed to see a noticeable difference. The enemy counter–attacked a number of times, and showed a high resistance. But the Franco–British forces pushed the enemy to retreat and to abandon much of its equipment. Heavy fighting in the villages of the Somme, and shelling on the outskirts of cities in the north of the Oise river continued without interruption during the month of August.

Between 15 August and 26 September, the regiment of Cholesky participated in the attacks on the Hindenburg line where the Germans fell back. This line passed through Lassigny, Ribécourt–Dreslincourt and Tracy–le–Mont. The details of the offensive were decided by Général Mangin:

> *The line marked by the following points: Pontoise, Cuts, Blérancourt, Mont-de-Grocq, Vézaponin, Tartiers, Cuizy–en–Almont, should be considered as an indication of minimum target. It is especially important to act in the best circumstances of the fighting for progressing as fast and as far as possible, so as to finally achieve: 18th Corps, the Oise river and the confluence with the Ailette river; 7th Corps, the Ailette river; 30th Corps, the line Crécy, Juvigny; 1st Corps, the northern edge of the ravine of Juvigny.*

The Germans had installed numerous fortifications and had arranged for quarries, command posts, observation posts, shelters, as well as cemeteries and hospitals. The 3rd Army aims to relieve Compiègne, then proceeds to the north, and moves forward to the hills of Thiescourt and the region of Plémont. On 10 August, Montdidier is taken while the second battle of Lassigny, launched on August 9, allows the French to retake Belval on 13 August, and the hills of Lassigny on 15 August. On 17 August, the 10th Army, to improve the starting conditions of the attack, clears by surprise, with the 7th and 30th Corps, the ravine of Autrêches and Hautebraye, and these two villages and takes 250 prisoners. On 18, the 10th Army attacks on the whole front between Tracy–le–Val and Fontenoy, and removes all the forward positions of the enemy (plateau of the farm des Loges, the farm of Tiolet) and arrives at the edge of the ravine Nampcel–Morsains; 2,000 prisoners. The next day it occupies Audignicourt, and prepares the attack of the main position. On 20, after an intense artillery preparation which lasted throughout the preceding day, it moves forward, at 7 am, with its four left wing Corps: 38th, 7th, 30th, and 1st. The attack is crowned by a complete success and, on a front of 25 kilometers stretching from the Oise to the Aisne rivers, all the main positions of the enemy were taken away. The left wing arrives at Carlepont and Mont–de–Choisy, the center front at Blérancourt and Selens, the right wing, gaining 5 to 6 kilometers, exceeds Vézaponin, and Tartiers Cuisy–en–Almont. The town of Soissons begins to be surrounded. The number of prisoners amounts to more than 7,000, with a hundred guns. The growth will continue non–stop for the night and the days to come. The cities of Lassigny, Carlepont, Mont–de–Choisy, and Cuts are delivered on 21 August. On 22, the 10th Army makes a new leap forward and reaches all the goals it was assigned by the original order of 11 August. Its left wing exceeds the forest of Carlepont and borders the Oise river, from Sempigny to Varesnes; the center is on the edge of the plateau overlooking the Ailette river; her right wing forms a tenaille with the Aisne river, from Pommiers to Pont–Saint–Mard .

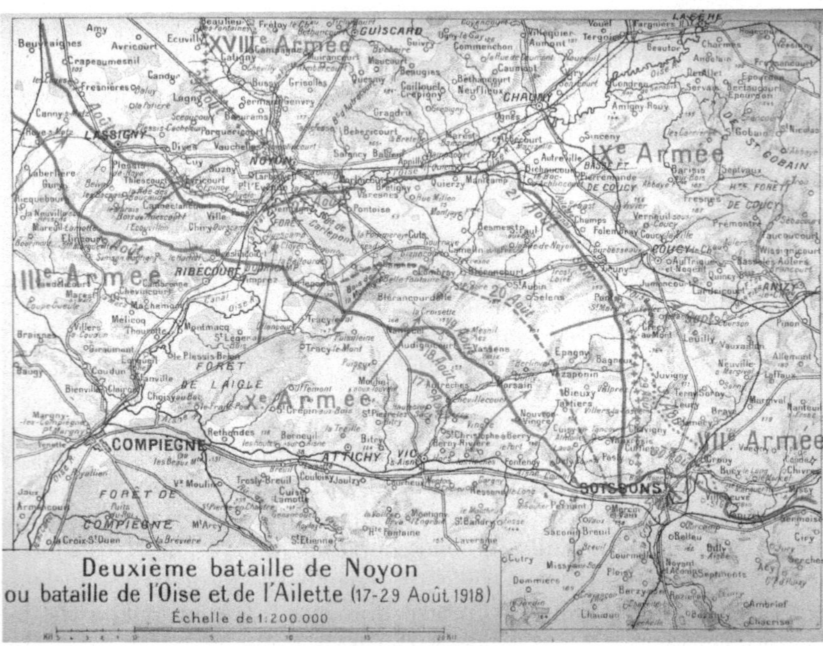

Map of the front between 17 and 29 August 1918 (from [90])
Bagneux is north of Soissons

The 202th Field Artillery Regiment is engaged in the fighting on the Ailette river, on 23 August, and Courson. On 25 August, Mangin's Army is ready to break the enemy front between the Aisne river and the village of Saint–Gobain. On 27 August, the town of Roye is occupied by British and French troops, and the next day the cities of Chaulnes and Nesle in turn are retaken during the advance to Noyon, recovered on the morning of 29 August.

In the journal *Le Miroir*, published during the war, a French soldier says [7]:

Monday 26 August. Between the Ailette and the Aisne rivers, we have progressed to the east of Bagneux, making a hundred prisoners. North of Roye, a raid of the enemy failed. We made about twenty prisoners.

Tuesday 27 August. Activity of the two artilleries in the region of Lassigny. Between the Ailette and the Oise rivers, we made further progress east of Bagneux and repelled the counter–attacks of the enemy west of Crecy–au–Mont. We made 400 prisoners during these actions. Two enemy raids in the Vosges obtained no result. We made prisoners.

Tuesday 28 August. On both sides of the Avre river, we executed two local operations that allowed us to take back Fresnoy–lès–Roye and Saint–Mard, despite the strong resistance of the enemy. We made over 1,100 prisoners, including 36 officers. Between the Oise and the Aisne rivers, a counter–attack of the Germans west of Chavigny failed. We made thirty prisoners. To the east of Bagneux, we have progressed from 1,200 meters. Cannonade between the Aillette and the Aisne rivers.

During the night of 29–30 August, the 10th Army extends from Champs, on the Ailette river, to Braine included, on the Vesle river. Its mission is to take over the plateau north of Soissons, in particular that of Laffaux. On 30 August, the progress to the Hindenburg line begins. East of Soissons, the forces of the 10th Army near the Aisne and Vesle rivers, from Sermoise to Braine, continue to face a vigilant and strong opponent during the night of 30–31 and in the morning of 31.

On Saturday 31 August 1918, André–Louis Cholesky died at 5 am in a quarry north of Bagneux (in the department of Aisne, about 8 km northwest of Soissons) from injuries received on the battlefield [8]. Lieutenant Marcel Desbrosses (born 15 July 1895 in Héricourt in the department of Haute–Saône) was also killed. Cholesky was buried in the military cemetery of Chevillecourt near Autrèches in the department of Oise (with André Léon as his first names!), about fifteen kilometers west of Soissons. On 24 October 1921, his body was transferred to the national cemetery of Cuts (in Oise, 10 km southeast of Noyon) grave 348, square A, where it still is. The cross bears the first name of René (the nickname he took). His name also appears on the memorial monuments dedicated to the soldiers *Morts pour la France* in the secondary school of Saint–Jean–d'Angély, in *École Polytechnique* where he had been a student, and in Saint–Martin–d'Ary near Montguyon. His name has now been added to the monument in Montguyon.

---

[7] see http://grande.guerre.pagesperso-orange.fr/aout18.html.

[8] On map No. 2611 at the 1:25,000 of the *Institut Géographique National* a ravine or a small rocky bank named *Mont Tantin* is mentioned.

The military cemetery of Chevillecourt

Cholesky was Officer of the Nicham Iftikhar (10 June 1907), *Officier d'Académie* (23 April 1908), *Chevalier de la Légion d'Honneur* (10 April 1915), *Croix de Guerre avec Palmes*, Officer of the Star of Romania with Swords (17 December 1917), and he was awarded the Order of Saint–Stanislas (6 August 1917) and the Nicham Medjidie.

Here is the military account as sent to the mayor of Montguyon on 22 January 1923:

> *The year 1918, on 31 August, at five in the morning, being north of Bagneux (Aisne) the death certificate of André–Louis Cholesky, commander of the 2nd group of the 202nd Regiment of field artillery, born October 15, 1875 at Montguyon (Charente–Inférieure), domiciled in last place in La Roche–Chalais (Dordogne)* Mort pour la France, *died north of Bagneux (Aisne) on 31 August 1918 at five in the morning on the battlefield from his injuries.*
>
> *Married to Miss Henriette Anne Brunet (wife Cholesky) based in La Roche–Chalais (Dordogne), son of André and Marie Garnier based in Montguyon (Charente–Inférieure).*
>
> *Compiled by me, Andrew Ducout, Captain temporarily commanding the the 2nd group of the 202nd Regiment of artillery, on the state-*

The grave of Cholesky
© Claude Brezinski

ment of Charles Céas, 22, as Second Lieutenant in the 2nd group of the 202nd Regiment of artillery, witnesses, who signed with me.

After reading (Approved six words scratched as null), signature of the witnesses: Cease, Deschavanne. The officer of civil status. Signed Ducout (I say a word scratched as null), seen by us Hautière Pierre Malo, his quartermaster. Signed: unreadable. Checked for the legalization of the signature of Mr. Pierre Malo Hautière: Paris, 7 April 1919, the Minister of War, by delegation, the Chief of the Office of administrative records. Signed: unreadable.

Mention of amendment: (Act of 18 April 1918). Commandant Cholesky was legally domiciled in Montguyon (Charente–Inférieure) and not last in La Roche–Chalais (Dordogne), the deceased was awarded the Croix de Guerre. Paris on 1 April 1919. The Minister of War, by delegation, the Chief of the Office of administrative records. Signed: unreadable.

The act above was transcribed on 11 August 1919 by us, Brault of Bournonville, Mayor.

The history of the 202th Field Artillery Regiment is told in a small booklet with an introduction by Général Noël Marie Joseph Édouard, vicomte de Curières de Castelnau (Saint–Affrique, 24 December 1851 - Montastruc–la–Conseillère, 19 March 1944), commander of the Second Army [4]. Let us give its translation corresponding to the period (5 June - 31 August 1918) when Cholesky was in this regiment:

We participate in several raids, the two main are those of 12 July on the farm Bel–Air and of 21 July on the farm of the Tuilerie, the prisoners made give interesting information.

The area is quiet, but we do not lack of work: the officers must be made familiar with new firing methods, the batteries with the speed of movement. Each battery is split in two sections occupying different locations, often even each piece [of artillery] *is independent, the structuring work is the double.*

A new organization of our defense lines on a large depth results in new work and new displacements, alternating infantry and artillery. All positions are fully explored, so as to allow a rapid occupation. The countryside is scored by basic reference nameplates.

Chef d'escadron Fauché, of the second group, is replaced by Chef d'escadron Cholesky, from the geographical service of the Romanian mission.

We are relieved on 7 August by the second American division. The regiment left Millery, Belleville and Atreville, where the officers had made a happy life, for quartering in Neuves–Maisons and embark, on 11 August, in the train stations of Ludres and Chaligny, for the Oise.

The Oise - The pursuit. *We disembark in Creil, on 12 August, and quarter some days in Rhuys and in the castle of Beaurepaire.*

Since our departure from Somme (in May), Ludendorff took the offensive on the Chemin des Dames. *His troops, which pierced the front by surprise and arrived breathless near Soissons and up to the Marne river, taking Fère–en–Tardenois and Château–Thierry, threaten Paris.*

His Army is about to encircle Compiègne, but Général Foch is watching; the city is saved by the attack of the troops of Mangin between Rubescourt and Saint–Maur (11 June); the shot is missed.

On 15 July, the enemy no longer looking for Amiens, or Calais, or Paris, tries to crush the Allied forces on the front of the Champagne, by attacking between Château–Thierry and La Main de Massiges. The disposal in echelon in depth we have experienced in Lorraine foiled this attempt; the enemy, after a heavy bombardment, can seize the small posts that were sacrificed in advance on the first line and find intact the bulk of the Army Gouraud established on the second line.

While our armies stand firm in Champagne, Mangin and Degoutte save the country by a flank attack on the German pocket of the Tardenois (18 July). They force the Germans to abandon the Marne river, Château–Thierry and Soissons. On 3 August, Général Foch orders to take the offensive *along the whole front. Montdidier, the Lassigny massif are taken. The Army of Général Mangin attacks for its part on the Ailette river; it is with it that our division has to take part in the action.*

With five nights of successive stages through Pierrefonds, Montigny–Lengrain, Morsain, in destroyed villages, looted, full of troops, the regiment takes its position in the night between 23 to 24 between Epagny and Vezaponin, relieving the 252th Field Artillery Regiment. Our infantry, pushed in line, assisted by tanks, immediately progresses towards the farm Montécouvé and in the woods of the Domaine. Enhanced by the 372th, the 240th and the 60th Field Artillery Regiments, we support it by a roaring fire; the railway track is reached.

The groups then change position and are going in the ravines of Bagneux and Mareuil to track the progress of the infantry. This is where Chef d'escadron Cholesky was to be killed with his orientating officer, Lieutenant Desbrosses. Commandant Cholesky showed in the first days of the attack an unmatched fearlessness. Carrying himself in the head of the liaison staff in the most perilous reconnaissances, searching for the observatories from which one could perform the most accurate settings.

His high value was immediately recognized by all his chiefs. His courage and boldness admired by of all those who approached him.

Lieutenant Desbrosses, successively aspirant, sous–lieutenant, lieutenant at the 25th battery, then assigned to the staff of the divisionary artillery and at the staff of his group, had managed to attract the esteem and sympathy of all. The same bombshell took us two eminent officers.

The Hindenburg line was broken on 2 September 1918, during the second battle of Arras. The armistice was signed on 11 November, the same year.

André–Louis Cholesky
© Cholesky's family

Chapter 2

Cholesky's family

This chapter gathers the information we were able to find on the family (past and present) of Cholesky. Most of it is due to the extensive research of Raymond Nuvet, the vice–mayor of Montguyon, in the municipal archives of all the villages around.

2.1 Cholesky's ancestors

The great great grandfather, Thomas Cholesqui (1) [1]

The first member of the family who has been found in the municipal archives so far is Thomas Cholesqui (**1**), with this spelling. He was married to a woman whose first name was Elouche.

It seems that he was issued from a noble Polish family whose blazon was Cholewa, but it is not known when they arrived in France. They were good patriots whose name, at the origin, was written Cholewski. This is probably why, on some documents in his possession, Cholesky himself spelt his name as Choleski. Cholewa is a common Polish name. In 1990, there were some 8,100 Poles by that name, 797 with the spelling Holewa (both pronounced the same), plus quite a few more with related names such as Cholewiak (175), Cholewka (761), etc. Cholewa appears all over the country, with the largest numbers (more than 300) living in the provinces of Bielsko–Biala (731), Katowice (1,015), Kielce (395), Krakow (1,069), Lublin (517), Opole (307), Radom (405), Tarnow (694), Warsaw (338). The name seems to be more common in Southcentral and Southeastern Poland than elsewhere. It is interesting to notice that Cholewa is a metonymic occupational name for a bootmaker, or a nickname for someone who habitually wore boots, from cholewa "boot upper" since, as we will see below, several members of

[1] The number in bold after each name refers to the family tree given at the end of the chapter.

the family will later undertake this profession. The coat of arms of this family can be found at http://en.m.wikipedia.org/wiki/File:POL_COA_Cholewa.svg.

It is also possible that Thomas Cholesqui came from Pozsony (Presburg), the capital of Hungary from the 16th to the 18th century, named Bratislava after 1919 when Czechoslovakia was created, then the capital of Slovakia, or that he emigrated there. It seems quite difficult to establish the exact origin of the family since, in fact, the pronunciation of "Ch" is the same, in Polish, as that of "H", and since "Ch" can also be written "Sz". Presently, many people wear the names Scholeszy, or Cholewski, or Holewski, or even Scholewski, which can explain the various spelling in the official acts of civil status of the family (see below).

The first partition of Poland took place on 5 August 1772. Russia, Prussia and Austria signed a treaty on 30 September, ratified by the Polish Assembly, which cut from Poland one third of its population, and 30 % of its territory. Russia received the territories east of a line delimited by the Dvina, the Prut, and the Dniepr rivers. Prussia obtained the rich region of Royal Prussia populated by 90% of Germans, with the northern part of Great–Poland (Wielkopolska), populated by Polish. Small–Poland (Malopolska), the south basin of the Vistula river, and Western Podolia were given to Austria. It may be these events that brought the family Cholesky to emigrate to Presburg. According to some internet sites, the family may have originated from Ukraine. On 26 December 1804, after the victory of Austerlitz, Napoléon imposed the treaty of Presburg to Austria, which has to yield Veneto, a part of Istria and Dalmatia to France, and Tyrol, Vorarlberg and Trentino to Bavaria. Need we see in Napoléon's presence in these places a connection with the arrival of Cholesky's family in France?

The great grandfather, Jean Cholesqui (2)

Thomas' son, Jean Cholesqui **(2)**, also called Inebriysorsky, was born around 1770–73, probably in Pozsony, Hungary, or, at least, he originated from there as confirmed in the act of his second marriage in 1825. He came to France around 1797 when he was 25 or 27, and settled in Saint–Martin in the district of Cognac (since, as stated in the act of his second marriage in 1825, he lived in Saint–Martin for more than twenty five years) [2]. In 1801, he is listed as a farmer and in 1814, as *bordier/journalier*, that is a day–sharecropper. He is surrounded by clog–makers and shoemakers, a business that he will eventually learn in order to become a master shoemaker himself. In 1814, he was living in the house of Mr. Vounage.

The second partition of Poland (1793) was the result of the application for aid to Russia by the Confederation of Targowica on 4 May 1792. Russia agreed and, as Prussia, sent troops. An agreement between these two countries led to the second partition which was also ratified by the Polish Assembly. Austria had its hands tied by the war declared by its former ally, France.

[2] The municipalities of Cognac, Saint–Brice, Verrières, Saint–Preuil and Juillac–le–Coq are in the department of Charente.

The third partition of Poland arose in 1795. The following year, the Holy Roman Emperor and the King of Prussia signed an alliance against the revolutionary France. The Polish Legions (French Army) were born from a rallying of soldiers behind Napoleonic France. Perhaps young people were enlisted and entered France as members of the Army. Was our Jean Cholesqui one of them?

Jean Cholesqui married Marie Vinante who will die on 19 April 1816 in Cognac. They had two sons, Jean **(3)** (in 1801) and Louis **(4)** (in 1814). At the birth of his second son Louis in 1814, and at the death of his wife, this Jean declared that he was unable to sign (demonstrating that it is not him who signed Jean Cholesky on the civil status certificates). On this last certificate, his name was spelled Jean Chaulski.

On 25 April 1825, when he was 52 years old, he remarried Marguerite Ménard, born in the village of Saint–Brice near Cognac, on 28 August 1778. Again, he did not sign his wedding certificate and it was his son Jean, issued from his first marriage, who was his witness and signed in his place. It is probably him whose name and signature are also found on the other certificates from Montguyon.

Jean Cholesky (3)

The first son of Jean Cholesqui and Marie Vinante is Jean Cholesky **(3)**, also called Inebriysorsky. He was born on 26 August 1801 in Saint–Preuil in the department of Charente–Inférieure (spelling very difficult to read). In another document, it is written that he was born in Juillac–le–Coq (Charente–Inférieure), in 1800. He was a mason and a stonecutter. On 25 June 1825, he married Jeanne Chapeau (Chapaud?) (indicated on the records, but the original documents were not found), born in Saint–Brice on 13 February 1799. She will die on 21 January 1833. They had a son, Victor **(5)**, born on 6 July (or April?) 1831, who could be the one found on the documents in the town hall of Montguyon.

Widower, Jean remarried, on 6 July 1835 at the age of 33, Marie Pignon, born in Jarnac Champagne (department of Charente–Inférieure), on 16 June 1809. On the document of this second marriage, he appeared as a stonecutter, born in Saint–Preuil on 26 August 1801, widower of Marie (?) Chapaud, son of Jean Choleski, said Inebriysorsky, and of (?). In 1860, he is presented as a wine merchant. He died in Cognac, on 6 August 1864. At this time, he was an innkeeper. Jean and his second wife had a daughter, Marie Cholesky **(6)**, born on 17 March 1838.

Victor Cholesky (5)

The son of Jean Cholesky and Anne (Marie? Jeanne?) Chapaud, Victor **(5)**, was born on 6 July (or April?) 1831. On 10 April 1860, he married Mathilde Prudence Honorine Séraphine Coubart, born in Cognac on 19 July 1836. She was a dressmaker and he was a shoemaker. Then, he became an innkeeper in 1871. At this time, they live in Boulevard du Nord in Cognac.

Their first son, Ernest **(14)**, was born on 4 September 1861 in Cognac where he died on 6 January 1884.

Their second son, Édouard **(15)**, was born on 18 January 1871 also in Cognac. He was exempted from military service after his review board, confirmed by the Prefect of Charente, on 1 August 1892. He worked as a painter and married, in Chateauneuf–sur–Charente on 25 June 1894, Georgette Maury, born on 14 December 1870 in the same town. She was the daughter of Jean Maury, born in 1822 in Le Bugue in the department of Dordogne and of Jeanne Bourland, born in 1838. Édouard had a daughter, Marguerite **(31)** who married a certain Levêque to whom she gave a son, Guy **(39)**. Let us mention that, on the internet, one can find a decree, dated 23 January 1992, on the appointment and the promotion in the reserve officers, of a reserve major of artillery named Guy Édouard Ernest Levêque (http://france.globe24h.com/lex/jorf/003/00345/0000345206.shtml). Is it him? On this branch of the family, see the site:
http://www.genealogie33.org/pmarteau/dat13.htm#17.
Victor died in Chateauneuf–sur–Charente at an unknown date.

Marie Cholesky (6)

Jean Cholesky and his second wife Marie Pignon had a daughter, Marie Cholesky **(6)**, born on 17 March 1838. She got married in Cognac, on 29 May 1855, with Jean Édouard Théodore Sauvage, a masonry contractor, born in Bordeaux on 13 March 1831. She worked as a laundress, and died in Cognac on 31 October 1857, at the age of 19.

The grandfather, Louis Cholesky (4)

The second son of Jean Cholesqui **(2)** and Marie Vinante is Louis, born in Cognac on 11 May 1814. On 4 March 1837, he married Marie Amélie Moreau (aged 25) in Montguyon. She was the daughter of Pierre Moreau (remarried with one named Anne), master shoemaker in Montguyon, and Jeanne Guit (wife Moreau?), deceased on 23 March (or August) 1820. This Marie Amélie will die on 26 April 1869. Louis worked as a shoemaker in the premises of his parents in–law. He was the owner of two houses in the town center of Montguyon with respectively 3 and 9 apertures. Around 1870, he demolished the two houses to reconstruct another one with 21 apertures that will be named *Café du Centre et Hôtel de l'Étoile*. Therefore, Louis became an innkeeper. We have the proof that, in 1900, André Cholesky was still running this hotel. Then, his name disappears. His name was always written with the spelling Scholesky. Throughout the collection of old bills available from the business houses of Montguyon, there is no invoice for the *Hôtel de l'Étoile* or for the *Café du Centre* with the name of Cholesky. Later, this settlement will become the *Hôtel du Cheval Blanc* (maybe located on the other side of the square). Louis Cholesky died on 25 December 1900.

From his union with Marie Amélie Moreau, seven children were born, all in Montguyon, probably at the *Hôtel de l'Étoile*.

The eldest son, André **(7)**, was born on 5 February 1837 and died on 2 May 1838 in Montguyon.

Café du Centre and *Hôtel de l'Étoile* on the left
© Public domain

The eldest daughter is Julie **(8)**, born on 14 (or on 4 according to her wedding certificate) October 1838. On 17 September 1868, in Montguyon, she married Théodore Naintré, carpenter, widower of Marie Blanc, born on 2 April 1833 in Buxeuil in the department of Vienne. She died in Montguyon, on 21 May 1885, at the age of 46, without known descendants.

Then, a second daughter came, Anne Agathe **(9)**, on 3 February 1840 (her surname and his father's are written Scholesky on the birth certificate). She was a milliner, then a mercer and, on 10 September 1860, she married François Pierre Laroche (21 October 1839 - 29 April 1890), bootmaker in Montguyon, born and died in this village. Let us mention that, on some documents, she signed Élodie Cholesky, wife Laroche, and she was probably called by this name in her family, which does not facilitate genealogical researches. But, at that time, it was a common practice to modify one's first name. Anne Agathe Laroche died in Montguyon on 26 January 1910.

They had five children: Marguerite **(26)** (born on 16 January 1864), married to Louis Lalande, a house–painter, born on 3 December 1854 in Guîtres in the department of Gironde, widower of Thérèse Godechaud (deceased on 23 March 1887), in Montguyon, on 7 January 1888; a second daughter, also named Marguerite **(27)** (Montguyon, 24 November 1866 – id., 12 July 1924) who was a laundress and married, on 12 January 1889, Jean Auguste Isidore Ceyral, born on 26 July 1860 in Cercoux in the department of Charente–Inférieure, ancestor of a well–known family in Montguyon; Pauline **(28)**, born on 14 December 1868 in Montguyon, a milliner, who married, on 8 April 1890, Jean Alexis Paul, born from an unknown father, on 6 October 1864 in Escala in the department of Hautes–Pyrénées, school teacher in La Billette, in the commune of Saint–Martin–d'Ary. It is he who purchased, at the departure of the Ursulines nuns of the Sacred–Heart in Pons in 1905, half of the large building which is the convent of Montguyon. Pauline died in Montguyon on 17 January 1957; Amélie **(29)**, born on 11 August 1873 in Montguyon, married, on 25 October 1902, Antoine Marie Léotade Tresseres, constable in Pessac, born on 22 October 1873.

These four sisters, the daughters of Anne Agathe Cholesky, lived together in a building which now belongs to Mr. Largeteau, a notary, in the center of Montguyon. This is why its facade bears the inscription *Aux quatre sœurs* which could still be seen, at some hours of the day, by standing at the right place in the street. They were running a haberdashery.

The last child of the family Laroche seems to be François André **(30)**, born on 6 Mai 1876 in Montguyon, and who died there on 30 May 1900.

The fourth child of Louis Cholesky and Anne Amélie Moreau is Augustin **(10)**, born on 25 May 1841 and deceased on 29 May 1843 in Montguyon. On his death certificate, his name is spelled Scholesky.

The next child is André Cholesky **(11)**, who was born on 8 October 1842. He is the father of our hero André–Louis **(18)**. He married Marie Garnier, also called *Lovely*, on 4 March (or on 11 November) 1872, in La Roche–Chalais, district of Saint–Aulaye, a small city of 2,300 inhabitants in the department of Dordogne. He has then no occupation and his father, Louis, is mentioned in the wedding certificate as an innkeeper in Montguyon (around 1,500 inhabitants at this time), whereas in all other settings he is presented as a master shoemaker. Anyway, in 1874, he became an innkeeper in Montguyon, at the *Hôtel de l'Étoile et du Café du Centre* on the town hall square, opposite to the covered market. They will have ten children.

Then came Isidore **(12)**, born on 3 April 1845 and deceased on 27 April of the same year, always in Montguyon.

Their last child is Paulin **(13)**, born on 30 May 1849 and died on 3 November 1862 in Montguyon.

Here is a story about André (the father of André–Louis), which can be found in the deliberations of the City Council of Montguyon:

> *The mayor informed the Council of the difficulties raised by Mr. Cholesky, regarding the construction of toilets in the garden of the building Lafargue, belonging to the municipality of Montguyon.*

The building *Aux quatre sœurs*
© Public domain

Mr. Cholesky claims that a part of the impasse behind his building with the main facade on the town hall square belongs to him.

However, he recognizes that he must give the right of way at all times and in any case, but that the existence of public toilets that the city will build, will worsen the right of way, as soon as the contractor would make the work to be performed.

In this case the mayor asks what he should do.

After consideration of the matter, the Council expressed the view that the city, owner of the building Lafargue, has the indisputable right to build the building that suits it to build, letting the workers and vehicles

André Cholesky, father of André–Louis

go by the ordinary way that leads to the stables and the bondage buildings of the building Lafargue, as, from an immemorial time, the owners of this building practiced at any time and in any case.

If Mr. Cholesky were to impede the work in any way, the Council decides that the mayor will bring a legal action to obtain the path that leads to the buildings of the city completely free.

The mayor said that, for this purpose, he will contact Mr. Filhol, lawyer in Jonzac.

Marie Garnier, mother of André–Louis
© Cholesky's family

2.2 The siblings

What about the ten children of André Cholesky **(11)** and Marie Garnier, all born in Montguyon?

The first child is Onésime **(16)** who was born on 8 August 1873 in the newly constructed Hôtel de l'Étoile, and died on 7 September 1874 in La Roche–Chalais.

The eldest daughter Marie–Suzanne **(17)** was born on 14 June 1874, and died when she was seven months old, on 12 January 1875 in La Roche–Chalais.

The third child is the one who interests us. André–Louis **(18)** was born on Friday 15 October 1875, at 1 pm, at the Hôtel de l'Étoile.

The *Hôtel du Cheval Blanc*
© Public domain

The fourth child of the couple was Sara Andrée **(19)**, born on 26 August 1877 at the Hôtel de l'Étoile. She married Venceslas Handelsmann Bronislawski (born in 1863), a medical doctor, on 27 December 1900. This doctor defended his thesis in 1900 in Bordeaux, with the title *Contribution à l'étude de l'amusie et de la localisation des centres musicaux* (that is *Contribution to the study of amusia and the location of the musical centers*), published by G. Gounouilhou. They had a son, Vitold, and a daughter, Yadia. Sara Andrée died in the 16th district of Paris, on 9 January 1959.

Then, Marie Maurice **(20)** was born on 5 January 1879. In 1903, he was living in Paris and was working as an electrical engineer. On 25 April 1923, in Paris, he married Lucienne Marguerite David, born in the 4th district of Paris on 10 April 1894. During World War I, she will serve as a nurse and will be decorated with the Legion of Honor. They had no children. After the years 1940, they lived until their death in Saint–Martin–d'Ary, near Montguyon, at La Billette. Marie Maurice died there on 31 August 1959. The couple is buried in the cemetery of Montguyon. Some residents of the city still remember him and his wife.

The next child is Onésime Louis **(21)** who was born on 2 February 1881 in Montguyon. Married to Pauline Blum, on 1 August 1905, he had a daughter Arlette **(36)** who married Commandant Jacques Peyron. They had three children, Yveline, Jean Louis and Jean Marc **(48)**. He died on 31 August 1913 in Leysin, in the canton of Vaud in Switzerland.

The next child is Ernest Léon **(22)**, born on 6 January 1883 in Montguyon. On bad terms with his father who wanted him to stay home to manage the family properties, he did not graduate like his brothers. He married Yvonne Fernande Laure Bouchez, on 18 July 1907 in Paris where they lived, 2 Avenue Bugeaud, in the 16th district (according to [193], Yvonne was born in 1888 in Bondy, and died in Iran in 1966). They had two daughters, Christiane Maximilienne **(37)** and Yveline **(38)**. He enlisted in the artillery, where he served for three or four years as a sergent (sergeant). But he was discharged on the eve of the Great War because of a shadow in the lung. Between the two wars, he made his fortune, first in the pharmaceutical industry, then in the food industry and salt provisions. He was one of the founders and managers of the famous French company GÉO. He was described by his grandson [193] as tall, bald and vaulted, with a mustache yellowed by tobacco. In 1952, he went to Iran with his wife to join their two daughters. He died in Tehran on 14 May 1957. First buried there, his ashes were then transported to the cemetery of Colombes, near Paris, where he rests in the family vault of his wife.

The central square of Montguyon
© Public domain

Then, Jean Alfred **(23)** was born on 3 April 1884 in Montguyon where he died on 18 April 1885.

The next child was Henriette Suzanne **(24)**, born on 12 June 1885 in Montguyon. She was married there, on 20 August 1903, with Jules Maurice Marcel

Heldt, a high school professor, born on 24 January 1876 in Bordeaux. They had two sons, whose youngest Frédéric, a partner of the French writer and film director Marcel Pagnol (Aubagne, 28 February 1895 - Paris, 18 April 1974), married Baroness Brun de Saint Hippolyte, and had a son, Eric Heldt.

The last son, Élie Charles (**25**), was born on 8 September 1886 in Montguyon where he died on 14 May 1889.

2.3 The next generation

André–Louis Cholesky (**18**) got married, on Friday 10 May 1907 in the town hall of La Roche–Chalais, with his first cousin Anne Henriette Brunet, born on 27 June 1882. She was the daughter of François Brunet, 52, owner of a farm, and of Anne Garnier, the sister of Marie Garnier, the mother of André–Louis (**18**). They had two sons, one posthumous, and two daughters: René (**32**) (Montguyon, 16 June 1908 - Paris, 23 May 2000), Françoise (**33**) (Montguyon, 30 November 1909 - Rabat, 17 May 1974), Hélène (**34**) (Montguyon, 4 June 1911 - Bordeaux, 6 July 2003) and André–Louis (**35**) (La Roche–Chalais, 19 January 1919 - Toulon, 7 May 1987).

René (**32**) served in military communications. His name can be found in the list of former students, section deck, of the 10th promotion 1928–1929 on the school ship *Jacques–Cartier* of the merchant navy [3]. He is also mentioned, as a lieutenant–colonel, among the officers of SHAPE [18]. He married Annie Plazy, widow Quiniou. They adopted a girl, Catherine (**40**), who had two children from her husband, a Mr. Daumain, and then divorced.

Françoise (**33**) married Oleg Popoff, a chief physician at the so–called *Hôpital Indigène* in Rabat (this is how were named, at that time, the hospitals where Moroccans were treated). They adopted two children, Anne and Nicolas (**41**). Anne, married a Mr. Meley, and had three sons, Nicolas, Serge and Daniel. Nicolas Popoff, married Marie–France Geisser and had three daughters, Katia, Alexia and Olga.

Hélène (wife Gross) (**34**) had six children. They are: Michel (**42**), born on 25 June 1932 in Bordeaux (it is with him that I classified the archives of his grandfather at *École Polytechnique*), François (**43**), born on 3 January 1935 in Casablanca, Jacqueline (**44**), born on 14 November 1937 in Casablanca, Philippe (**45**), born on 23 August 1940 in Casablanca, and who has ten grand–children, Françoise (**46**), born on 12 June 1948 in Bordeaux, and Christian (**47**), born on 1 January 1952 in Bordeaux. Hélène Gross got the bronze Medal of the French Family which was delivered to her by the former French Prime Minister Jacques Chaban–Delmas (Paris, 7 March 1915 - Paris, 10 November 2000).

André–Louis (**35**), the posthumous son who bears the names of his father (**18**), was born in La Roche–Chalais. Because his father was killed during the war, he was adopted by the Nation after a judgement of the civil court of Ribérac

Henriette Cholesky, the wife of André–Louis
© Cholesky's family

(Dordogne), dated 14 October 1920. He embraced a military career in the Navy and became an officer. He belonged to the last class of students of *École Navale* which hurriedly left Brest on the battleship Jean–Bart to take refuge in Casablanca where it arrived on 22 June 1940. Lieutenant in the Navy, he commanded the submarine *La Créole*, called *La Locomotive*, from 18 January 1954 to 18 July 1955 (http://sous-marin.france.pagesperso-orange.fr/Q193.htm). He was knighted in the Legion of Honor, on 28 December 1954. When he was 49 years old, he married, on 28 December 1968 in Revest–Les–Eaux (department of Var), Anna Browislawa Romanow (53 years old), born in Germany, the daughter of Valentin Romanow, minor, and Antonia Romanow, born Pollednick. She was divorced from Georges Jean Baptiste Secher. By marriage, Mrs. Antonia Cholesky took French nationality. André–Louis Cholesky died at Toulon on 4 (or 7?) May 1987. This death was recorded in La Roche–Chalais on 11 May 1987.

Currently no further descendent bears the surname of Cholesky.

Yveline **(38)**, the youngest daughter of Ernest Léon **(22)**, probably born in Paris, married there in 1937 an eminent iranian ophthalmologist, the professor

Jimmy Farahmand. She will follow him in Iran where they settled. She died in Nogent–sur–Marne.

They had eight children **(51)**: a daughter, Christiane (called Riri), born on 5 December 1937 in Paris, then Philippe, born on 7 October 1941 in Paris, Jean Yves, born on 12 September 1943 in Paris, Jacques, born on 10 September 1945 in Paris, Laure–Lovely, born in Tehran in 1950 and who died there in 1952, Patricia, born in Tehran in 1951, and finally twins, Jean–Marie and Marie–Christine, also born in Tehran in October 1954. Jean–Marie died in Grenoble in 1998.

The fate of the other daughter of Ernest Léon, Christiane Maximilienne **(37)**, is more surprising. She was born in Paris on 18 December 1914. She was a very beautiful woman, red–haired, and, in Deauville, she received the first prize for the nicest legs from the hands of Marlene Dietrich.

Reza Khan Mir Panj (Alasht, 16 March 1878 - Johannesburg, 26 July 1944), known as Reza Shah Pahlavi, reigned as Shah of Iran from 1925 to 1941 when, accused of Germanism, he had to cede the throne to his eldest son Mohammed Reza Pahlavi (Tehran, 26 October 1919 - Cairo, 27 July 1980), who ruled the country until the Islamic Revolution in 1979. The latter, to symbolize the entry into the war of his country against the Axis Powers, sent his younger brother Ali Reza Pahlavi (Tehran, 1 March 1922 - Youch, Iran, 26 October 1954) to join, as a captain, the headquarters of Général Jean Joseph Marie Gabriel de Lattre de Tassigny (Mouilleron–en–Pareds, 2 February 1889 - Paris, 11 January 1952). At a reception at the Iranian Embassy, rue Fortuny in Paris, Christiane Maximilienne, who had accompanied her sister and her husband, Professor Jimmy Farahmand, met the Prince Ali Reza Pahlavi. They fell in love and were married on 20 November 1946 at the town hall of the 16th district of Paris. The Shah opposed the marriage of his brother, but Christiane Maximilienne was pregnant. Their son, Patrick **(50)**, was born on 1 September 1947. She already had another boy, Joachim Christian Philippe Pahlavi **(49)**, born on 15 September 1941 in Paris, the son of a German medical doctor, Joakim Martens (Birkenwerder, north of Berlin, 18 December 1914 - Kartschinskaja/Tschir, west of Stalingrad, 18 January 1943), who had been working in a hospital installed in the house of the Cholesky family in Montguyon during the second World War. Patrick was declared under the family name of Cholesky, but Ali Reza adopted him on 13 November 1946, one week before his marriage with his mother in Paris, and named him Joachim Christian Philippe Pahlavan Nassan Pahlavi–Pahlavi.

Despite the hostility of his family to his wedding, Ali Reza brought his young wife to Iran in 1952. Her position as wife of Prince Ali Reza and mother of the heir to the throne of Iran earned Christiane Maximilienne a very rough life. In Iran, she was called *Princess Pahlavi*. Her parents, Ernest Léon Cholesky **(22)** and his wife, left their apartment near Place Victor Hugo in Paris to follow their daughter to Tehran. Two years after her marriage to the Prince, she was forced to divorce. She died in 1995 in Nice, three days after her 81st birthday. Ali Reza died in a plane accident on 26 October 1954 near Youch in the Alborz mountains.

The Shah, who had married Soraya (Ispahan, 22 June 1932 - Paris, 25 October 2001) and had no child from her, considered for some time nominating Patrick Ali (he obliged him to have Ali as his middle name) as his heir. You can find his autobiography on the Internet [194]. The following is a translation of some parts:

> *My name is Patrick Ali Pahlavi. I came to the world in Paris on 1 September 1947, at quarter past one in the morning. But my story starts before that.*
>
> *In 1941, my grandfather, Reza Shah, was exiled by the British and the Americans, under the pretext of having flirted with Hitler. His son Mohammad–Reza Pahlavi succeeded him as Shah of Iran. The latter, in action of allegiance to the allies, sent his younger brother and heir, Prince Ali–Reza, who would become my father, fight the Nazis, incorporated into the headquarters of Général de Lattre de Tassigny. In 1945, he entered with him in liberated Paris.*
>
> *The Iranian embassy then gave a reception in his honor. Prince Ali Reza Pahlavi appears in the uniform of an officer of the French Army.* Il avait un charme fou *(he had a tremendous charm), told me my mother, who was invited to the reception. A few days later he contacted her. They meet again and fall in love of each other. Their love seems to have been very strong. There are signs that are unmistakable. For example, she had a son from a previous marriage and he owned him, giving him his name even before he married my mother...*
>
> *On 20 November 1946, Christiane Maximilienne Cholesky married Ali–Reza Pahlavi, at the town hall of the 16th district of Paris.*
>
> *The Shah, learning the union of his brother, the heir Prince, with ... this French woman, threatened to cut off funds if he does not come back to Tehran. Never mind, Prince Ali Reza who has a beautiful voice is preparing himself for a career as a singer at the opera. The Shah who is afraid of scandals backtracked.*
>
> *– All right, you married her without my permission, you made me very badly, but I forgive you. I ask you, at least, not to have children.*
>
> *No luck, I'm already training in the womb of my mother. The fact that she is pregnant gets out and ends back to the ears of the sovereign. Then, a bunch of emissaries descends on the apartment of my French grandparents, Léon Cholesky and Yvonne Bouchez, at no. 2 Avenue Bugeaud in Paris. My mother told me later the state of stress and anxiety she experienced throughout her pregnancy and how the fear to see the man she loved go, tortured her...*
>
> *Shah finally wins. Some weeks after my birth, my father, saying he was going out to buy a journal, leaves to Iran, never to return. Four*

years later, however, and no longer containing himself, he wrote to my mother. He asks her to join him incognito in Iran. Thing said is done, my mother takes the passport of a cousin who looks like her, adds the photos of my brother and me, and we went to the faraway East.

Iran, it changes the atmosphere, especially at that time. We live at the bottom of a large shady park, the park Amine Dowlé, and in the evenings, my mother go and join her husband, secretly in the palace.

My French grandparents were with us and their other daughter, aunt Yveline, who is also married to an Iranian and has already a dozen of children, lives with us...

We all lived in this huge house from the Ghadjar time. Colonnade, small windows, roof in iron sheets. In the morning, all the children were playing in the park. There was my brother, Christian, and I, plus all the children of aunt Riri (at that time she was almost an adult and did not play), Philippe, Jean–Yves, Jacquot and Patricia. The twins, Jean–Marie and Marie–Christine, were not yet born...

Our lives went by in a happy and peaceful Iran in peace, where religion was not yet a problem. Everything seemed to smile on us, until in the fall of 1953, when fate decides otherwise.

One morning, I discovered a lot of people in the room of my grandmother, Yvonne, where we usually had breakfast. Men and women I did not take the time to identify. I recognized the father Toulemonde, he was holding in his arms my mother crying. My aunt, mom Yveline, as she was called, then came to me. She took me in her arms as if to comfort me, she also was crying. My father had just been killed in a plane and I was seven...

In addition, if at home it was a drama, the Shah in his palace seemed not at the party. He indeed lost, in the person of his younger brother, his Crown Prince. However, the Iranian constitution imposed on him to have a permanently designated heir, so that if anything happened to him, the Pahlavi dynasty remains. The problem that confronted the king was that his half–brothers having the Ghadjar blood (previous dynasty) could not constitutionally reign. The Shah had either to repudiate Soraya who could not give him a child, or to abdicate. But the Shah was very much in love with Soraya, and he relied on the progress of medicine for seeing her, one day, give birth. What to do?

It is the Minister of the Court, Mr. Alâ, who blew out the solution.

– Your Majesty, do not worry, there is a remedy. The son of your late brother can reign.

– What? The little French? But we do not even know where he is, or whether his mother will agree!

– Your Majesty, the Constitution only prohibits the Ghadjar blood, remember - you, your first wife, Fowziah, was not Iranian. As for the child, he is here in Tehran. His mother hides with him, but we know where, and she will be much obliged to obey us.

We stop here for a moment to savor the irony of fate. This child, born from a French woman, and that the Shah forbade his brother to procreate seven years ago, today he pulled him away from his mother to save both his throne and his love...

The problem of succession was eventually submitted to a council of wise men who proposed that the Shah take a second wife, which was allowed by Islamic law. But Soraya refused the idea, and the couple divorced in 1958 for reasons of State. The following year, the Shah married Farah Diba (born in Tehran, 14 October 1938) which will give him four children and an eldest son, Reza Pahlavi, born 31 October 1960 in Tehran, Imperial Prince, a birth that moved Patrick Ali away from the throne. Thus, no descendant of Cholesky family reigned over Iran.

However, Reza Pahlavi having committed suicide in Boston on 4 January 2011, and having only daughters, Patrick Ali is back as the heir of the throne of Iran, fifty years later.

Patrick Ali married Sonja Laumann in 1972, and they had three sons **(53)**, their Highnesses the Imperial Princes Davoud Pahlavi, born on 7 July (or October) 1972, Hud Pahlavi, born on 26 November 1973, and Mohamed Younes Pahlavi, born on 19 May 1976 (see http://fr.wikipedia.org/wiki/Ali_Patrick_Pahlavi).

Joachim Christian Philippe Pahlavi **(49)**, the first son of Christiane Maximilienne **(37)** and half–brother of Patrick Ali, followed her mother in Iran. He married Jana Broz (born in 1938 in Moscow) and they had 4 sons **(52)**, Xavier, Cyril, Axel, and Niels. He wrote a biography [193] where many details on this branch of the family are given.

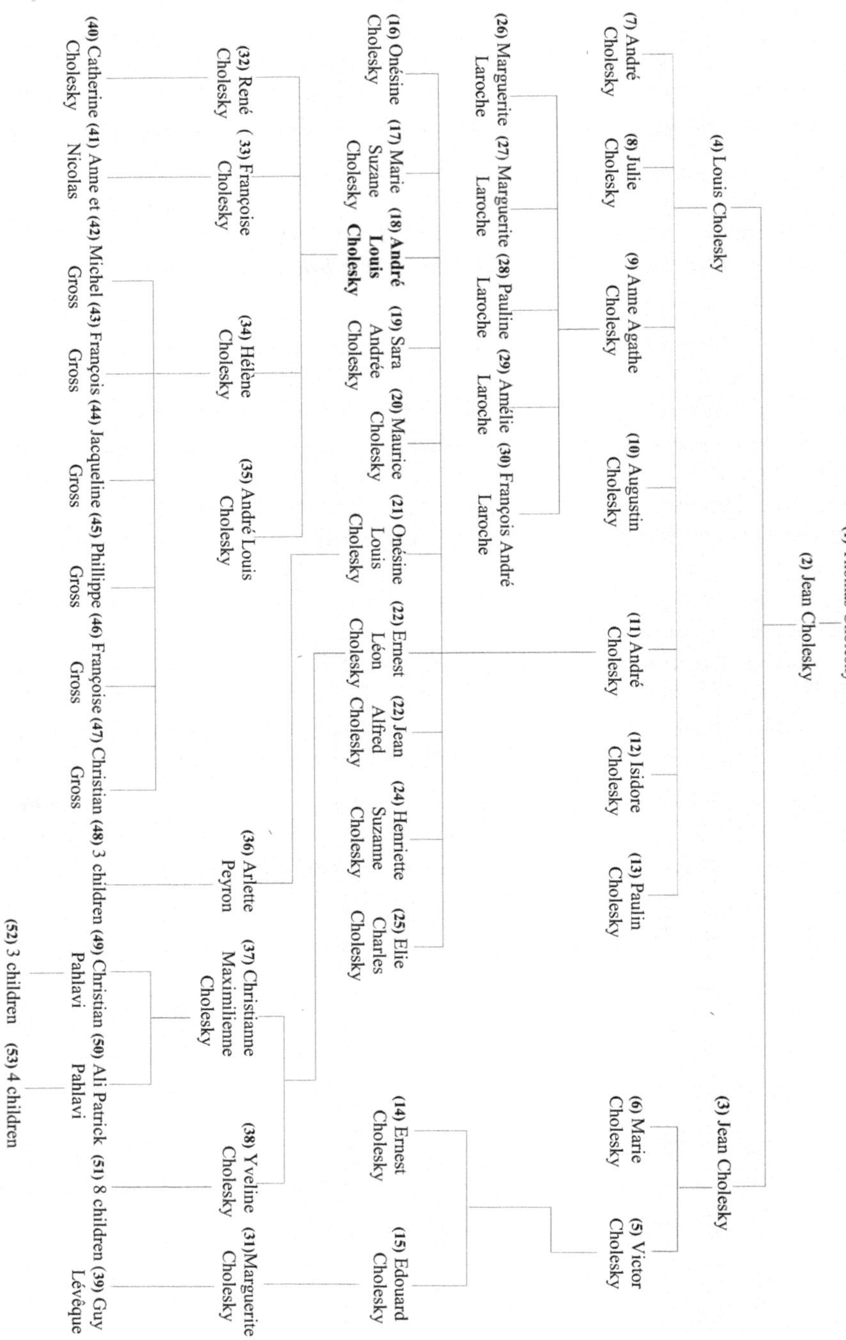
(1) Thomas Cholesky
(2) Jean Cholesky
(4) Louis Cholesky
(3) Jean Cholesky
(7) André Cholesky
(8) Julie Cholesky
(9) Anne Agathe Cholesky
(10) Augustin Cholesky
(11) André Cholesky
(12) Isidore Cholesky
(13) Paulin Cholesky
(6) Marie Cholesky
(5) Victor Cholesky
(26) Marguerite Laroche
(27) Marguerite Laroche
(28) Pauline Laroche
(29) Amélie Laroche
(30) François André Laroche
(16) Onésine Cholesky
(17) Marie Suzane Cholesky
(18) André Louis Cholesky
(19) Sara Andrée Cholesky
(20) Maurice Cholesky
(21) Onésine Louis Cholesky
(22) Ernest Léon Cholesky
(22) Jean Alfred Cholesky
(24) Henriette Suzanne Cholesky
(25) Elie Charles Cholesky
(14) Ernest Cholesky
(15) Edouard Cholesky
(32) René Cholesky
(33) Françoise Cholesky
(34) Hélène Cholesky
(35) André Louis Cholesky
(36) Arlette Peyron
(37) Christianne Maximilienne Cholesky
(38) Yveline Cholesky
(31)Marguerite Cholesky
(40) Catherine Cholesky
(41) Anne et Nicolas
(42) Michel Gross
(43) François Gross
(44) Jacqueline Gross
(45) Phillippe Gross
(46) Françoise Gross
(47) Christian Gross
(48) 3 children
(49) Christian Pahlavi
(50) Ali Patrick Pahlavi
(51) 8 children
(39) Guy Lévêque
(52) 3 children
(53) 4 children

Chapter 3

On topography

In this chapter, we will give some notions of topography and describe the work that a topographer has to perform on the ground. For more details, see [57] and the many other books that exist.

Topography is a difficult topic since it consists in representing on a plane sheet of paper a part of a surface that is not plane. The Earth is not a sphere but roughly an ellipsoid flattened at the poles. The vertical at each point is defined by a plumb bob. It is the only absolute geometrical data that is independent of the neighboring points. The point where the vertical meets the celestial sphere is the zenith, and the plane perpendicular to it defines the horizontal. It is given by the surface of a liquid at rest. The angle between the horizontal and a celestial object is called its *azimuth.* The Earth's theoretical surface should be, at each point, perpendicular to the vertical. But there are infinitely many such surfaces, parallel to each other. Thus, the one closest to the level of the sea along the coastline is chosen. But, the level of the sea changes with time. Its mean level has first to be determined. But, this mean level depends on its location on the Earth. All seas and oceans do not have the same mean level. Thus, an origin has to be chosen and the level is defined step by step by each vertical. The level surface obtained in that way, which is as close as possible to the surface of the sea, is called the geoid. It is the equipotential surface of the rotating gravity field. It has roughly the shape of an ellipsoid. In 2003, during construction of the bridge of Laufenburg across the Rhine river, a height difference of 54 centimeters was found between the bridge built from the Swiss side and the roadway of the German side. The reason of the error was that Germans refer to the sea level of the North Sea and the Swiss to the Mediterranean Sea. These few explanations are sufficient to show the difficulties of the problem.

The work of a topographer consists in taking measurements on the surface of the Earth for establishing maps. A map is governed by its scale which is the ratio between the distances between two points on the map and their exact distances on the Earth. For representing a large portion of its surface, the shape of the Earth has to be taken into account while, for a small portion of it, the Earth is approximated by its tangential plane at a certain point. But, anyway, the Earth is represented on a plane surface which means that it has to be projected on it, each point of it corresponding to a point on the map. There exist several types of projections according to the projecting surface (a cone or a cylinder for example), and bearing the names of Mercator, Lambert, Bonne, etc. For more details, see [57].

3.1 Triangulation

The method of *triangulation* is used for positioning points on a map and establishing it. It seems difficult to know exactly who devised it. Thanks to the compass, invented by the Chinese and transmitted to Europe by the Arabs, sailors were able to deviate from the coast and keep their course. Knowing the direction and the distance of travel (thanks to the log, a hemp rope weighted, graded by nodes and which is allowed to spin), it was possible to obtain maps of the coasts.

In the second edition of his book *Cosmograficus liber Petri Appiani (Apianus)*, Gemma Frisius (Dokkum, Netherlands, 9 December 1508 - Leuven, Belgium, 25 May 1555) gives an appendix *Libellus de locorum describendorum ratione* of 16 pages where he founded modern geodesy and explains how networks of triangles allow us to survey areas as large as desired. Four years later, he describes how to build an instrument, derived from the astrolabe, for measurements on the ground, the goniometer. His nephew, Gualterus Arsenius (Clèves ?, ca. 1530 - Leuven, ca. 1580), will improve it in 1572, by adding a compass. Triangulation was perfected in 1556 by the Italian mathematician Niccolò Tartaglia (Brescia, ca. 1499 - Venice, 14 December 1557), and used by Jacob Roelofs, called Jacob van Deventer (Kampen or Deventer, ca. 1500-1505 - Cologne, May 1575) for mapping the Netherlands between 1557 and 1573. It is thought that the Danish astronomer Tycho Brahé (Knudstrup, 14 December 1546 - Prag, 24 October 1601) used it in 1578 to connect the island of Hven, where he maintained his observatory of Uranienborg, to the coast of Denmark. Johannes Kepler (Weil der Stadt, Wurtemberg, 27 December 1571 - Regensburg, 15 November 1630) also used it. Triangulation was popularized by the Dutchman Willebrord Snell Van Royen (Leiden, 13 June 1580 - Leiden, 30 October 1626), called Snellius. In 1621, he measured by triangulation the arc of $1°11'$ from Berg–op–Zoom to Alkmaar with 33 triangles. The 1,230 meters long baseline was measured with a land–chain and verified by a ruler in wood. He observed the azimuth in Leiden. He obtained 10,004 km for the quarter of the terrestrial meridian. Triangulation required a complete mastery of angular measures. It really became known only at the beginning of the 16th century with the arrival of the plane table in England and Germany.

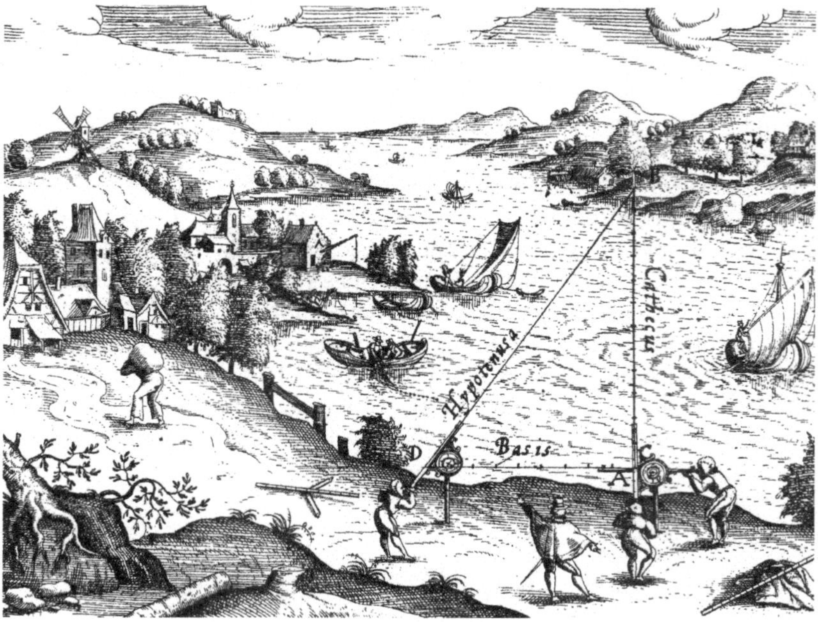

Triangulation (from a treatise by Levinus Hulsius (1546–1606))
© Public domain

By triangulation, the region to be mapped is covered by a chain of adjacent triangles. For refinement, one can use networks of smaller and smaller triangles. One begins by measuring the length of a side of the first triangle (called the *baseline*) and then, standing at each corner of the adjacent triangles, to measure only angles. It is easier and less subject to errors to quantify angles than distances. Vertical angles have also to be determined for bringing the map to a horizontal plane, an operation called *leveling*. Then, the usual trigonometric formulae give the lengths of the sides of all triangles (see Appendix A). For short distances, the errors due to the shape of the Earth are small, and formulae due to Adrien Marie Legendre (Paris, 18 September 1752 - Paris, 10 January 1833) allow one to easily correct them. For long distances spherical trigonometry has to be used. Thus, the topographer obtains a network of adjacent triangles whose dimensions and angles are known. Obviously, the three angles of each triangle (instead of two) could be measured for safety, and a second baseline also, thus leading to more equations than unknowns. But one has to take into account the various sources of errors. This is called *adjustment*. The adjustment of geodetic networks is due to Carl Friedrich Gauss (Braunschweig, 30 April 1777 - Göttingen, 23 February 1855), and it will be explained more precisely in Chapter 4.

There are two types of leveling: direct and indirect. In direct leveling (also called geometrical or precise leveling), vertical distances are measured directly by using leveling instruments. Elevation is carried out from a known point to an unknown one. In this method, a horizontal sight is taken on a graduated staff and the difference in the elevation between the line of sight and the ground on which the staff is held is measured. Knowing the height of the line of sight, the difference in the elevations of the station and the staff is obtained. On the other hand, indirect leveling consists in measuring, with a theodolite, the angle between the horizontal or the vertical and the point whose elevation has to be obtained, and then to use trigonometrical relations. This technique is particularly adapted to rough, uneven terrain where direct leveling is impracticable or too time consuming. Barometric leveling can also be used, as it was, for example, by Horace–Bénédict de Saussure (Conches, Switzerland, 17 February 1740 - Conches, 22 January 1799) during his famous ascension of Mont Blanc on 3 August 1787. Direct leveling is the most precise procedure.

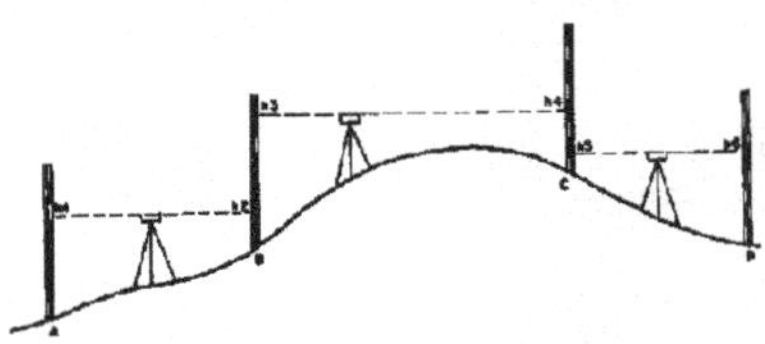

Direct leveling
© Public domain

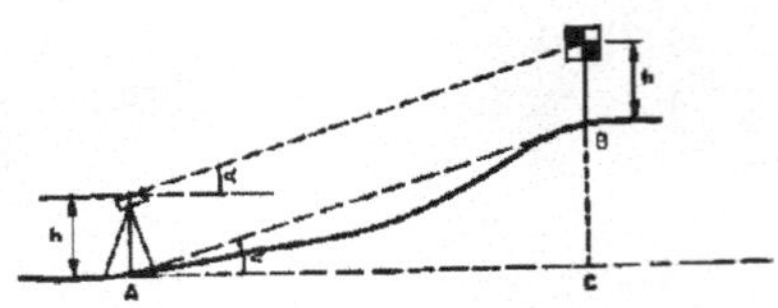

Indirect leveling
© Public domain

When a region has been triangulated, it has to be connected with the triangulation of its neighboring regions. A point of the triangulation is chosen as the fundamental point. By astronomical measurements, its longitude, its latitude, and the direction of the meridian are determined. Then, a direction of reference is deduced by measuring angles. The astronomical coordinates of the fundamental point allow us to define the position of the ellipsoid in space: it is tangent to the geoid, which means that the normal (the perpendicular) to the ellipsoid at this point coincides with the vertical, and that the meridian of the ellipsoid coincides with the astronomical meridian. But it must be clearly understood that this position in space is only a local one, and that points common to two triangulations could then have slightly different coordinates in each of them. Within the same network, there might be differences between the geodetic and the astronomical coordinates of the points. Only the astronomical coordinates are absolute, the geodetic coordinates are only relative. As we already explained, these differences are due to the deviation of the vertical which is the angle between the vertical and the normal to the ellipsoid.

The work of the surveyor on the ground is divided into triangulation and leveling. It is then necessary to transfer the measurements between charts or other drawings which are not at the same scale in order to prepare the production of the map itself.

First, it is necessary to choose which primary geodetic control points will be used for the triangulation. They have to be easily identified such as summits, towers or churches. Some of them could be inaccessible for installing instruments. The points obtained by triangulation are plotted on a *grid chart* and form a *network* (different levels of networks are distinguished, from 1 to 4, depending on their approximation). For measuring the length of the baseline, that is the side of the first triangle, its two ends must be, as much as possible, located on a flat and smooth piece of ground. For this, rules in wood and then surveyor chains in metal were used. Later, more specific processes were introduced, like those of Jean–Baptiste Joseph Delambre (Amiens, 19 September - Paris, 19 August 1822) and Pierre François André Méchain (Laon, 16 August 1744 - Castellón de la Plana, Spain, 20 September 1804) who were involved in the measurement of the Paris meridian during the French revolution in order to define the length of the meter. The points of the triangulation must be distributed as homogeneously as possible in order that, firstly, they can be used to build a finer network, and, on the other hand, to facilitate the verification of the measures.

After measuring the base, the angles between the various points of the network are measured using an alidade or a goniometer installed on a plane table. The plane table, or goniograph, is a key instrument in topography. It is a wooden board mounted on a tripod, so that it can rotate around a vertical axis, on which a straightedge and an instrument (the *alidade*) for measuring angles are fixed. The alidade is a sight instrument used to aim and draw the directions. It was invented by Archimedes in the 3rd century B.C. It consists of a rule with two vanes which pivots on a graduated circle and is mounted on the plane table. The alidade is directed toward the points to be positioned on the framework, and pencil rays are drawn along the straightedge toward them. The procedure is then repeated from another point, and the points where the rays intersect on the plane table give the positions of the various sites on the map. In the holometric alidade, invented in 1667 by the French Adrien Auzout (Rouen, 28 January 1622 - Rome, 23 May 1691) and Jean Picard (La Flèche, 21 July 1620 - Paris, 12 October 1682), this is performed with a telescope equipped with a clinometer. It is used for surveys of medium and large scales. These instruments require a set of delicate adjustments to ensure a maximum of accuracy, in particular the horizontality of the plane table has to be very carefully checked. All these instruments are extensively described by Cholesky in his various writings.

From the length of a side and its two adjacent angles, trigonometry tells us that it is possible to calculate the length of the two other sides of the triangle and the third angle (see Appendix A). Gradually, we thus obtain the angles and sides of all triangles. To reduce the uncertainties, too acute angles or too obtuse ones have to be avoided because a small error in their value causes a significant error on the length of the sides. The most favorable triangles are those which are nearly

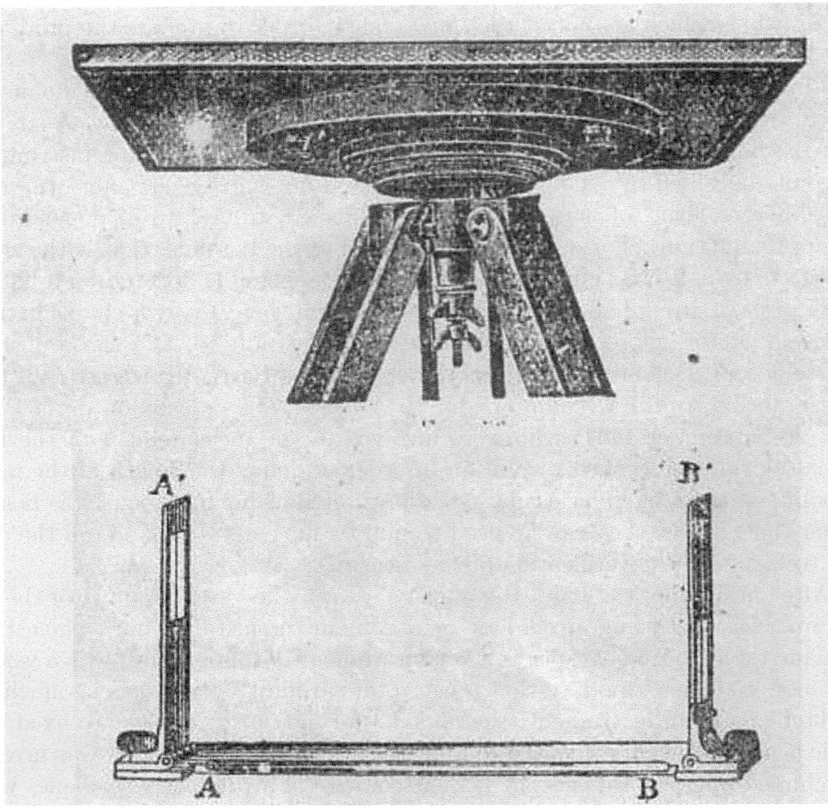

Plane table and alidade

equilateral. The density of the points has also to be uniform for the homogeneity of the network.

To verify the measures, several baselines are often used. The differences between their lengths measured and calculated are then used to evaluate the accuracy of the work, and they help to minimize the errors.

In each triangle, it is often possible to measure more than two angles and the length of one side. Thus, instrumental and measurement errors can be minimized and the accuracy is increased. In this way, we obtain overabundant measures. If the triangles subject to these measures do not close, the point that best fits all the measures is chosen for each vertex, and the corrections to be made on the angles measured are determined by computation. As we already mentioned, this operation is the adjustment of the network . Pierre Simon de Laplace (Beaumont–en–Auge, 23 March 1749 - Paris, 5 March 1827) was the first to submit these questions to a rigorous mathematical analysis in his *Théorie analytique des probabilités* (Analytic theory of probabilities), a work published in 1812. Then, it was Gauss who, in 1826, showed how to compensate networks. By adjustment, a linear system with more equations than unknowns (or reversely according to the unknowns chosen) is obtained. It is solved by the least squares method which leads to a system of linear equations whose solution can be obtained by the method Cholesky invented for this purpose (see Chapter 4).

The concern of geodesists and, more generally, surveyors is to obtain by adjustment, a geometric network that is very close to the observed one with only very slight deformations, and with unique coordinates for each point. The method of least squares is aimed at determining a final network which minimizes the sum of the squares of the corrections brought to the observations. It consists of several steps: calculation of the approximate coordinates of the points of the network, writing of the observation relations or of the condition equations, computation of the coefficients of the normal equations, and solution of the symmetric system by a numerical method. The condition equations are based on the mathematical relationships between the elements of the triangles. They have the disadvantage of not being linear, except those concerning the closing at 180° of the triangles. They are only suitable for simple cases because they are difficult to write down. The only condition equations which are easy to write are, in altimetry, those related to the closure to zero of the loops of direct leveling.

The observation relations corresponding to sights, measures of distances, and azimuthal determinations, are relatively easy to settle. They do not need the measure of all the vertices of the triangles. Each observation leads to a relation of the form

$$\textit{residual} = \textit{approximate value} + \textit{corrections} - \textit{observed value}.$$

The unknowns are the corrections and the residuals. They are lengths in the meaning of a *dimensional formula*, and they can be expressed graphically.

For a network of triangulation, we obtain n observation relations linking p unknowns which are the corrections to the approximate values of the initial unknowns (coordinates of the new points, constants of orientation in the closure of the horizon...), and n unknown residuals. Thus, a linear system of n observation relations with $p + n$ unknowns, with $p > n$, is obtained. This system admits infinitely many solutions. Its solution by the least squares method consists in imposing that the sum of the squares of the residuals of the observation relations be minimum, which gives the p equations which are missing to the system formed by the n observation relations. This system of $p + n$ equations, called the *normal equations*, is symmetric (and positive definite, a notion that Cholesky does not seem to have been aware of) and it has a unique solution. Its computation is the last step in the treatment of adjustments.

Following this procedure, the Belgian cartographer of French origin Guillaume Adolphe Nerenburger (Amsterdam, 23 April 1804 - Brussels, 19 March 1869) solved a system of 60 equations in 22 unknowns. Friedrich Wilhelm Bessel (Minden, 22 July 1784 - Königsberg, 17 March 1846) treated a system of 31 equations with 70 unknowns, and Johann Jakob von Baeyer (Müggelsheim, near Berlin, 5 November 1794 - Berlin, 10 September 1885) 47, and then 86 condition equations. For solving such systems, geodesists and mathematicians began to seek for effective methods, fast and easy to use without having an advanced training in mathematics. After a number of attempts, the method of solution that imposed itself was that of André–Louis Cholesky invented around 1903.

All these measurements lead to a network of points both on the map, where their positions are plotted by their coordinates, and on the ground, where they are indicated by various benchmarks such as pillars, concrete ground marks, bolts let into the soil, or wooden pegs. If a more accurate map is needed, the network has to be refined by measuring smaller triangles inside the primary network. From the first measured points (50 km from one another), an internal triangulation, called of second order (20 - 30 km), is established, then a triangulation of the third order (5 - 10 km), and so on until the desired accuracy for the details of the map has been achieved.

Leveling benchmark on the town hall of Montguyon
© Claude Brezinski

As the vertices of the triangles are not located at the same height, the triangles are tilted. It is then necessary to reorient them horizontally by measuring the angle made by each side with the vertical. This is the *leveling* which can be performed using zenithal measures.

3.2 Leveling

Leveling (or altimetry) consists in measuring the elevation of points with respect to a horizontal surface taken as reference. Often, it is the geoid in order to be able to draw level curves , also called contour lines. In fact, only the difference of elevation between two points is measured. These points are usually identified by rods located on each side of a sight device made perfectly horizontal by using a water level. The graduated vertical rods are observed through a telescope. By moving the telescope and the rods and repeating the procedure, the differences in height can be accurately measured. The operation can only be conducted on a small distance between each point (since an uneven terrain and/or the vegetation that may interfere with the target have to be taken into account), and it must be repeated as many times as necessary. The topographer has therefore to pay a particular attention to the accuracy of each sight and to the instrumental errors, and even to the aerial refraction of the light and other physical parameters that can affect the results. In the case of the most precise work, the error may be kept to less than a centimeter over a distance of 100 kilometers. To achieve this accuracy great care has to be taken.

But such measures do not provide the altitude above the geoid because the sum of the height differences depends on the path followed. For finding this altitude, it is necessary to appeal to the notion of *potential* of the gravitational field, and to perform the sum of the products of the elevation differences by the value of gravity starting from a point of zero altitude, located in France at the tide gauge in Marseille. As we will see, leveling was improved by Cholesky.

Leveling goes back to antiquity. The plumb bob is indispensable. The Old Testament says that, faced with an attack threat of the Assyrian King Sennacherib (reigned 705–680 B.C.), King Hezekiah of Judah (741 B.C. - 688 B.C.) decided to block the water sources that were outside the city, and to construct a reservoir and an aqueduct to bring water into it. This aqueduct was entirely underground and it connected the source of the Gihon to the reservoir of Siloam, built in 701 B.C. in Jerusalem, to supply the city with water in the case of a siege. S–shaped, it was 512 meters long for a difference of level of 10 meters. Circa 550 B.C., the Greek architect Eupalinos of Megara built a tunnel 1036 meters long to bring water from the mountains to the city of Samos. Two teams of two men advanced one to the other and the feat is that they met. The Romans built many aqueducts as well as other sewers, thus proving their mastery of technical leveling.

The water level long remained the preferred instrument, if not the unique one, for leveling. It consists of two vertical transparent open tubes, connected by a rigid horizontal tube. It is filled with a colored liquid and the line of two outcrops indicates the horizontal. One can also use the spirit level which is a cylindrical glass tube, very slightly curved. It contains a liquid that does not fill it completely, leaving a visible air bubble which comes exactly between two marks on the graduated glass when it is perfectly horizontal. Other types of levels also exist.

Once the general plan has been obtained, before drawing the map, one has to carry back to the offices the measures (angles and lengths) obtained on the ground on a sheet of paper, according to the scale of the map. Then, the map has to be checked. The best procedure is to conduct a new survey. But, since the time is doubled, only the most important points are verified and this is usually performed simultaneously to the first survey. If there is an error, its cause must be immediately found. One cannot rely on the accuracy of a map if these checks have not been performed.

Obviously, the methods of measurement have greatly evolved with technical developments. They no longer have to be performed manually, and they are now almost all automated. However, Cholesky's method is always the method of choice to solve on a computer the systems of linear equations which are obtained.

3.3 Cholesky's double–run leveling

Cholesky is well known among topographers for a leveling method of his own named *double–run leveling* (*méthode de double cheminement*). Let us quote the report (35 pages) [16] written by Cholesky on the surveying campaign 1910–1913 in Algeria and Tunisia since it contains information on this new leveling procedure:

> p.4 - *In May 1912, the Capitaine Cholesky was ordered to study a leveling process which allowed to carry out the leveling in Morocco faster than those in Algeria and Tunisia, while maintaining a sufficient accuracy so that the results could be immediately be exploited in the study of railways and also to possibly enter into the general scheme of all the railway lines to be later leveled in Morocco.*
>
> *The method and the general working conditions were first studied in the office, and then tested on the ground at the polygon in Vincennes by the administration officers Marchal and Legay and the non–commissioned officers Chassagne and Busquet made available at Cholesky's disposal. These four operators, then trained, left for Casablanca in early July 1912, and worked in Morocco until January 1913 under Commandant Perret, head of the operations of the reconnaissance map at 1:200,000th of Morocco...*
>
> *The working method used is not significantly different from that adopted by the Service du Nivellement général de la France (General leveling Service of France) (1).*

Each leveling is measured twice, as much as possible by two different operators, the two operations being separated by a reversal of the telescope on its collar and of the level of the telescope. Each level line of the first fundamental order is measured twice, in opposite directions and, as much as possible, in different working sessions.

(1) See, for the working method that has been discussed above, the special instructions written for the brigades of Morocco in June 1912. The results appeared to be satisfactory. However it would be premature to give a final conclusion, no polygon having been closed with this way of working.
The question would deserve to be studied in detail, it would allow to choose between the traditional method used so far and this new method which seems to allow a better use of the staff available to the Geographical Service while maintaining a comparable, or even better, accuracy with a significant saving.

It seems that Cholesky introduced this new procedure in 1910. It is still taught today (see, for example, [117] and [134]). As he explained, it consists in conducting simultaneously two separate survey traverses, very close to each other, by placing the leveling rod successively at two distinct points located behind the station and then at two points in front of it, and so on. Then, the two traverses are calculated separately and the results obtained are compared. Its purpose is to limit the effects of some instrumental defects. For completeness, it is reported that, to distinguish the two rod–men moving in parallel, one of them has to wear red trousers so that the operator could not be confused!

A few pages later (p. 29), he wrote:

Determination of the errors - Adjustment - *The calculations for determining the accidental and systematic errors are in progress, as the whole adjustment of the fundamental network of Tunisia, in which only a part of the section H_1N, in front of the train station of Sers to Graseria by Maktar and Kairouan, is missing.*

A comprehensive report will be presented later on the conduct of these calculations and the results obtained.

Was this report about the method for solving systems of linear equations that he had invented some years earlier?

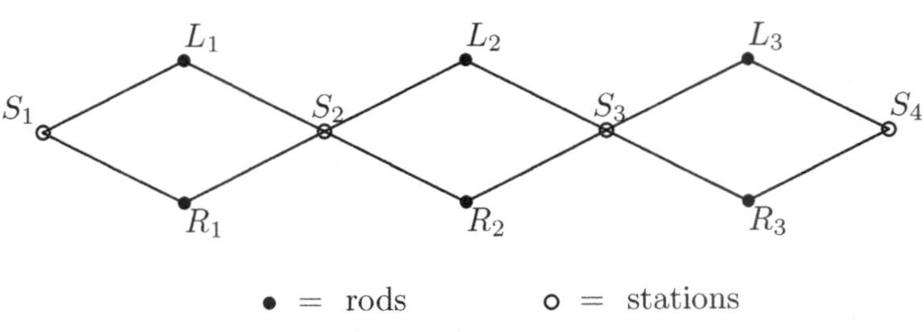

Double–run leveling

It is in his book *Levés d'Études à la Planchette* (pp. 124–126) that Cholesky presented his method of double–run leveling. Let us give a rough translation of this part of the text:

> *The points of the grid are usually at a distance of 100 to 150 meters. If the difference of level between the two consecutive positions of the plane table is not too substantial, it can be determined by using only one station of the leveling instrument placed approximately at equal distance of the two extremities of the grid. We can then operate as follows.*
>
> *The operations of planimetry being completed at point A, an assistant goes to the next station with the plane table and the rod. The operator with the assistant carrying the leveling instrument (or carrying it himself) walks in the direction of point B until where he wants to install it. During this time, the assistant remained at point A installs his rod there for the leveling measurement.*
>
> *The operator installs, as quickly as possible, the leveling instrument... Then, he sights successively the rods placed in A and B.*
>
> *This operation allows to determine not only the difference of level between the points A and B, but also the plane of the leveling instrument, that is the elevation of the horizontal plane passing by the optical axis of the telescope...*
>
> *During this time, the assistant who was carrying the plane table to the point B and who had laid it nearby in order to go with his rod exactly on the leveling plate of the point B of the grid, can install his rod and the plane table or, at least, to rough the installation.*

To end this chapter, let us give a translation of [134, pp. 308–309] to show that this method is still explained nowadays:

> *Two distinct traverses are simultaneously carried on by placing the rod consecutively in two neighboring but distinct points B and B', then in two points C and C' and so.*
>
> *The traverse $ABCDR$ can be calculated, on the one hand, and, on the other hand, the traverse $AB'C'D'R$ and thus compare the two elevations for the benchmark R; if there is a slight discrepancy, R will be taken as the average of the two elevations obtained.*
>
> *To perform this type of traverse, two rod–men are generally used, each with a rod that they place successively at the points B, B' - C, C', etc.*

It is added later in this book that the same technique can be used for round–trip traverses.

In a paper dated 2008 [183], we can read that:

> *The double–rodded level circuit is a time–saving leveling technique that Carl Parsons, a party chief from West Virginia, described to me many years ago. Since that time, I have been in many suitable situations to use the technique and am now happy to describe it to other surveyors for their benefit. In doing research for this article, I have found the best text book reference to this technique in a surveying book entitled* Surveying Theory and Practice, *by John Clayton Tracy Ph.D., C.E. (1947 by John Wiley & Sons, Inc.). In his description of the procedure Mr. Tracy provides an alternate name, "Double–Line Leveling", which is a more accurate description of the technique because the procedure does not employ two rods as the more common name, the one that I have heard, implies. Mr. Tracy also provides two note forms for taking notes with the technique that are different than what I am showing in this article; the note form in this article appears to be the simplest to use and understand.*
>
> *Basically, this leveling procedure involves running two separate level circuits between two points and keeping notes for both simultaneously.*

The method is also explained in the *Glossary of the Mapping Sciences* edited by the American Society of Civil Engineers.

André–Louis Cholesky

Chapter 4

The method of Cholesky for linear systems

This chapter is devoted to the method of Cholesky for solving systems of linear equations with a symmetric (and positive definite) matrix. In Section 4.1, we will introduce the method of least squares which is used for treating data generated by a topographical survey. These data have to be adjusted as explained in Section 4.2. Then, a system of linear equations is obtained. Its solution can be obtained by various methods which were in use before Cholesky's discovery. They will be reviewed in Section 4.3. Section 4.4 is devoted to the manuscript of Cholesky. Then, it will be analyzed in Section 4.5. In Section 4.6, we will discuss other methods for linear systems which were introduced after Cholesky's. Finally, its diffusion in the scientific community will be the subject of Section 4.7.

4.1 The least squares method

Consider a system of m linear equations with n unknowns $Mx = c$, where M is a rectangular matrix with m rows and n columns, and c a known vector of dimension m. The unknown vector x has dimension n. There are two cases to discuss according to the respective values of m and n. The method for solving such a system is called the *method of least squares*, and the solution x is said to have been obtained in the *least squares sense*. This method is very useful and is much used in many branches of applied mathematics (geodesy, astronomy, statistics, etc.) for the treatment of experimental data and fitting a mathematical model to them. For an extensive treatment of the various cases that can occur, see [47,165].

If $m > n$, there are more equations than unknowns and the system is called *overdetermined*. In general, it has no solution x satisfying exactly the equations. Then, we try to solve them at best, that is in order to minimize the Euclidean norm $\|r\|_2^2$ of the residual vector $r = c - Mx$, that is the sum of the squares of its components. Writing down that the partial derivatives of $\|r\|_2^2$ with respect to the components x_i of the vector x have to vanish leads to the *normal equations* $M^T Mx = M^T c$. The matrix $A = M^T M$ is symmetric and, if the columns of M are linearly independent, positive definite. The vector x, unique solution of the normal equations $Ax = b$, with $b = M^T c$, is the least squares solution of the system $Mx = c$. Solving the normal equations $Ax = b$ amounts to searching the vector x such that

$$\|b - Ax\|_2^2 = \sum_{i=1}^{n} [b_i - (Ax)_i]^2 = 0.$$

Gauss rewrote this sum as a sum of other squares by eliminating one of them at each step. Expressed in terms of linear algebra, this *Gaussian elimination*, as it is called, is his *pivot method* for solving a system of linear equations [67].

Let us now consider the case $m < n$. When there are more unknowns than equations, the system $Mx = c$ is called *underdetermined* and it has infinitely many solutions. Among them, one looks for the solution minimizing the sum of the square of the unknowns $x_1^2 + \cdots + x_n^2$. It is exactly this case that appears in the adjustment of geodetic networks treated by Cholesky. As advocated by Gauss, the most plausible adjustments in term of probability theory, are those minimizing the sum of the squares $\|x\|_2^2 = x_1^2 + x_2^2 + \cdots + x_n^2$. For solving this problem, Lagrange introduced m new variables $y = (y_1, \ldots, y_m)^T$, the *Lagrange multipliers*, and then sought to cancel the partial derivatives of the function $f(x) = \|x\|_2^2 - 2(y, r)$ with respect to the unknowns x_i. One has $\partial f(x)/\partial x_i = 2x_i - 2(y, M_i^T) = 0$ where M_i^T is the ith row of the matrix M^T, which leads to $x = M^T y$. Replacing x by this expression in the first system, gives $MM^T y = c$. The matrix $A = MM^T$ is symmetric and positive definite, and the system $Mx = c$ is said to have been solved in the least squares sense.

Therefore, in both cases, the least squares method leads to a system with a symmetric positive definite matrix.

As long before as 1746, Ruggero Giuseppe Boscovich (Ragusa, Dubrovnik Republic, now Croatia, 18 May 1711 - Milan, 13 February 1787) determined the orbit of a comet from three observations of its position by minimizing the sum of the absolute values of the errors. He was followed in this way by the German astronomer Johann Tobias Mayer (Marbach, 17 February 1723 - Göttingen, 20 February 1762) in 1748 and 1760 while studying the libration of the moon (a slow apparent oscillation). Laplace, in Book III, no. 39, of his *Traité de Mécanique Céleste* which appeared in 1802, proposed to find out the unknowns so that the greatest error be minimum [162]. He applied this idea to determine the ellipse which deviates the least from experimental data. In [163] and [164], in 1810 and 1811, he derived the least squares method from what is now called the central limit theorem, and he applied it to the computation of the mean value of a series of observations [116].

The least squares method was published for the first time by the French mathematician Adrien–Marie Legendre in the Appendix of a book in 1806 [166]. Its interpretation as a statistical procedure was given by Gauss, first in 1809 [110] and then in 1810 in his Memoir on the asteroid Pallas discovered by Heinrich Wilhelm Olbers (Arbergen, 11 October 1758 - Bremen, 2 March 1840) on 28 March 1802 [112]. This work was immediately recognized as a major contribution. Gauss acknowledged that the method appeared in Legendre's book, but he claimed that he already used it in 1795, which cannot be verified. But, he certainly used it in 1801 to determine the orbit of the asteroid Ceres discovered by Giuseppe Piazzi (Ponte in Valtellina, 16 July 1746 - Naples, 22 July 1826) on 1 January 1801 [47,118] from the observatory of Palermo in Sicily. Ceres was observed for 41 days during which time it covered an angle of 9 degrees before it disappeared in the rays of the Sun. Piazzi's observations were published in June 1801 by János Ferenc Zach (Pozsony, Slovakia, 13 June 1754 - Paris 2 September 1832), director of the Seeberg observatory and editor of *Monatliche Correspondenz*, the main German astronomical periodical at that time. The challenge among astronomers was to predict when and where Ceres will reemerge. The problem was that no one knew the shape of its orbit, and it could only be assumed to be an ellipse with eccentricity between 0 and 1. But a similar problem had been already solved by Gauss for the Moon. He was able to determine an orbit as long as it was assumed to be a conic section. He adopted Kepler's assumptions that the motion of a celestial object is only determined by its orbit which, if it does not pass extremely close to another body in the solar system, is a conic section with its focal point at the center of the Sun. With these assumptions, the orbit is only determined by 5 parameters which specify its form and its position in space. In addition, a single time measure of the object at a specific point of the orbit is needed. Gauss had a total of 22 observations. He first determined a rough approximation of the orbit from 3 observations, and then refined it to a higher degree of precision again and again. The procedure required him over 100 hours of calculation. In September 1801, Zach published several forecasts of the prospective orbit. Gauss's predictions were quite different from the others. On 7 December, Zach, and then Olbers on 31 December, located Ceres very close to the positions predicted by Gauss. Let us mention that, in a footnote to a paper published in 1806, Olbers asserted that Gauss had shown him the method in 1802. See [242] for a more detailed account on this work.

A priority dispute about the method arose between Legendre and Gauss. In May 1809, Legendre wrote to Gauss (translation given in [196]):

> *It was with pleasure that I saw that in the course of your meditations you had hit on the same method which I had called* Méthode des moindres quarrés *in my memoir on comets... I confess to you that I do attach some value to this little find. I will therefore not conceal from you, Sir, that I felt some regret to see that in citing my memoir p. 221 you say* principium nostrum quo jam inde ab anno 1795 usi sumus *etc. There is*

no discovery that one cannot claim for oneself by saying that one had found the same thing some years previously; but if one does not supply the evidence by citing the place where one has published it, this assertion becomes pointless and serves only to do a disservice to the true author of the discovery. In Mathematics it often happens that one discovers the same things that have been discovered by others and which are well known; this has happened to me a number of times, but I have never mentioned it and I have never called principium nostrum *a principle which someone else had published before me. You have treasures enough of your own, Sir, to have no need to envy anyone; and I am perfectly satisfied, besides, that I have reason to complain of the expression only and by no means of the intention.*

It seems that it was not the first time Legendre experienced an insufficient recognition of his own work by Gauss. The dispute greatly hurt Legendre who fought for many years to have his priority recognized. Indeed, in a second supplement to his 1805 book, he came back to his claim for the method of least squares [220, 222]. According to [66], the numerical results reported by Gauss could not have been obtained by the method of least squares or by any other approach he mentioned.

In 1821–1823, Gauss put the least squares method on a firm theoretical basis and he showed that it leads to the best possible combination of the observations whatever be the probability law of the errors [113] (see [218] for an English translation). However, this result was somehow neglected until it was rediscovered by Andrei Andreyevich Markov (Ryazan, 14 June 1856 - Petrograd, now St Petersburg, Russian Federation, 20 July 1922) in 1912 [174]. According to the Gauss–Markov Theorem, consider a system $Mx = \widehat{c} = c + \varepsilon$, where the $m \times n$ matrix M has rank n, and where ε is a random vector whose components are uncorrelated and follow a normal distribution with a zero mean and the same variance. Then, the best linear unbiased estimator of any linear functional $c^T x$, where c is a vector, is $c^T \widehat{x}$, where $\widehat{x}$ is the least squares estimator obtained by minimizing the Euclidean norm of the residual $r = \widehat{c} - Mx$ [47, p. 4]. It is important to mention that, in 1809–1810, Gauss applied the method of least squares to overdetermined systems, but that it was only in 1826 that he applied it to underdetermined systems [114]. Gauss's approach was popularized in 1838 by the book of Friedrich Wilhelm Bessel (Minden, Westphalia, now Germany, 22 July 1784 - Königsberg, Prussia, now Kaliningrad, Russian Federation, 17 March 1846) and Johann Jacob von Baeyer [45]. Cartographic offices adopted it, and then Gauss's method became part of the mathematics curriculum for geodesists [120]. The next improvement of the method will be Cholesky's.

Gauss' works are analyzed in detail in [26,121,211,217]. For his contributions to geodesy, see [212]. For the history of the least squares method, see [118, pp. 209–224], [180], [213], and many other references. For French readers, [40] contains much informations.

Carl Friedrich Gauss (1777 - 1855)

The method of least squares is related to the reduction of a quadratic form to canonical form, a topic treated by Joseph–Louis Lagrange (Turin, 25 January 1736 - Paris, 10 April 1813) in 1759 [152]. At this occasion, he gave elimination formulae similar to those of Gauss. The presentation of Gauss himself was in terms of quadratic forms. It seems that the matrix interpretation of Gauss is due to Carl Gustav Jacob Jacobi (Potsdam, Prussia, 10 December 1804 - Berlin, 18 February 1851) in a posthumous paper dated 1857 [141].

Let us mention that the American mathematician of Irish origin Robert Adrain (Carrickfergus, Ireland, 30 September 1775 - New Brunswick, NJ, USA, 10 August 1843), at the occasion of a problem in topography, wrote a paper in 1808 (only published in 1809) where he presented the method of least squares [21]. According to some authors, he can be credited for its discovery [19,93]. This work was completely unnoticed in Europe. In 1818, Adrain applied this method to the determination of the flattening of the Earth from the measures of the length of the meridian, and gave an estimation of the axes of the terrestrial ellipsoid [22]. On this problem, see [147, pp. 226–230]. Some history about methods for adjusting observations in land surveying is given in [223] and in the discussion in [27].

Friedrich Robert Helmert (Freiberg, 31 July 1843 - Potsdam, 15 June 1917) can be considered as the founder of modern geodesy. He made important contributions to the theory of errors. In his book [130], whose first volume appeared in 1872, he used the method of least squares in the examination of measuring instruments, and the adjustment of geodetic networks. He introduced a coordinate transformation named after him. This transformation depends on parameters which he determined by the least squares method. In this book, Helmert also derived Student's distribution of the standard deviations of samples from an indefinitely large normal population 30 years before Student in 1908 [251]. Let us mention that Student is the pseudonym of William Sealy Gosset (Canterbury, 13 June 1876 - Beaconsfield, 16 October 1937), an employee of the Guinness brewery near Dublin who was not allowed to publish his researches under his own name. On the conditions of application of the method of least squares in geodesy, see [145]. On the history of fitting of linear relationships between 1750 and 1900, see [103]

4.2 The adjustment of networks

Let us now explain the use of the method of least squares in topography for the adjustment of networks. For French readers, we recommend [195].

For establishing a map, the surveyor realizes a triangulation of the land. According to the scale of the map, it is necessary to take into account the exact form of the Earth. The angles and the lengths are compelled to satisfy the condition equations which express the fact that the sum of the angles of a triangle must be equal to some precise value (greater than 180 degrees since the Earth is, more or less, a sphere), that, at each point, the sum of the angles must be 360 degrees, and that the lengths must be invariant whatever be the order in which the measures are performed. But some points can only be observed from a distance, and the measurement instruments cannot be installed on them. This is the case of bell towers, tops of tall structures or fireplaces. Cholesky writes [69, p. 264]

> *Each time a triangulation is made, it is advantageous to make also an adjustment by computation. Then, one is led to write down a certain number of equations expressing the geometrical relations between the various elements of the triangulation drawing and since, generally, there are more unknowns than equations, the indeterminacy is removed by writing that the sums of the squares of the corrections is minimum.*

Thus, a system of linear equations with more unknowns than equations is obtained. It is always possible to modify the values of the angles in order to satisfy these condition equations. This is called *adjustment* by Gauss.

Let us explain in detail how this adjustment of geodetic networks is performed. Assume that n measurements $l_1, \ldots, l_n$, eventually reduced to a plane representation, of N quantities $X_1, \ldots, X_N$ have been made. In general, they are measures of angles or distances. The unknowns $X_1, \ldots, X_N$ are the coordinates, in the Lambert conformal conic projection, of the new points. They can be compensated, that is corrected from the errors which could affect them due to the precision of the instruments and the experimental uncertainties. If there are no errors, the measures l_i are linked to the unknowns by relations of the form

$$l_i = f_i(X_1, \ldots, X_N), \quad i = 1, \ldots, n,$$

where the f_i's are nonlinear functions. Let v_i be an estimate of the error on the measure l_i. They are random variables which follow a normal Gaussian distribution centered at zero. Thus, we have the following *observation relations*

$$l_i + v_i = f_i(X_1, \ldots, X_N), \quad i = 1, \ldots, n.$$

This is a nonlinear system of n equations with $N + n$ unknowns $X_1, \ldots, X_N$ and $v_1, \ldots, v_n$. It is necessary to replace it by a linear system.

Let $l_i^* = f_i(X_1^*, \ldots, X_N^*)$ be the measures computed from the approximate values X_i^* of the unknowns. These approximate values are obtained by an arbitrary adequate process, by using only a part of the measurements. For example, for compensating a network of 5 points, only 3 measures will be used and the solution obtained will serve as an approximate solution. In order to linearize the preceding system, a Taylor expansion at the first order of the functions f_i is realized, which gives

$$l_i - l_i^* + v_i = \frac{\partial f_i}{\partial X_1^*}(X_1 - X_1^*) + \cdots + \frac{\partial f_i}{\partial X_N^*}(X_N - X_N^*), \quad i = 1, \ldots, n.$$

Let $x_i = X_i - X_i^*$ be the *correction of the adjustment.* Denoting by A the matrix of the coefficients $a_{ij} = \partial f_i / \partial X_j^*$ for $i = 1, \ldots, n$ and $j = 1, \ldots, N$, by b the vector with components $b_i = l_i - l_i^*$ for $i = 1, \ldots, n$, by x the vector with components x_i for $i = 1, \ldots, N$, and by v the vector with components v_i for $i = 1, \ldots, n$, the preceding system becomes

$$Ax = b + v.$$

Let us now see how to solve it by the *method of least squares.*

Each measure l_i is a random variable characterized by its *standard deviation* σ_i. The number $p_i = 1/\sigma_i^2$ is the *weight* of this measure. It can be proved that the residuals v_i are independent normal random variables centered at zero, and that the most probable solution of the system of the observation relations $Ax = b + v$ is the one which minimizes the quantity

$$\varepsilon^2 = \sum_{i=1}^{n} p_i v_i^2.$$

It is the *least squares condition.*

This quantity is the smallest possible when the partial derivatives with respect to the x_j's are all zero. We have

$$\frac{\partial \varepsilon^2}{\partial x_j} = \sum_{i=1}^{n} p_i \frac{\partial v_i^2}{\partial x_j}.$$

But $\partial v_i^2/\partial x_j = 2v_i \, \partial v_i/\partial x_j$. According to the system of the observation relations, we have $\partial v_i/\partial x_j = a_{ij}$, and the least squares condition therefore leads to

$$\sum_{i=1}^{n} a_{ij} p_i v_i = 0, \quad j = 1, \ldots, N.$$

Denoting by P the diagonal matrix with elements $p_1, \ldots, p_n$, this condition is

$$A^T P v = 0.$$

This is a system of N equations with n unknowns and we finally obtain a system of $N + n$ equations with the same number of unknowns

$$\begin{cases} A^T P v = 0 \\ Ax - v = b. \end{cases}$$

But $v = Ax - b$ and, substituting in the first system, we obtain

$$A^T P A x = A^T P b.$$

This system, called *normal equations*, only contains the unknown vector x of adjustments. It has dimension N and its matrix is symmetric positive definite since the weights p_i are strictly positive. It is this system which has to be solved.

4.3 Methods for linear systems

It is well known that many results are rediscovered independently by several scientists, sometimes at distances of several years and thousands of miles. This is the case of methods for solving systems of linear equations. In the sequel, we will only mention some of the main developments of such methods. For their complete history, we refer the interested reader to [120, 121].

Consider the $n \times n$ linear system $Ax = b$. In 1750, the Swiss mathematician Gabriel Cramer (Geneva, 31 July 1704 - Bagnols–sur–Cèze, 4 January 1752) showed that the solution can be obtained by means of determinants [73]. The ith component x_i of the vector x is given by the formula

$$x_i = \det(A_i)/\det(A), \quad i = 1, \ldots, n,$$

where $\det(A)$ is the determinant of the matrix A, and where A_i is the matrix obtained by replacing the ith column of A by the right hand side b.

According to the historian of mathematics Carl Benjamin Boyer (Hellertown, 3 November 1906 - New York, 26 April 1976) [51, p. 431], Colin Maclaurin (Kilmodan, Scotland, February 1698 - Edinburgh, 14 June 1746) probably knew the same method as early as 1729, and he described it in his book [172] published two years after his death.

Of course, Cramer's formulae give, in theory, the solution but, from the practical point of view, they cannot be used. Computing determinants by their minors requires $\mathcal{O}(n^2 n!)$ arithmetical operations for obtaining x which is impracticable. Indeed, on a computer performing 10^7 operations per second, it requires 37 seconds when $n = 10$, 2 hours when $n = 12$, more than one day for $n = 13$, almost one year for $n = 15$, 17 years when $n = 16$, and... 10^{10} years if $n = 23$ which is more than the age of the Universe! This is why scientists rapidly turn to elimination methods such as Gauss's one.

The method proposed by Gauss in 1810 [112] consists in an elimination technique. By a linear combination of the first and the second equations of the system (an operation which does not change its solution), the first unknown x_1 can be eliminated from the second equation. Then, by a linear combination between the first and the third equations, x_1 can be eliminated from the third equation. Continuing the same procedure until the last equation, x_1 is eliminated from all equations except the first one. This is the end of the last step. Then, by a linear combination of the second and the third equations of the system obtained after the first step, the second unknown x_2 is eliminated from the third equation. Combining the second and the fourth equations yields to the elimination of x_2 from the fourth equation. And so on, until x_2 has been eliminated from the last equation. This is the end of the second step. The equations 2 to n of the new system do not contain x_1, and the equations from 3 to n do not contain x_1 and x_2. The same procedure can be continued and, after $n-1$ steps, the final system contains all the unknowns in its first equations, $x_2, \ldots, x_n$ in its second equation, $x_3, \ldots, x_n$ in its third equation, up to the last equation which only contains x_n. Thus, this last equation immediately gives the value of x_n. The penultimate equation contains only x_n and x_{n-1}, but x_n is known and thus the value of x_{n-1} is obtained, and so on until the first equation which furnishes the unknown x_1. The method requires $\mathcal{O}(n^3)$ arithmetical operations. Gauss did not present his method with matrix notations, while Cholesky did. Gauss's method is valid for any regular matrix but when the matrix is symmetric (and positive definite) it does not take into account the symmetry, it does not benefit of it and, thus, it needs too many (and unnecessary) arithmetical operations contrarily to the method of Cholesky which became so popular for that reason.

Let us mention that it has been shown very recently how to implement Cramer's rule in $\mathcal{O}(n^3)$ arithmetical operations, which makes it comparable to other direct methods for solving linear systems [128].

The idea of solving a system of linear equations by eliminating successively the unknowns by linear combination of the equations seems to have been used by many mathematicians of the 18th century [120]. Let us give an example. José Anastácio da Cunha (Lisbon, 11 May 1744 - Lisbon, 1 January 1787) is a Portuguese mathematician who studied and taught at the University of Coimbra, and then, from 1781, at the Real Casa Pia in Lisbon. There, he began to write his *Principios Mathematicos*, a book he finished while correcting the proofs one day before he died in 1787. This book was translated into French and published in 1811 by João Manuel D'Abreu (Valença do Minho, ca. 1757 - Azores, 1815), a colleague of him, a member of the Academy of Sciences in Lisbon [78]. On page 151, Da Cunha wrote (translation from French):

> *... from which we deduce* $2x + z + y = 78$*;* $3z + x + y = 126$*;* $2y + x + z = 90$*. Operating as we see here*
>
> $$\begin{array}{rcrcrcl} 2x & + & z & + & y & = & 78 \\ x & + & 3z & + & y & = & 126 \\ x & + & z & + & 2y & = & 90 \\ \hline 0x & + & 5z & + & y & = & 2\times 126 - 78 = 174 \\ & & z & + & 3y & = & 2\times 90 - 78 = 102 \\ \hline & & 0 & + & 14y & = & 5\times 102 - 174 = 336 \end{array}$$
>
> *we will find* $y = 24$*;* $z + 3\times 24 = 102$*, or* $z = 30$*;* $x + 30 + 2\times 24 = 90$*, or* $x = 12$.

More examples are also given in the following pages. This is, obviously, the Gaussian elimination method. Let us mention that the word *elimination* was introduced in 1797 by the French mathematician Sylvestre François Lacroix (Paris, 28 April 1765 - Paris, 24 May 1843) who wrote [150, p. 114]:

> *When a question contains as many distinct conditions as unknowns, each of these conditions gives one equation, where it often arises that the unknowns are mixed together, as we already saw it in problems with two unknowns; but if these unknowns are only in the first degree, we can, as we did in the previous numbers* [i.e. section]*, take in one of the equations the value of one of the unknowns, as if the remainder was known, and substitute this value in all the other equations, which then will only contain after that the other unknowns.*
>
> *This operation, by which we expel one of the unknowns, is called* elimination*. By this means, if we have three equations with three unknowns, we will deduce two equations with two unknowns, that we will treat as above; and having obtained the values of the last two unknowns, we will substitute in the expression of the first one.*

The term *matrix* was first introduced in 1850 by James Joseph Sylvester (London, 3 September 1814 - London, 15 March 1897), and matrix algebra was developed by Arthur Cayley (Richmond, England, 16 August 1821 - Cambridge, 26 January 1895) around the same time. As showed by Paul Sumner Dwyer (8 December 1901 - Mackinaw City, MI, 1 September 1982) and Henry Jensen (7 October 1915 - 10 August 1974) in 1944 [96, 142], Gauss's method consists, from the theoretical point of view, in decomposing the matrix A into a product $A = LU$, where L is a lower triangular matrix with a unit diagonal (that is with all the elements above the diagonal being zero and those of the diagonal being 1), and where U is an upper triangular matrix (that is with all the elements below the diagonal being zero). Then, the system is written $LUx = b$, that is $Ly = b$ if we set $y = Ux$. Solving the system $Ly = b$ gives the vector y, which is the right–hand side of the second system $Ux = y$, and the solution x is finally obtained after solving two triangular systems, an easy task.

A variant of Gauss's method, in which the system is transformed into a diagonal one, is due to the German surveyor Wilhelm Jordan (Ellwangen, 1 March 1842 - Hanover, 17 April 1899) who exhibited it, in 1888, in the third edition of one of his books [144]. It was the first appearance of the method in a book, which led to sometimes denominate it as the Gauss–Jordan's method. This Jordan must not be confused, as it is often the case, with Camille Jordan, who had Cholesky as a student at *École Polytechnique*, and who was, with Eugenio Beltrami (Cremona, Lombardy, Austrian Empire, 16 November 1835 - Rome, 18 February 1900), at the origin of the singular value decomposition of a matrix; see [29] for a discussion. Let us notice that some authors attribute the Gauss–Jordan's method to the abbot Bernard–Isidore Clasen (1829 - 1902) [70], a priest in Echternach and a professor in Luxembourg.

The United States Coast Survey was created by the Congress in 1807, following a bill of the President Thomas Jefferson (Shadwell, Virginia, 13 April 1743 - Monticello, Virginia, 4 July 1826) for the survey of the coast. Progress moved ahead slowly and, in its early decades, it was only responsible for charting the coastlines. Triangulation work began in the vicinity of New York City in 1816, and the first baseline was measured and verified in 1817. In 1818, the U.S. Army and the U.S. Navy were made responsible for the survey work, and various superintendents developed their activities in astronomy, cartography, meteorology, geodesy, geology, geophysics, hydrography, navigation, oceanography, exploration, pilotage, tides and topography. After the acquisition of Alaska in 1867, the 1871 law required the Coast Survey to carry geodetic surveys into the interior of the country. Thus, in 1878, the U.S. Coast Survey became the U.S. Coast and Geodetic Survey.

On 9 November 1878, Myrick Hascall Doolittle (Addison, Vermont, USA, 17 March 1830 - Linden, Maryland, 27 June 1913), a mathematician of the Computing Division of the U.S. Coast and Geodetic Survey in Washington, presented

Myrick Hascall Doolittle circa 1862
© Courtesy of Antiochiana, Antioch College

a method for solving the normal equations coming out from problems of triangulation [89]. His method consisted of canceling step by step the elements of the matrix in order to transform it into an upper triangular matrix. It is equivalent to decompose A into a product $A = LU$ by a succession of n intermediate steps, where n is the dimension of the system, with $L = L_1 + \cdots + L_n$ and $U = U_1 + \cdots + U_n$. The matrix L is lower triangular and U is upper triangular but with a unit diagonal, contrarily to Gauss's method. Doolittle also showed how it is possible to solve an enlarged system, obtained by adding new equations and new unknowns, without having to repeat all calculations. This technique is therefore similar to the *bordering method* which is used for solving the normal equations by enlarging the system [165, pp. 122–124]. According to him, this was one of the main advantages of his method. Doolittle had no calculating machine at his disposal and he only used multiplication tables. He said that he solved, with the help of J.G. Porter, a system of 41 equations containing 174 side coefficients in five and a half days [89].

Let us give some details about Doolittle's life [102]. He graduated from Antioch College, Ohio, in 1862, and was a professor of mathematics there for the following year. Then, his physical condition preventing enlistment, he went to Harvard to study mathematics with Benjamin Peirce (Salem, Massachusetts, 4 April 1809 - Cambridge, Massachusetts, 6 October 1880) who will be the director of the U.S. Coast Survey from 1867 to 1874. Afterwards, Doolittle joined his wife who was a volunteer nurse in the Army hospitals in Washington, and he served a short time in the Army sanitary commission. He was for many years an active member of the Philosophical Society of Washington. Further details of his life were given in a letter published by his wife in the January 1895 issue of *The Antiochian*:

> *In 1864 he came to Washington to enter upon Astronomical work in the Naval Observatory. His special work was observing Meridian Zenith Distances. He found the night work very wearing and after six years resigned and went into the Patent Office, where he stayed three years, examining applications for patents for steam boilers. In 1873, he entered upon work in the Coast Survey Office. This was more congenial, and consisted of the least square adjustments of Triangulations. In this department he still remains.*

Doolittle resigned on 25 September 1911 because of ill health. He died at his home on 27 June 1913 at Linden, Md., after a short illness. As written in the account sent by the Survey to the newspapers *"In the details of the work as well as in general principles he displayed a rare talent which has left a lasting impression on the work of the office and made his name well known among computers the world over."*

Doolittle's method was quite successful and it was used, with variations, for many years in geodesy. For example, Thomas Wallace Wright (Galloway, Scotland, 3 August 1842 - Schenectady, N.Y., 13 September 1908), a professor of Applied Mathematics and Chairman of the Department of Physics and then Physics from 1895 to 1904 at Union College (Schenectady, N.Y.), described it in the Section entitled *Combination of the direct and indirect methods of solution* [247, pp. 167–174] of his 1884 book based on the methods he had developed in his surveying works during the decade 1873–1883 when a civil engineer with the U.S. Survey [216]. In the second edition of this book, written in 1906 with the cooperation of John Fillmore Hayford (Rouses Point, New York, 19 May 1868 - Evanston, Illinois, 10 March 1925), Chief of the Computing Division and Inspector of Geodetic Work, U.S. Coast and Geodetic Survey, the authors said [248, p. 114]:

> *In it* [Doolittle's method] *there is a combination of improvements on the Gaussian method of substitution. Its advantage lies mainly in the*

arrangement of the work in the most convenient form for the computer. This makes the solution more rapid than by the other method, the gain in speed being the more marked the greater the number of equations.

These two books were standard references for many years. Doolittle's method was also explained in 1912 in the book of Charles Jean–Pierre Lallemand (Saint–Aubin–sur–Aire, 7 March 1857 - Vecqueville, 1 February 1938), the elder brother of Nicolas Arthur Lallemand, then a member of the French Academy of Sciences and head of the *Service du Nivellement Général de la France* (General leveling Service of France) [157]. Cholesky had a copy of this book which was, however, published after he discovered his method.

Let us quote an anecdote about Doolittle, reported by Commander Harry A. Seran (New Philadelphia, Ohio, 28 January 28 1887 - ?); see http://www.history.noaa.gov/stories_tales/seran.html:

About 8:00 o'clock in the evening of June 30, 1907, direct from a fresh water college in Ohio, I reached Washington, D.C., for the first time. An appointment as Aid in the Coast and Geodetic Survey and instructions to report for duty on July 1st were carefully folded in my pocket.

Upon reporting to the Assistant Superintendent and taking the oath of office, I was ordered to duty in the Computing Division. The Computing Division in those days was an entirely different division from what it is to–day. Among others was Mr. Doolittle, a famous computer and the originator of the Doolittle method of solving equations. He was at this time quite old and in order to know when to stop work in the afternoon, he had an alarm clock in his office which sounded off at 4:30. Needless to say, this alarm could be heard throughout the Division. It may have been the forerunner of the signal which is given on the Auto–call today, which tells those who have not already done so to put on their hats, coats and galoshes preparatory to trying to work their various ways home through the jam of traffic, red and green lights, busses, street–cars and trucks. Mr. Doolittle always wore carpet slippers in the office. That type of footwear must be extinct as the dodo bird to–day for it is never seen except in fancy dress costume.

Doolittle's method seems to have been the standard method used by the engineers of the *U.S. Coast and Geodetic Survey* for a number of years, as exemplified in the small booklet [20] by Oscar Sherman Adams (Monroe Township, Ohio, 9 January 1874 - 5 March 1962). This method was independently rediscovered by Henry Agard Wallace (a farm near Orient, Adair County, Iowa, 7 October 1888 - Danbury, Connecticut, 8 November 1965), the 33rd Vice President of the United States. A farmer at the origin, he became a self–taught practicing statistician, and published an influential paper with George Waddel Snedecor (Memphis, Tennessee, 20 October 1881 - Amherst, Massachusetts, 15 February 1974) on computational methods for correlations and regressions. Let us quote [104]:

Henry Agard Wallace (1888 - 1965) served as U.S. Secretary of Agriculture 1929–1932 and as Vice President 1941–1945. (Indeed, he might have succeeded Franklin Delano Roosevelt as President later in 1945 but for the fact that his colleagues thought him too liberal, so he was replaced by Harry Truman when Roosevelt decided to run for a fourth term.)...

However, he [Wallace] *was a practicing statistician who published an influential monograph coauthored by George Waddel Snedecor [240]* [1]*, which has not been properly acknowledged. This pamphlet contained a very useful computational technique which was known to geodesists but not to statisticians at the time, and which apparently Wallace had rediscovered independently...*

At first sight, it seems unlikely that Wallace and other practicing statisticians would not have known of a computational procedure that had been in the public domain for 44 years. But Doolittle's paper was published in an unfamiliar journal in 1881, four years too late for inclusion in the extensive list of writings on the Method of Least Squares [179] compiled by Mansfield Merriman (1848 - 1925) in 1877...

Thus, if we can fully credit the recollections of the 72–year–old Wallace... then it is clear that Wallace's contribution to the 1925 monograph was prepared independently of Doolittle's work.

On these developments, see [123, pp. 159–166].

In 1907, Otto Toeplitz (Breslau, Germany, 1 August 1881 - Jerusalem, under the British Mandate at the time, 15 February 1940) proved that any Hermitian positive definite matrix can be factorized into a product LL^* with L lower triangular and L^* its conjugate transpose, but he did not give a method for obtaining L [227]. This is what Cholesky did.

4.4 Cholesky's manuscript

Let us consider the system of linear equations $Ax = b$, where the matrix A is real and symmetric positive definite (Cholesky only knew it was symmetric). Cholesky's method consists in decomposing A into the product $A = LL^T$, where L is a lower triangular matrix (that is a matrix where all the elements above the diagonal are zero) and the terms on the diagonal are strictly positive (Cholesky only knew they were different from zero). Then, the system becomes $LL^Tx = b$. Setting $L^Tx = y$, the system becomes $Ly = b$. It can be easily solved, which gives the vector y. Finally, we solve $L^Tx = y$. The elements of the matrix L are obtained by identifying the elements of A with the corresponding elements of the product LL^T.

[1] Our bibliography.

For any regular matrix, Gauss's method consists in decomposing A into the product $A = LU$, where L is a lower triangular matrix with all the elements on its diagonal equal to 1, and U an upper triangular matrix, that is a matrix whose elements under the diagonal are zero. Of course, Cholesky's method can appear as a variant of Gauss', but it is not. Gauss presented an elimination method for solving a system of equations which was much later interpreted as corresponding to a decomposition of this form (see Chapter 4.3). On the contrary, in Cholesky's method, and this is the way he presented it, the elements of the matrix L are directly obtained by identification of the elements of A with the corresponding elements of the product LL^T.

Let us set

$$L = \begin{pmatrix} l_{11} & & & \\ l_{21} & l_{22} & & \\ \vdots & \vdots & \ddots & \\ l_{n1} & l_{n2} & \cdots & l_{nn} \end{pmatrix}.$$

Multiplying by L^T and identifying with the corresponding terms of A, we get

$$\begin{aligned} l_{11}^2 &= a_{11} \\ l_{21}l_{11} &= a_{21} \\ &\vdots \\ l_{n1}l_{11} &= a_{n1} \end{aligned}$$

which gives

$$\begin{aligned} l_{11} &= \sqrt{a_{11}} \\ l_{i1} &= a_{i1}/l_{11}, \qquad i = 2, \ldots, n. \end{aligned}$$

Then we have

$$\begin{aligned} l_{21}^2 + l_{22}^2 &= a_{22} \\ l_{31}l_{21} + l_{32}l_{22} &= a_{32} \\ &\vdots \\ l_{n1}l_{21} + l_{n2}l_{22} &= a_{n2} \end{aligned}$$

and it follows that

$$\begin{aligned} l_{22} &= \sqrt{a_{22} - l_{21}^2} \\ l_{i2} &= (a_{i2} - l_{i2}l_{2i})/l_{22}, \qquad i = 3, \ldots, n, \end{aligned}$$

and so on. As we will see below, these formulae are exactly those given by Cholesky.

Cholesky described his method in an unknown and unpublished manuscript entitled *Sur la résolution numérique des systèmes d'équations linéaires* (that is *On the numerical solution of systems of linear equations*). It contains 8 pages (with only 8 lines on the last page), and it is written on sheets of paper of size 21.8 cm × 32 cm. It is dated 2 December 1910 and since, contrarily to the other manuscripts of Cholesky, very few words are crossed–out, it is almost sure that the method was obtained some time before, and that the paper is a rewriting or a new copy of a preceding text. As we will see, Cholesky wrote that he used his method in Algeria. He was sent there first in 1902, and then in 1911. Since his paper is dated 2 December 1910, he can have found his method, maybe, as early as 1902.

We have a letter, dated 25 September 1905 and written from Montellier, to an unknown capitaine where Cholesky explains:

> *The commandant submitted me, the day before yesterday, his scruples on the adjustment of stations (angles) as it is currently practiced. After I soaked in the book of instructions, I tried to solve the problem in the most general perspective, that is to say to determine the compensated directions I, II, III, IV, etc. by means of a formula linearly containing the angles observed directly (1.2) (1.3)... (2.3), etc.*
> *Last night I reached the last conclusion of my researches which is as follows:*
> *Adjustment as it is done is an absolutely useless complication. The application of least squares leads to a single formula (I-II)=angle (1.2) compensated* $= 2(1.2) + [(1.3) - (2.3)] + [(1.4) - (2.4)] + \cdots$
> *And if you compute so the compensated angles (I-II), (I-III) etc., you can easily check that you can deduct the angle (K-L) by difference, the result being none other than the weighted value of the angle (k-l) obtained by the first formula, that is by introducing in it all the differences or sums 2 by 2 that ought to reproduce the angle considered.*
>
> *To compensate a first–order station, it is sufficient to make the weighted averages for all angles observed, and we will have as a check the formation of the differences of the* $(n-1)$ *first angles, alone to be independent and fundamental.*
>
> *The commandant wrote to the colonel the obtainment of this result which will probably cause some emotion, but which will simplify adjustment in the future. I wanted to warn you at the same time so that you can think about it and you do not continue the research you have undertaken to modify the method.*

Unfortunately, this is the only letter we have, and it is difficult to conclude from it if Cholesky was speaking of his method for solving systems of linear equations although the relation 2(1.2)+[(1.3)-(2.3]+[(1.4)-(2.4)]+$\cdots$ looks like an elimination procedure.

Let us now give a translation of this unpublished manuscript of Cholesky on his method. The original layout and the structures of the sentences have been preserved as much as possible. The original text in French is given in Appendix B.

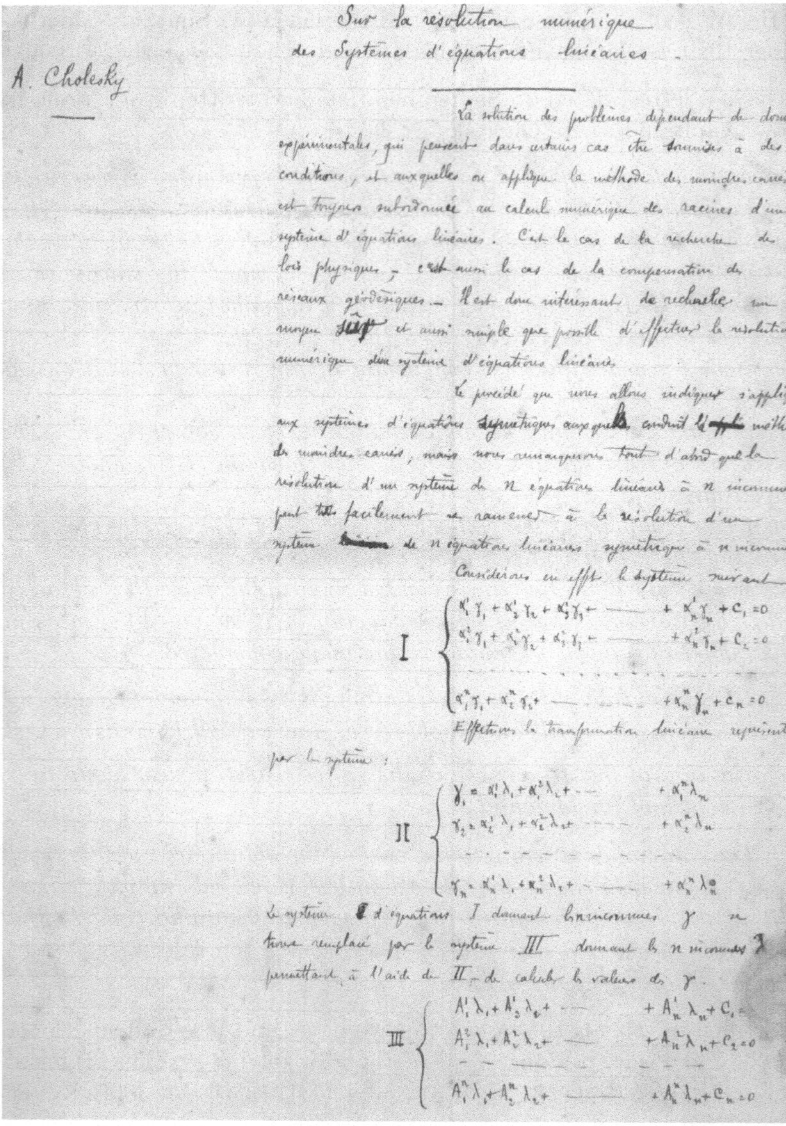

Sur la résolution numérique des systèmes d'équations linéaires.

A. Cholesky

La solution des problèmes dépendant de données expérimentales, qui peuvent dans certains cas être soumises à des conditions, et auxquelles on applique la méthode des moindres carrés, est toujours subordonnée au calcul numérique des racines d'un système d'équations linéaires. C'est le cas de la recherche des lois physiques, c'est aussi le cas de la compensation des réseaux géodésiques. Il est donc intéressant de rechercher un moyen sûr et aussi simple que possible d'effectuer la résolution numérique d'un système d'équations linéaires.

Le procédé que nous allons indiquer s'applique aux systèmes d'équations symétriques auxquels conduit la méthode des moindres carrés, mais nous remarquerons tout d'abord que la résolution d'un système de n équations linéaires à n inconnues peut très facilement se ramener à la résolution d'un système de n équations linéaires symétrique à n inconnues.

Considérons en effet le système suivant

$$\text{I}\quad\begin{cases} \alpha_1^1 y_1 + \alpha_2^1 y_2 + \alpha_3^1 y_3 + \cdots + \alpha_n^1 y_n + C_1 = 0 \\ \alpha_1^2 y_1 + \alpha_2^2 y_2 + \alpha_3^2 y_3 + \cdots + \alpha_n^2 y_n + C_2 = 0 \\ \cdots \\ \alpha_1^n y_1 + \alpha_2^n y_2 + \cdots + \alpha_n^n y_n + C_n = 0 \end{cases}$$

Effectuons la transformation linéaire représentée par le système :

$$\text{II}\quad\begin{cases} y_1 = \alpha_1^1 \lambda_1 + \alpha_1^2 \lambda_2 + \cdots + \alpha_1^n \lambda_n \\ y_2 = \alpha_2^1 \lambda_1 + \alpha_2^2 \lambda_2 + \cdots + \alpha_2^n \lambda_n \\ \cdots \\ y_n = \alpha_n^1 \lambda_1 + \alpha_n^2 \lambda_2 + \cdots + \alpha_n^n \lambda_n \end{cases}$$

Le système d'équations I donnant les inconnues y se trouve remplacé par le système III donnant les n inconnues λ permettant à l'aide de II de calculer les valeurs des y.

$$\text{III}\quad\begin{cases} A_1^1 \lambda_1 + A_2^1 \lambda_2 + \cdots + A_n^1 \lambda_n + C_1 = 0 \\ A_1^2 \lambda_1 + A_2^2 \lambda_2 + \cdots + A_n^2 \lambda_n + C_2 = 0 \\ \cdots \\ A_1^n \lambda_1 + A_2^n \lambda_2 + \cdots + A_n^n \lambda_n + C_n = 0 \end{cases}$$

First page of the manuscript

On the numerical solution of systems of linear equations
A. Cholesky

The solution of problems depending on experimental data, which, in certain cases, can be subject to conditions, and to which we apply the method of least squares, is always subject to the numerical calculation of the roots of a system of linear equations. This applies to the search of the laws of physics; it is also the case for the adjustment of geodetic networks. It is therefore interesting to find a safe and as simple as possible process for the numerical solution of a system of linear equations.

The process that we will specify applies to symmetrical systems of equations to which leads the method of least–squares; but we will first notice that the solution of a system of n *linear equations in* n *unknowns can easily be reduced to the solution of a system of* n *symmetric equations in* n *unknowns.*

Indeed, let us consider the following system:

$$I\left\{\begin{array}{lll} \alpha_1^1\gamma_1+\alpha_2^1\gamma_2+\alpha_3^1\gamma_3+ & \cdots & +\alpha_n^1\gamma_n+C_1=0\\ \alpha_1^2\gamma_1+\alpha_2^2\gamma_2+\alpha_3^2\gamma_3+ & \cdots & +\alpha_n^2\gamma_n+C_2=0\\ \cdots\cdots\cdots\cdots & \cdots & \cdots\cdots\cdots\cdots\\ \alpha_1^n\gamma_1+\alpha_2^n\gamma_2+ & \cdots & +\alpha_n^n\gamma_n+C_n=0. \end{array}\right.$$

Let us perform the linear transformation represented by the system:

$$II\left\{\begin{array}{l} \gamma_1=\alpha_1^1\lambda_1+\alpha_1^2\lambda_2+\cdots+\alpha_1^n\lambda_n\\ \gamma_2=\alpha_2^1\lambda_1+\alpha_2^2\lambda_2+\cdots+\alpha_2^n\lambda_n\\ \cdots\cdots\cdots\cdots\cdots\cdots\\ \gamma_n=\alpha_n^1\lambda_1+\alpha_n^2\lambda_2+\cdots+\alpha_n^n\lambda_n. \end{array}\right.$$

The system of equations I giving the n *unknowns* γ *is replaced by the system III giving the* n *unknowns* λ *allowing, using II, to compute the values of the* γ*'s.*

$$III\left\{\begin{array}{l} A_1^1\lambda_1+A_2^1\lambda_2+\cdots+A_n^1\lambda_n+C_1=0\\ A_1^2\lambda_1+A_2^2\lambda_2+\cdots+A_n^2\lambda_n+C_2=0\\ \cdots\cdots\cdots\cdots\cdots\cdots\\ A_1^n\lambda_1+A_2^n\lambda_2+\cdots+A_n^n\lambda_n+C_n=0. \end{array}\right.$$

Generally, we have

$$IV)\quad \begin{array}{l} A_p^p=\sum_{k=1}^{k=n}(\alpha_k^p)^2\\ A_p^q=\sum_{k=1}^{k=n}\alpha_k^p\alpha_k^q. \end{array}$$

The coefficient A_p^q *is obtained by the product of the coefficients of the rows* p *and* q *of the system I which are in the same column and effecting the sum of the* *products thus obtained in the n columns, which can be symbolically expressed by saying that A_p^q is the product of the row p by the row q.*

As the order of the factors can be reversed in each product, we immediately see that

$$A_p^q = A_q^p$$

in the determinant of the system III the terms symmetric with respect to the diagonal are equal, i.e. the system of equations with λ is symmetric.

Therefore, we propose to solve a system of equations of the form III. Notice, from what precedes, that the system of equations II, if the unknowns γ were assumed to be known, will be a system of equations in the λ's, equivalent to the system III. It would therefore be a way to solve the system III if it was possible to find a system I allowing to easily compute the γ's.

This is what happens if, in the system I

the first equation only contains γ_1
the 2nd γ_1 *and* γ_2
the 3rd γ_1, γ_2 *and* γ_3

and so on. We can indeed compute similarly all the γ's successively from γ_1.

The problem is thus reduced to the search of the system

$$V)\left\{\begin{array}{ll}\alpha_1^1\gamma_1 & +C_1=0\\ \alpha_1^2\gamma_1+\alpha_2^2\gamma_2 & +C_2=0\\ \alpha_1^3\gamma_1+\alpha_2^3\gamma_2+\alpha_3^3\gamma_3 & +C_3=0\\ \dots\dots\dots\dots\dots\dots & \dots\dots\\ \alpha_1^n\gamma_1+\alpha_2^n\gamma_2+\alpha_3^n\gamma_3+\cdots+\alpha_n^n\gamma_n & +C_n=0.\end{array}\right.$$

This system being indeed found, the problem becomes very easy since the system II is replaced by the system IV which allows to compute the λ's step by step from λ_n.

$$VI)\left\{\begin{array}{llllll}\alpha_1^1\lambda_1+ & \alpha_1^2\lambda_2+ & \cdots & \cdots & +\alpha_1^n\lambda_n-\gamma_1=0\\ & \alpha_2^2\lambda_2+ & \alpha_2^3\lambda_3 & \cdots & +\alpha_2^n\lambda_n-\gamma_2=0\\ & & \alpha_3^3\lambda_3 & \cdots & +\alpha_3^n\lambda_n-\gamma_3=0\\ & & \cdots & \cdots & \dots\dots\dots\\ & & & & \alpha_n^n\lambda_n-\gamma_n=0.\end{array}\right.$$

We will easily compute the coefficients α starting from the coefficients A of the system III, by applying the general relations IV) to the system V. This shows that one can compute row by row all the coefficients of the system VI

$$1^{th}\ row\ \left\{\begin{array}{lll} A_1^1 = (\alpha_1^1)^2 & \textit{from which} & \alpha_1^1 = \sqrt{A_1^1} \\ A_2^1 = \alpha_1^1 \alpha_1^2 & & \alpha_1^2 = \dfrac{A_2^1}{\alpha_1^1} \\ \cdots\cdots\cdots & & \\ A_p^1 = \alpha_1^1 \alpha_1^p & & \alpha_1^p = \dfrac{A_p^1}{\alpha_1^1} \\ A_n^1 = \alpha_1^1 \alpha_1^n & & \alpha_1^n = \dfrac{A_n^1}{\alpha_1^1} \end{array}\right.$$

2nd row
the α_1 are already known by the computation of the 1th row

$$\left\{\begin{array}{ll} A_2^2 = (\alpha_1^2)^2 + (\alpha_2^2)^2 & \alpha_2^2 = \sqrt{A_2^2 - (\alpha_1^2)^2} \\ A_3^2 = \alpha_1^2 \alpha_1^3 + \alpha_2^2 \alpha_2^3 & \alpha_2^3 = \dfrac{A_3^2 - \alpha_1^2 \alpha_1^3}{\alpha_2^2} \\ A_p^2 = \alpha_1^2 \alpha_1^p + \alpha_2^2 \alpha_2^p & \alpha_2^p = \dfrac{A_p^2 - \alpha_1^2 \alpha_1^p}{\alpha_2^2} \end{array}\right.$$

p^{th} row
All the α with a lower index smaller than p are known from the computation of the preceding rows.

$$\left\{\begin{array}{l} A_p^p = (\alpha_1^p)^2 + (\alpha_2^p)^2 + (\alpha_3^p)^2 + \cdots + (\alpha_{p-1}^p)^2 + (\alpha_p^p)^2 \\ \alpha_p^p = \sqrt{A_p^p - (\alpha_1^p)^2 - (\alpha_2^p)^2 - (\alpha_3^p)^2 - \cdots - (\alpha_{p-1}^p)^2} \\ A_p^q = \alpha_1^p \alpha_1^q + \alpha_2^p \alpha_2^q + \alpha_3^p \alpha_3^q + \cdots + \alpha_{p-1}^p \alpha_{p-1}^q + \alpha_p^p \alpha_p^q \\ \alpha_p^q = \dfrac{A_p^q - \alpha_1^p \alpha_1^q - \alpha_2^p \alpha_2^q - \cdots - \alpha_{p-1}^p \alpha_{q-1}^q}{\alpha_p^p} \\ q > p. \end{array}\right.$$

As for the computation of the γ's, it is easily obtained using the equations V. We obtain

$$\begin{array}{lcl} (-\gamma_1) & = & \dfrac{C_1}{\alpha_1^1} \\ (-\gamma_2) & = & \dfrac{C_2 - \alpha_1^2(-\gamma_1)}{\alpha_2^2} \\ \cdots & \cdots & \cdots\cdots\cdots\cdots\cdots\cdots \\ (-\gamma_p) & = & \dfrac{C_p - \alpha_1^p(-\gamma_1) - \alpha_2^p(-\gamma_2) - \cdots}{\alpha_p^p} \end{array}$$

which shows that the coefficients $(-\gamma)$ which appear in the table of the equations VI) are computed with respect to the constant terms C of the table III exactly in the same way as the coefficients α with respect to the A's.

The computations can be presented in a convenient way into a single table. The given equations being symmetric, it is sufficient to write, in the table, the coefficients in a single side of the diagonal, above for example.

The transformed equations of system VI can then be displayed under the diagonal symmetrically to the given equations, each new equation occupying a column below the diagonal.

The writings are restricted to the replication of the coefficients α; indeed, the computation of a coefficient α of the form $\frac{\sum mn}{K}$ is realized on a calculator by a sequence of multiplications which are automatically and algebraically added on the machine, the algebraic sum being immediately divided by K. Thus, the most complicated operation that the machine is able to perform is used, then the maximal output of this machine is obtained, while staying as much as possible away from the errors [deleted and replaced by "faults"], *frequent in the transcription of the digits read.*

Simplifications in the application of the formulae or precious indications are found in the layout of the computations, and in the use of the calculator as well. Let us give one single example: the calculators of the "Dactyle" type record the quotient in white or red digits according if the division is performed by turning [translator: the crank] *in the direction of addition or substraction, and it follows that every coefficient α being the result of a division on the machine, its sign is indicated by the color of the digits that compose it; sign errors are thus easily avoided.*

It seems useless to dwell on the rest of the resolution, that is on the computation of the λ's from the system VI). Indeed, we immediately see how we can turn back from λ_n to λ_{n-1}, then to λ_{n-2}, and so on until λ_1.

We can highlight the advantages of this method of solving linear systems in terms of the approximation with which the results are obtained.

Every solution method must necessarily lead to a system of equations of the type of the system VI allowing to obtain directly one of the unknowns and to determine successively all the others. Assume that we have been led to the system:

$$
VII \left\{
\begin{array}{lllr}
\delta_1^1\lambda_1 + & \delta_1^2\lambda_2 + & \cdots & +\delta_1^n\lambda_n - \varepsilon_1 = 0 \\
 & \delta_2^2\lambda_2 + & \cdots & +\delta_2^n\lambda_n - \varepsilon_2 = 0 \\
 & & \cdots & \cdots\cdots\cdots\cdots\cdots \\
 & & & \delta_n^n\lambda_n - \varepsilon_n = 0
\end{array}
\right.
$$

We can put in correspondence this system with a second system giving the values of the ε's, that is:

$$
VIII \left\{
\begin{array}{lr}
\beta_1^1\varepsilon_1 & +C_1 = 0 \\
\beta_1^2\varepsilon_1 + \beta_2^2\varepsilon_2 & +C_2 = 0 \\
\cdots\cdots\cdots\cdots\cdots\cdots & \cdots\cdots \\
\beta_1^n\varepsilon_1 + \beta_2^n\varepsilon_2 + \cdots + \beta_n^n\varepsilon_n & +C_n = 0.
\end{array}
\right.
$$

Formulae IV are then replaced by the following ones:

$$A_p^p = \sum_{k=1}^{k=p} (\beta_k^p \delta_k^p)$$
$$A_p^q = \sum_{k=1}^{k=p} (\beta_k^p \delta_k^q).$$

From what we can conclude that the unique system of the coefficients α that we previously used is replaced in all the other modes of resolution by a double system of coefficients β and δ such that we always have $\beta_p^q \delta_p^q = (\alpha_p^q)^2$.

But the calculations are performed necessarily with a limited accuracy and we are led, to avoid errors [deleted and replaced by "faults"] *and make the computation as simple as possible, to compute all the numbers used with a fixed number of figures. It follows that the numbers α, β, δ are affected by an error η depending on the neglected figures and independent of the magnitude of the number calculated.*

The use of the quantity $(\alpha_p^q)^2$ in the computations corresponds to the introduction of an error $2\alpha_p^q \eta$.

The use of the equal quantity $(\beta_p^q \delta_p^q)$ corresponds to the introduction of the error $(\beta_p^q + \delta_p^q)\eta$.

We know that the product $\beta_p^q \delta_p^q$ being constant, the sum of its two factors attains its minimum when they are equal. Thus, the smallest error that could be introduced is

$$2\alpha_p^q \eta.$$

As a result, the mode of resolution of linear systems which has just been exposed appears as the one which provides the best approximation of the computations.

This property allows to reduce to its strict minimum the number of figures to be used in the computations, it widely compensates the small inconvenience of using the square root in the solution of linear equations.

Especially since the square root can be easily and rapidly obtained with a calculator by the following procedure which is completely different from the processes usually indicated by calculators' manufacturers.

Let us extract the square root of a number N.

Assume that a number n close to the required root r is known.

As to fix the ideas

$$\begin{aligned} r &= n + \varepsilon \\ N &= r^2 = (n+\varepsilon)^2 = n^2 + 2n\varepsilon + \varepsilon^2. \end{aligned}$$

If ε^2 has its last digit one order less than what we want to compute, we have the right to write

$$N = n^2 + 2n\varepsilon = n(n + 2\varepsilon)$$

that is, dividing N by n, we have as the quotient

$$n + 2\varepsilon.$$

2ε represents the excess of this quotient on the divisor n, and we obtain r by adding the half of this excess to n.

Practically, it is advantageous to have at one's disposal a table of squares which gives at first sight the square root of any number with 3 exact significant figures.

$$\frac{\varepsilon}{n} \text{ is then smaller than } \frac{1}{10^2}$$

$$\frac{\varepsilon^2}{n^2} \qquad \frac{1}{10^4}.$$

The first division gives the square root with 5 significant figures.

$$\frac{\varepsilon^4}{n^4} \text{ is smaller than à } \frac{1}{10^8}$$

thus the 2nd division made with the root having 5 exact figures would give 9 exact significant digits and so on.

We can state a simple rule assuming that the number has a decimal point at the right of the first digit on the left. Under these conditions each division doubles the number of decimal places of the root.

Using this method, an exercised operator gets in a few seconds the square root of a number of 5 digits with the same number of exact figures.

The method for solving linear systems which has just been exposed was supplemented by an adaptation of the verification procedure indicated by Gauss under the name of proof by sums.

The verification is obtained as follows: we juxtapose to the constant term C_p of the equation of rank p a term V_p given by the relation $-V_p = A_1^p + A_2^p + \cdots + A_n^p + C_p$.
Under these conditions the sum of the numbers listed in the row p of the system of equations is zero. If we treat V_p in the resolution the same way as C_p, this linear relation will be preserved and it will still be true for the coefficients α.

Moreover, it will still be possible to verify the computation of the λ's from the system of equations VI), since if we replace in the equations III the constant terms C by the checking terms V, the operation is equivalent to changing λ into $(1-\lambda)$; the computation of the unknowns made with the V's therefore gives the values λ' such that $\lambda_p + \lambda'_p = 1$.

It is possible, operating as just said, to succeed for sure and in a short time the resolution of very complex systems of equations.

The solution of a system of 10 equations with 10 unknowns can be obtained with 5 exact figures in 4 to 5 hours, including the verification of the equations and the computation of the residuals.

By this method, several systems over 30 equations were solved and, in particular, a system of 56 equations. This last case is part of an adjustment calculation of the altitudes of the primordial chains of the triangulation of Algeria. Because of the importance of calculations and to avoid congestion, we had to adopt a special arrangement, but the computations have been conducted exactly as has been said.

Vincennes 2 December 1910

Signature

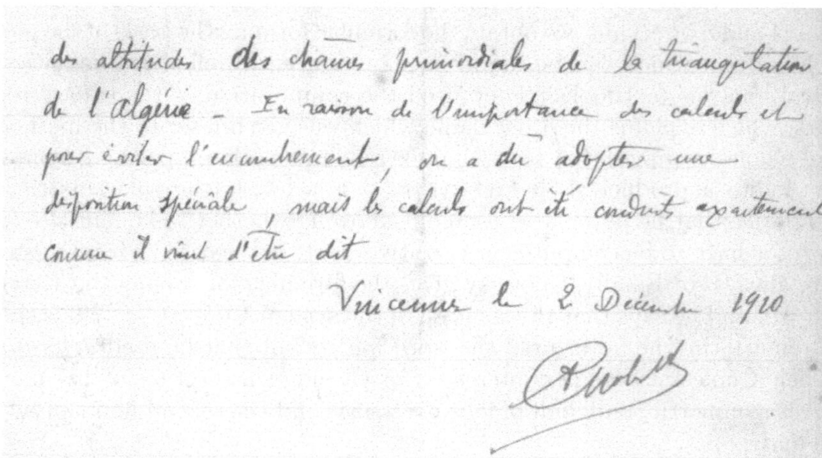
des altitudes des chaînes primordiales de la triangulation
de l'Algérie – En raison de l'importance des calculs et
pour éviter l'encombrement, on a dû adopter une
disposition spéciale, mais les calculs ont été conduits exactement
comme il vient d'être dit

Vincennes le 2 Décembre 1910

Last page of the manuscript

4.5 Analysis of the manuscript

Let us now follow, step by step, the presentation given by Cholesky. He begins by considering the square linear system I: $\alpha\gamma + C = 0$, where α is a $n \times n$ matrix, and γ and C are vectors of dimension n. Then, he sets II: $\gamma = \alpha^T \lambda$. Thus I becomes III: $A\lambda + C = 0$. Next, he gives the formulae IV which allow one to compute the elements of the matrix A: the element of A which is at the intersection of column p and row q is the scalar product of the rows p and q of the matrix α of the system I. He notices that $A = \alpha\alpha^T$ and that, since the order of the factors can be inverted in the product, A is symmetric.

Cholesky therefore proposes to solve a system of the form III. He observes that, if γ is known, then the system II is equivalent to III, but with the unknown λ.

Thus, the system III can be solved if it is possible to find a system I allowing one to easily compute γ.

This is what happens if the matrix α of the system I is a lower triangular matrix (that is all the elements above the main diagonal are zero). Indeed, the first equation only contains γ_1, the second one only contains γ_1 and γ_2, and so on. Thus, one has to find a system V: $\alpha\gamma + C = 0$, with α lower triangular. Having obtained γ, the system II becomes the system VI: $\alpha^T\lambda - \gamma = 0$, which can be solved gradually from λ_n since α^T is an upper triangular matrix.

Now, it remains to compute the elements of the lower triangular matrix α. For that purpose, one has to use the formulae IV which give the elements of A in terms of those of α by identification of the corresponding elements in the matrices A and $\alpha\alpha^T$. Thus, we obtain the formulae forming the basis of the method of Cholesky and which can be found in many current applied mathematics and numerical analysis textbooks. They need the computation of the square root of numbers, which explains the other name which was given later to the method. In passing, Cholesky proves that his technique amounts to decomposing a symmetric matrix A into a product $A = \alpha\alpha^T$ where α is a lower triangular matrix. We have to notice that he never worries about knowing whether the quantities whose square roots have to be computed are positive. But, in the case he is interested in, they are always so. Finally, Cholesky gives the formulae for solving the system V: $\alpha\gamma + C = 0$, and he says that the solution of the system VI: $\alpha^T\lambda - \gamma = 0$ is similar. Let us remark that he never used the word *matrix* but that he used *determinant*.

Then Cholesky becomes interested in the implementation of his method. Since A is symmetric, only half of it is necessary, and the second half can be used for the matrix α.

The computation of the elements of the matrix α needs an algebraic sum of products. This sum is automatically performed on a pinwheel *Dactyle* calculating machine, using all its capabilities. On the other hand, this machine is free of sign errors for the operator since the result appears in white or red according to its sign. These *Dactyle* calculators were built in Foncine–le–Haut in the Jura massif by the company *Château Frères*, 125 Boulevard de Grenelle in Paris, until the early 1950. They were invented by the Swedish engineer Willgodt–Theophil Odhner (Dalby, Värmland, Sweden, 10 August 1845 - Saint Petersburg, 15 September 1905, Gregorian calendar) around 1878. The patent having fallen into the public domain in 1906, many copies were manufactured all over the world, some including improvements.

Then, Cholesky discusses the advantages of his method from the point of view of numerical precision. He considers a general system $A\lambda + C = 0$, and he replaces it by the successive solutions of $\beta\varepsilon + C = 0$, where β is a lower triangular matrix, and of the system $\delta\lambda - \varepsilon = 0$, where δ is an upper triangular matrix. Thus $A = \beta\delta$, and then, by identification, the formulae giving the elements of these two matrices. The product of the elements of β and δ located in row p and in column q is equal to the square of the corresponding element of α, that

A Dactyle calculating machine
© Freddy Haeghens - Particular collection
http://users.skynet.be/Fredscalculators/p7.htm

is $\beta_{pq}\delta_{pq} = \alpha_{pq}^2$. The computations are necessarily done with a limited precision (fixed point arithmetic). Thus, these numbers are affected by an error η. The computation of α_{pq}^2 introduces, at the first order, an error $2\alpha_{pq}\eta$ and that of $\beta_{pq}\delta_{pq}$ an error $(\beta_{pq} + \delta_{pq})\eta$. Since $\beta_{pq}\delta_{pq} = \alpha_{pq}^2$, we thus have $(\beta_{pq} + \delta_{pq})\eta = (\beta_{pq} + \alpha_{pq}^2/\beta_{pq})\eta$. Differentiating this expression with respect to β_{pq}, we see that this error is minimum when $\beta_{pq} = \alpha_{pq}$. Thus, we must have $\delta_{pq} = \beta_{pq}$, and Cholesky concludes that his method, where the two matrices are transposed one from each other, is the method leading to the least numerical error. We see that Cholesky did a real numerical analysis work, in the modern sense of the word.

And he proceeds. His method needs the computation of square roots. He proposes a procedure which differs from the one recommended by the calculator's manufacturers. Let us consider the computation of $r = \sqrt{N}$, and let n be an approximation of this square root. He sets $r = n + \varepsilon$, and thus

$$N = r^2 = (n + \varepsilon)^2 = n^2 + 2n\varepsilon + \varepsilon^2 \simeq n(n + 2\varepsilon),$$

at the first order. Then, he deduces that

$$\varepsilon \simeq \frac{1}{2}\left(\frac{N}{n} - n\right),$$

and that a better approximation of r is obtained by adding this quantity to n. Indeed, we obtain

$$r \simeq n + \frac{1}{2}\left(\frac{N}{n} - n\right) = \frac{1}{2}\left(\frac{N}{n} + n\right),$$

and the process can be iterated. This is nothing else than the method of Hero of Alexandria (a byproduct of Newton's method for solving the equation $r^2 - N = 0$).

Then, Cholesky discusses the speed of convergence of this method for extracting the square root. Assume we have at our disposal a numerical table giving the square root with 3 exact figures. Thus ε/n is smaller than 10^{-2} and its square is smaller than 10^{-4}. Consequently, the first iteration leads to 5 exact significant figures. Therefore, $(\varepsilon/n)^2$ is smaller than 10^{-8}, and the second iteration furnishes 9 exact figures. Cholesky concludes that, if the numerical table gives the square root with only one exact significant figure, then the number of exact figures is doubled at each iteration. In modern terms, the method has order 2, it converges quadratically.

Next, Cholesky explains a procedure for verifying the correctness of the computations. For that, he considers the system $A\lambda' - V = 0$, where $V = Ae + C$ with $e = (1, \dots, 1)^T$. Solving this system by the same method leads to $\alpha\gamma' - V = 0$, whose solution gives the vector γ'. Then λ' is the solution of the system $\alpha^T\lambda' = \gamma'$. Cholesky expresses this property by asserting that this linear relation remains valid and is still true for the coefficients α. Thus, it holds that $A\lambda' - V = A\lambda' - Ae - C = A(\lambda' - e) - C = 0$, which shows, by comparison with the initial system $A\lambda + C = 0$, that, for all p, $\lambda_p + \lambda'_p = 1$. Thus, this is a procedure for verifying the computation at each stage, step by step.

Cholesky ends his manuscript by reporting the computing time needed for solving various systems by his method.

To conclude this analysis, let us say that this manuscript of Cholesky constitutes a perfect piece of research in numerical analysis, quite remarkable for the time (and even for ours when we see some papers submitted for publication to journals): motivation of the problem, presentation and theoretical justification of a new algorithm, study of its numerical stability, explanations about its implementation by hand and on a calculator, study of the numerical problems due to the finite precision of the computations, method for extracting the square root and study of its convergence, procedure for verifying the results, commentaries on the computing time. If this paper was submitted today to a numerical analysis journal, it would be recommended for publication without any hesitation. Nowadays, Cholesky's method still has a major importance.

In all the works of Cholesky, constant care is taken to propose not only good solutions, often even new ones, but which are also easy to use by people without an advanced knowledge in mathematics. For that purpose, he also developed verification processes, showed how to construct tables easy to fill up, and improved the manipulation of the arithmetical operations on hand calculators.

4.6 After Cholesky

The Dutch mathematician, physicist and astronomer Willem de Sitter (Sneek, The Netherlands, 6 May 1872 - Leiden, 19 November 1934) described a method for solving the normal equations in his inaugural lecture entitled *The New Methods in Celestial Mechanics* at the University of Leiden in 1908, and it was currently used at the Astronomical Laboratory of Groningen. This method was indeed easy to use even by people inexperienced in numerical calculation. Less than three hours were needed for solving a system of dimension 6. It seemed to consist mainly of some practical arrangements for implementing the method of Gauss. De Sitter also provided a procedure for verifying the solution obtained. In 1915, he again exposed this method in [82, Appendix, pp. 160–173].

A simplified form of the Doolittle's method is due to Frederick Vail Waugh (1898 - Arlington Hospital, 17 February 1974) in 1935 [241]. Variants of the Gaussian elimination method were also studied by Alexander Craig Aitken (Dunedin, New Zealand, 1 April 1895 - Edinburgh, 3 November 1967) in 1932 [25], Prescott Durand Crout (28 July 1907 - Switzerland or Lexington, 25 September1984) in 1941 [75, 76], Harold Hotelling (Fulda, Minnesota, USA, 29 September 1895 - Chapel Hill, North Carolina, USA, 26 December 1973) in 1943 [135], and finally Paul Sumner Dwyer in 1944 [94, 95].

In 1938, the Polish astronomer Tadeusz Banachiewicz (Warsaw, 13 February 1882 - Kraków, 17 November 1954) proposed a *square root method* [35–37] very similar to that of Cholesky where a decomposition procedure for an arbitrary matrix is given. However the language used was that of *Krakowians*, mathematical objects invented by Banachiewicz which are quite similar to matrices but with a different law of multiplication. He was the first to formulate the fact that elimination methods correspond, in fact, to the factorization of the matrix A into a product of two matrices. But he was preceded by Cholesky in this interpretation for symmetric matrices. It can be seen from the paper [30] of the Belgian astronomer Sylvain Julien Victor Arend (Robelmont, Luxembourg, 6 August 1902 - Brussels, 18 February 1992), dated 1941, that the square root method amounts to Cholesky's factorization. In his obituary written in 1992, after explaining that Arend went to Kraków to study Krakowians with their founder, its author wrote [87]:

> *Back to Belgium, he diffused the cracovian calculation in our country and in neighboring countries without much success, given the regrettable incomprehension of the scientific world - still present - about the practicality of this method.*

A comparison of Cholesky's, Banachiewicz's and Gauss's methods is given in [127]. Historical remarks on these methods and a discussion about priority can be found in [210].

In 1941, Dwyer provided an abridged version of the Doolittle's method and linked it to other resolution methods [95]. In 1944, he gave the matrix interpreta-

Tadeusz Banachiewicz
© Public domain

tion of the method of Doolittle [96]. He also showed that $L = DU^T$, where D is a diagonal matrix, and he noticed that it would be more interesting if the matrices L and U^T were the same in order to perform only half of the computations. For this, it would be enough to take the square roots of the diagonal terms, that is to say, of the matrix D, and he noticed that the method obtained in that way will be similar to that of Banachiewicz. This is indeed Cholesky's method.

In 1951, Paul Sumner Dwyer wrote a book on linear computations [98] where, Section 6.5, pp. 113–118, he described the square root method. He wrote:

> *The square root method has been worked out independently by a number of different authors. Banachiewicz published it in 1938 in Poland [35, 37]. It has more recently been advocated by Dwyer [97], Duncan and Kenney [91], and Laderman [151] as an excellent method for studying least squares, correlation, and regression problems. Articles [108, 131,*

136, 232] have traced the method to the earlier work of Cholesky and Schur. It is doubtful that such a simple method was not considered by authors prior to Cholesky. Before the days of computing machines, the method would probably have been discarded as impractical by any author who considered it.

Dwyer gave no reference to Cholesky or Schur. According to Derrick Shannon Tracy (India, 1 July 1933 - 27 December 1998, New Delhi airport, India), a Ph.D. student of Dwyer, he was not willing to credit Cholesky with his procedure since he did not publish it in his lifetime [2].

As explained in [121], the mathematician Ewald Konrad Bodewig (Bruckhausen, Hamborn, 22 July 1901 - ?) followed Jensen's approach [143] (see below) in a five–part paper that summarized, in matrix notation, all methods for solving linear equations that were known until 1947. He emphasized that Cholesky's method corresponds to a factorization of A into the product of a lower triangular matrix by its transpose [3].

However, none of these contributions, except that of Banachiewicz in 1938, is similar to Cholesky's method. Moreover, they are all posterior to it.

For more details about the other methods for solving systems of linear equations and the use of the least squares method in topography, see [54] and [67, Chap. 9]. On the chronology of these various works, see [120, 121]. An account on the history of geodesy, topography and cartography is given in [53, 57].

A bibliography on Cholesky, Doolittle and Crout factorizations can be found at http://math.fullerton.edu/mathews/n2003/cholesky/CholeskyBib/Links/CholeskyBib_lnk_3.html.

4.7 Diffusion of Cholesky's method

Cholesky's method was first made public and discussed in a paper dated 1924 [42], that is 6 years after its author's death, by a Commandant Benoît, of the *Artillerie Coloniale*, a former geodesist officer of the Geographical Service of the Army, and of the Geographical Service of Indochina, member of the French National Committee of Geodesy and Geophysics.

In his paper, Benoît begins by explaining that, in the adjustment of networks, one has to solve a linear system with more unknowns than equations (the condition equations) and that, to distort as little as possible the triangles that have been observed, one has to look for the solution with the minimum Euclidean

---

[2] Mail by Richard W. Farebrother.

[3] It seems that almost nothing is known of Bodewig's life except that, in 1931, he defended a thesis at the Rheinische Friedrich–Wilhelms Universität Bonn with the title *Die Stellung des Hl. Thomas von Aquino zur Mathematik.* Then, he went to Leipzig. It seems that, on 20 February 1936, he defended (in Basel?) an habilitation: *Biologische Massenerscheinung in mathematischer Beleuchtung.* He translated into German the book of L.E. Dickson on number theory. In 1954, he is found in The Hague, Netherlands, and worked for the Geodetic Commission. See http://bbf.dipf.de/kataloge/archivdatenbank/digiakt.pl?id=p86122.

norm. As already explained, this minimization problem is solved by setting to zero the partial derivatives of a special function which is expressed using the Lagrange multipliers. We thus arrive at the least squares method of Legendre and at the normal equations whose unknowns are these Lagrange multipliers. Benoît explains how, for solving these normal equations, Cholesky factorizes the matrix into the product of a lower triangular matrix by its transpose. Identifying the corresponding terms, he obtains the same formulae as Cholesky, but his presentation is somewhat different. Then Benoît explains how to display the computation into one single array in order to simplify them. Finally, he describes the completion of the calculations and the verification of the results as they were presented by Cholesky, and he illustrates the method by a complete numerical example.

As stated in [120, 121], it seems that, after its discovery, Cholesky's method was only known by geodesists and topographers. In a book written in 1912 by Charles Lallemand [157], whose brother was the head of the Section of Geodesy during Cholesky's campaign in North Africa in 1911–1912, the normal equations are solved by an elimination method similar to Gauss' which consists in a linear combination of the equations. The name of Cholesky is not mentioned. In Sweden, Cholesky's method was used by Tryggve Rubin (Stockholm, 1 March 1874 - Stockholm, 26 July 1946) in 1926 who supplemented it to produce other statistically relevant quantities [202]. In [24], the geographer Hans Jakob Konrad Wilhelmsson Ahlmann (Karlsborg, 14 November 1889 - Hägersten, 10 March 1974) and the cartographer Ludvig Rosenbaum explain in 1933 that:

> *The normal equations have been solved by the Cholesky–Rubin method, which offers the advantage that the solution is easily effected on a calculating machine (Cholesky) and that the most probable values and their mean errors are derived simultaneously (Rubin).*

In 1944, the Danish geodesist Henry Jensen, who will become the first professor of Geophysics at University of Copenhagen in 1965, wrote, after describing the method [143]:

> *The method is very quick and easy in practice. It ought to be more generally used than is the case. It is due to Cholesky (Bull. Géod., No. 2) and was later indicated by Banchiewicz...*
>
> *Cholesky's method seems to possess all advantages.*

The American geodesist Charles Arthur Whitten (Redfield, South Dakota, 2 October 1909 - Silver Spring, Maryland, 12 July 1994) of the Coast and Geodetic Survey initiated, in 1950, the use of computers for solving large (for that time) systems of linear equations in a variation of the coordinate method and introduced the Cholesky's method as an improvement of the standard Gauss–Doolittle's method.

The decisive spreading of Cholesky's method is due to John (Jack) Todd (Carnacally, Ireland, 16 May 1911 - Pasadena, 21 June 2007) who taught it in his numerical analysis course at King's College in London in 1946, and made it known. With his wife, the mathematician Olga Taussky (Olmütz, Austro–Hungarian Empire, now Olomouc, Czech Republic, 30 August 1906 - Pasadena, 7 October 1995), they relate [226]:

> *In 1946 one of us (J.T.) offered a course at Kings' College, London (KCL) on Numerical Mathematics. While we had some wartime experience in numerical mathematics, including characteristic values of matrices, we had had little to do with the solution of systems of linear equations. In order to see how this topic should be presented, we made a survey of Math. Rev. (at that time easy!) and found a review (MR7 (1944), 488), of a paper by Henry Jensen [143], written by E. Bodewig. Jensen stated* Cholesky's method seems to possess all advantages. *So, it was decided to follow Cholesky and, since the method was clearly explained, we did not try to find the original paper. Leslie Fox, then in the newly formed Mathematics Division of the (British) National Physical Laboratory (NPL), audited the course and apparently found the Cholesky Method attractive, for he took it back to NPL, where he and his colleagues studied it deeply [108, 232]. From these papers, the Cholesky (or sometimes Choleski) Method made its way into the tool boxes of numerical linear algebraists via the textbooks of the 1950s.*
>
> *In 1987 we complained to A.W. Tucker and G.H. Golub on the spelling of Cholesky, and Golub got R.W. Cottle to send us his file on Cholesky, of whom he was an admirer. This included his translation of an obituary of Cholesky [41], which we hoped he would publish. Now, in 1992, we have something to add to the story; when on a search on a different topic and turning the pages of MacDuffee [171], the following sentence hit us:* O. Toeplitz proved that, if H is hermitian, there exists a matrix P with 0's above the diagonal such that
>
> $$H = PP^{CT''}.$$
>
> *This paper [227] was presented by Hilbert on 23 February 1907 and therefore most likely before the work of Cholesky, who entered the French Army in 1905, was killed in action in 1918, but whose work was only published in 1924 [42]. It is necessary to discuss and compare the work of Toeplitz and Cholesky in a little more detail [4]. Naturally, the discussions go back to Gauss as geodesist, and as the founder of the theory of quadratic forms.*

As its title indicates, *Mathematical Reviews* (Math. Rev.) is a journal which publishes reviews of mathematical papers and books. Here is the review by Henry Jensen:

[4] See [224] for this discussion. This was their last paper.

MR0015921 (7,488d) 65.0X

Jensen, Henry

An attempt at a systematic classification of some methods for the solution of normal equations.

Geodætisk Institut, København, Meddelelse 1944, (1944) No. 18, 45 pp.

Explaining and comparing the usual methods for a direct solution (that is, noniterative) of linear equations, the author arrives at the following conclusions. The determinantal solution by Chiò's rule and the method of equal coefficients are only practicable in the case of few equations. Gauss's method gives a very plain solution for all systems. Banachiewicz's method of Cracovians is not superior to the Gaussian algorithm. Boltz's and Krüger's methods are useful in special systems, while the best of all methods seems to be Cholesky's. The explanations are illustrated by two examples which are computed by all methods or a part of them.

Reviewed by E. Bodewig

Thus, after the lectures given by John Todd, Cholesky's method was known by numerical analysts. Let us quote Todd again [226, p. 257]:

In early postwar years, local and national centers for applied mathematics, and in particular for computing, were organized in many countries.

I returned to King's College, London, and gave my first course in numerical mathematics in 1946. We had two Marchand ACT 10M machines for that class. In this course I introduced the Cholesky method as the preferred one for the solution of positive definite systems $Ax = b$. This was taken up by L. Fox, who analyzed it deeply with his National Physical Laboratory colleagues H.D. Huskey, J.H. Wilkinson , and Turing. It has indeed become one of the workhorses of numerical linear algebraists.

In a paper sent for publication on 14 October 1947, Leslie Fox (Dewsbury, 30 September 1918 - Oxford, 1 August 1992), Harry Douglas Huskey (born in Bryson on 19 January 1916) and James Hardy Wilkinson (Strood, 27 September 1919 - Teddington, 5 October 1986) [108] discuss, among others, Cholesky's method. They wrote:

Another method, due to Choleski, for the reciprocation of a symmetric matrix was pointed out to us recently by John Todd. This method is so simple in theory and straightforward in practice that its neglect in this

John Todd in 1977

country is surprising. The method, consists simply in expressing the symmetric matrix A in the form

$$A = LL',$$

where L is a lower triangular matrix and L', its transpose, an upper triangular matrix. Then

$$A^{-1} = L'^{-1}L^{-1}$$

and the problem is solved.

On 4 November 1947, Alan Mathison Turing (London, 23 June 1912 - Wilmslow, 7 June 1954), the well–known pioneer in computer science, sent to the same journal a paper on the propagation of rounding errors in methods for solving systems of equations and inverting matrices [232]. He generalized Cholesky's method to any matrix by factorizing it as $A = LDL^T$, where L is a lower triangular matrix with a unit diagonal and D a diagonal one. He named it the *unsymmetrical Choleski method* (notice, in these two papers, the error in the spelling), and studied rounding errors arising when programming these algorithms on a computer. When A is symmetric, the elements of D are positive and the factorization $A = (LD^{1/2})(LD^{1/2})^T$ is exactly Cholesky's. However, if A is not positive definite, $D^{1/2}$ *"may involve some purely imaginary numbers, but no strictly complex ones"*. For a general matrix, the factorization $A = L(DL)$ is exactly Gauss' [225]. Another seminal paper on the mathematical stability of a problem and the propagation of rounding errors due to the computer's arithmetic when solving a system of linear equations by an elimination technique was published in 1947 by Herman Heine Goldstine (Chicago, 13 September 1913 - Bryn Mawr, 16 June 2004) and John von Neumann (Budapest, 28 December 1903 - Washington, 8 February 1957) [239]. See [122] and [132, pp. 186ff. and pp. 224ff.] for a detailed analysis and the history of the work done on the propagation of floating point errors in Gaussian elimination and Cholesky's method.

Let us quote the journal M.T.A.C. [5], 2 (1946–1947) 289–290:

> *In 1942, in order to use more efficiently the scientific staff available in the Admiralty, the Director of Scientific Research set up, within the branch directed by Dr. J.A. Carroll, an Admiralty Computing Service to centralise, where possible, the computational and mathematical work arising in Admiralty Experimental Establishments.*
>
> *Mr. John Todd undertook the organisation and the supervision of the Service.*

In Germany near the end of the war, Todd prevented Allied forces from turning to rubble the Mathematical Research Institute in Oberwolfach (a well know congress center for mathematicians), where the University of Freiburg was sheltering its mathematicians. Todd successfully claimed the building for the Admiralty. It was *"probably the best thing I ever did for mathematics"*, he later said. For the complete history of this saving, see the notes he himself wrote: http://www.mfo.de/math-in-public/prizes/john-todd-award-1/article-oberwolfach-1945. John Todd was very influential in the development of numerical analysis during the war and the post–war periods both in the United Kingdom and in the United States where he later emigrated with his wife [173]. He and Olga Taussky were trained as pure mathematicians but, during the war, they had to perform calculations for physicists and, as he wrote:

> *This was rather frustrating: physicists were doing elementary computing badly and mathematicians like me were trying to do physics. I thought that I could see a way to improve this mismatching.*

[5] *Mathematical Tables and other Aids to Computation*, which preceded *Mathematics of Computation*.

In the journal *Mathematical Tables and other Aids to Computation*, 5 : 34 (April 1951), one can find the following:

> *37. The Square Root Method for Linear Equations.–In a letter dated 7 Feb. 1950 Mr. H.F. Rainsford of Colonial Surveys, Bushy Park, Teddington, England, commenting on the article entitled* The square root method for solving simultaneous linear equations *by J. Laderman in MTAC, v. 3, pp. 13–16, points out that this method was not* probably first discovered by Banachiewicz in 1938 *but goes back at least to Cholesky whose treatment of the problem was described by Benoît* [1] *in 1924. Can any reader supply an earlier reference to this method?*
>
> [1] *Benoît,* Note sur une méthode de résolution des équations normales provenant de l'application de la méthode des moindres carrés à un système d'équations linéaires en nombre inférieur à celui des inconnues. Application de la méthode à la résolution d'un système défini d'équations linéaires, *International Geodetic and Geophysical Union, Association of Geodesy, Bulletin Géodésique, No. 2, 1924, p. 67–77. An English translation of this article has been kindly supplied by Mr. Rainsford and is available in the UMT File.*

We were unable to find this translation.

In a famous paper published in 1953 [107], George Elmer Forsythe (State College, Pennsylvania, 8 January 1917 - Stanford, 9 April 1972), the founder and head of the Computer Science Department at Stanford University, wrote (the numbers into brackets refer to our bibliography):

> *The best known methods are based on systematic elimination of unknowns from equations in the system (1)* [that is $Ax = b$] *in the fashion of highschool algebra, as described by Gauss [114]. The elimination amounts to triangularizing the matrix A by premultiplying it by a triangular matrix, as Banachiewicz [35] and Turing [232] point out. The process can be rendered very efficient numerically by consolidating operations; see, for example, Benoît [42], Dwyer [98], and Turing [232]. When A is positive definite, the method is equivalent to the successive orthogonalization of the unit vectors in the A metric by the Gram–Schmidt process [201].*

According to William M. Kahan, Professor Emeritus at the University of California, Berkeley (USA) [6]:

> *As a graduate student at the University of Toronto starting in 1954, I recall seeing an old Friden electro–mechanical calculator with a square–root key used by statisticians (who were still part of the Mathematics department at that time) at least as early as the late 1940s. The calculator would have been ideal for Cholesky's method.*

[6] Private communication to me.

Carl Mauritz Fredrick Friden (Alvesta, Sweden, 11 April 1891- 29 April 1945) was a young Swedish engineer who arrived in the USA (via Australia) towards the end of the 1914–18 war. He established his own Friden Calculating Machine Company in 1933. Although he died in 1945, his company went on and, in 1952, produced the Model SRW square–root calculator [7]. This machine had *"the amazing ability to extract a square root at the touch of a single key... To extract a square root, the user could simply enter the number on the keyboard and then press the square root key at the base of the keyboard that corresponded to the position of the decimal point. The machine then produced the square root using a clever algorithm within a few seconds"* [8].

Another well known method for solving systems of linear equations is due to Alston Scott Householder (Rockford, Illinois, 5 May 1904 - Malibu, California, 4 July 1993). It consists in decomposing the matrix M of the system $Mx = b$ into the product of an orthogonal matrix Q (that is such that $Q^TQ = I$ which means that $Q^{-1} = Q^T$) by an upper triangular matrix R. Thus, the system $Mx = b$ reads $QRx = b$. Setting $y = Rx$, we obtain $Qy = b$ which gives $y = Q^Tb$. Then, solving the upper triangular system $Rx = y$ produces x. Householder related this QR–decomposition to Cholesky's method as follows. Since Cholesky has to solve a rectangular system, he multiplied his matrix A by its transpose, thus obtaining $M = A^TA$. A necessary and sufficient condition that M can be decomposed into LL^T by Cholesky's method is that $A = QL^T$. Indeed $M = A^TA = (LQ^T)QL^T = LL^T$ [138].

4.8 Early books

It is interesting to begin by quoting the preface of the book by William Edmund Milne (Pendleton, 1890 - 1971), a professor of mathematics at Oregon State College, which appeared in 1949 [182]. He wrote:

> *The growth of computational facilities in the period since World War I has already been phenomenal, and the possibilities in the near future are beyond imagining. The number of excellent calculating machines now available in almost every office or laboratory, to say nothing of the amazing Sequence Controlled Calculator, or the equally marvelous Electronic Numerical Integrator and Calculator, makes it possible to solve whole categories of problems that only yesterday were prohibitively difficult. In consequence the subject of numerical analysis is surely destined to make enormous strides in the decades to come.*
>
> *On the other hand our traditional courses in college and graduate mathematics too often turn out students poorly trained in the art of*

[7] See http://home.vicnet.net.au/ wolff/calculators/Friden/Friden.htm.
[8] Extracted from http://www.hpmuseum.org/srw.htm.

translating theoretical analysis into the concrete numerical results generally required in practical applications. As a physicist friend of mine once said, "You mathematicians know how to solve this problem but you can't actually do it".

Let us now quote Householder [139, pp. 59–97] on numerical analysis:

Numerical analysis, as a recognized discipline, may be said to have originated in the year 1947. In that year there occurred two events, quite independently of each other, which were to have considerable importance in the development of the subject. The first of these occurred in July. It was the establishment, by the National Bureau of Standards, of the National Applied Mathematics Laboratory, one branch of which was to be known as the Institute for Numerical Analysis. The second event, which occurred in November, was the publication of the now classical paper by von Neumann and Goldstine [239] [9] *entitled* Numerical Inverting of Matrices of High Order.

To the best of my knowledge, the phrase numerical analysis *had not been in common use prior to the establishment of INA. I think that the first book to use the phrase in the title was Hartree's [129] which appeared in 1952, and I think the second was my own [137] which appeared a year later. If you look among Hartree's references, you will find listed a* Numerical Analysis *by Milne, dated 1950, but there is no such book. There is, however, a* Numerical calculus *by Milne, [182] published in 1949. In 1947 there were two standard references in English. One of these was called* Numerical Mathematical Analysis *by Scarborough [209]; the other was the well-known classic by Whittaker and Robinson [243] called* The Calculus of Observations.

Let us point out that none of these references, except Hartree's book, mentions Cholesky. Householder continues and devotes a section to *"try to characterize what he* [I] *consider to be the business of a practicing numerical analyst"*. Then, he dedicates a large section to error analysis where he writes that *"Wilkinson and Gastinel, independently, have carried through the error analysis for the Gaussian inversion of a general matrix"*. Noël Gastinel (Le Muy, 25 December 1925 - Grenoble, 11 September 1984) was a pioneer, with Jacques–Louis Lions (Grasse, 3 May 1928 - Paris, 17 May 2001), of numerical analysis in France. Gastinel (my supervisor) defended his thesis [109] in 1960 under the supervision of Jean Kuntzmann (Pont–à–Mousson, 1 June 1912 - Grenoble, 19 December 1992), who created the first course of applied mathematics for engineers in France as soon as 1945 [148], and was the founder of the *Laboratoire de Calcul* in Grenoble. Then, practical sessions on calculators were introduced by Jean Marcel Laborde (1912 - 1997) in 1947 [149]. Although the power of computers had been multiplied by 10^6 between

[9] Our bibliography.

1939 and 1945, numerical mathematics were either completely ignored in France or considered as too technical, and it was common to consider that serious mathematics could only be axiomatic. Also in the United States, *"the social status of computation among theoretical mathematicians was low; this despite the fact that famous mathematicians has worked on numerical problems related to navigation, astronomy, geodesy, tides, etc."* [81]. On the development of numerical analysis, see [61].

André–Louis Cholesky, 2nd row, 5th from the left
© Cholesky's family

We tried to find out the earliest books which refer to Cholesky's method.

In Chapter 1, Section 6, of his book [182] Milne presented the solution of symmetrical systems of linear equations. His method consisted in a modification of Gauss's method where *"the labor of computing the solution is considerably reduced"*. It corresponds to the factorization $A = LDL^T$. His book also contains an annotated bibliography on early textbooks on numerical mathematics.

In the book of Rudolf Zurmühl (Soest, 14 September 1904 - Berlin, 27 October 1966) [253], published in 1950, the square root method is explained in Section 23.7, and the name of Benoît (with the corresponding reference) is quoted. Cholesky is also mentioned several times in [254]. In a later book [255], he wrote (posted by Hans J. Oser, Joint Policy Board for Mathematics, Washington, DC (USA) in NA Digest, Sunday, February 18, 1990, Volume 90, Issue 07):

The method has been used any number of times. For symmetric matrices it is mentioned first by M.H. Doolittle (U.S. Coast and Geodetic Report, 1878, pp. 115–120), later by Cholesky (op. cit.), and then with an explicit first-time reference to machinery and general systems, by T. Banachewicz (Bull. Internat. Academ. Polon. Sci., Ser. A, 1938, pp. 393–404)... The connection between Cholesky and the Gaussian elimination algorithm was not discovered until later.

It seems that the first book on numerical analysis, at least with this title, was written in 1952 by Douglas Rayner Hartree (Cambridge, 27 March 1897 - Cambridge, 12 February 1958) who also coined the expression [129]. Section 8.4, pp. 164–167, is devoted to Cholesky's method but, in fact, the method which is described is Gauss'. Let us quote him:

In the elimination method explained above the number of quantities written down in the process of solution of a system of n equations is of the order n^3. When n is more than about 6 or 8, there may be advantages in arranging the work so that the number of intermediate quantities recorded is considerably smaller, of order n^2 rather than of order n^3. The method usually ascribed to Choleski, of which there are several variants, is one in which the amount of recording of intermediate results is reduced in this way.

This book was quite influential for the development of numerical methods. Hartree was basically a theoretical physicist, and he first worked on the numerical integration of differential equations arising from atomic wave functions (the Hartree–Fock equations). During World War II, he used ENIAC (Electronic Numerical Integrator Analyser and Computer), the first entirely electronic and programmable computer, to calculate trajectories of projectiles. He also applied numerical methods to the physics of the atmosphere and to hydrodynamics.

Mario George Salvadori (Rome, 19 March 1907 - New York, 25 June 1997) was a professor of civil engineering at Columbia University. In his book [206], published in 1952, Section 1.8, pp. 23–28, is devoted to Cholesky's method. He wrote, in a footnote on p. 23:

Cholesky's method was used by A.L. Cholesky in France before 1916 in connection with symmetrical systems, was given in matrix form in Poland by Th. Banachiewicz in 1938, rediscovered and adapted to machine computations in the United States by P.D. Crout in 1941, studied again in England by A.M. Turing in 1948 and in Germany by A. Zurmühl in 1949.

On p. 25, he gave, in fact, a variant of Gauss's method with an upper triangular matrix with a unit diagonal, and he added:

The operations in Eqs. (1.8.1) and (1.8.2) can be performed on a calculating machine without writing any of the intermediate steps, that is, each element l_{ij}, t_{ij} is obtained in a single matrix operation.

Four years earlier, Salvadori already published a book for engineers [205] where he described Gauss's method but not Cholesky's. Thus, it may be conjectured that he learned it only after 1948.

In 1955, Andrew Donald Booth (Weybridge, 11 February 1918 - 29 November 2009), at that time a reader in computational methods at the University of London, wrote in his book [50]:

> *A third method of solution is usually known as the Choleski's process; it depends upon the reduction of a matrix A to the product of a lower triangular matrix L, with unit diagonal coefficients, and an upper triangular matrix U...*
>
> *It is stated than an advantage of the Choleski method lies in the fact that few intermediate results have to be written down; it is our experience, however, that the method is too complicated for* occasional *use, and is not suitable for use on an automatic computer.*

Obviously, he described Gauss's method and not Cholesky's, and his view was not quite prophetical. Let us mention that he is credited for the invention of the magnetic drum memory for computers.

In his book of 1956 [49], Ewald Konrad Bodewig devoted Section 2.5.3, pp. 110–112, to Cholesky's method. The name was also written Choleski at some places.

Cholesky is mentioned in a footnote of the book by Stephen Harry Candrall (born 2 December 1920), a professor of mechanical engineering at MIT, in 1956 [74].

The book of Richard Arthur Buckingham (17 July 1911 - 13 August 1994) [62], dated 1957, contains a description of the square root, or Choleski, method. He was the first and only director of the London University Institute of Computer Science, and then professor of Computer Education at Birkbeck College in London. He recognized before many of his colleagues that numerical computing would become increasingly important.

It seems that the first book in French where Cholesky's method (spelled Choleski) is described is due to Émile Victor Durand (Paris, 2 June 1911 - Toulouse, 15 January 1999). It was published in 1960–1961 [92, Vol. 2, pp. 74–77]. A physicist at the origin, he was interested in electronic optics which needed much numerical calculations. At the University of Toulouse, where he was a professor, he created in 1957 the *Institut de Calcul Numérique.*

However, it took some time before the method of Cholesky regularly appeared in numerical analysis textbooks, and was clearly attributed to him. The book of Vera Nikolaevna Faddeeva (Tambov, 20 September 1906 - Leningrad, 15 April 1983), published in Russian in 1950 and translated into English in 1959 [101], presented the square root method and attributed it to Banachiewicz. In the book of Francis Begnaud Hildebrand (1 September 1915 - 29 November 2002), a Ph.D. student of Crout at the Massachusetts Institute of Technology in 1940 and then a professor of mathematics there from 1940 until 1984, published in 1956 [133], it is only mentioned in a footnote on the Crout reduction that *"similar methods are attributed to Doolittle, Banachiewicz, Cholesky, and others"*. In his book of 1956 [161], Cornelius Lanczos (Székesfehérvár, Hungary, 2 February 1893 - Budapest, 25 June 1974) stated without any reference (p. 136) that *"any symmetric and positive definite matrix can be split into the row–by–row product of a triangular matrix by itself"*.

In the first volume of the *Bulletin Géodésique* in 1957, there is a Report of Study Group No. 2 of the International Association of Geodesy, J.E.R. Ross, President, Geodetic Survey of Canada, Ottawa. On page 9, it is written:

> *The availability of electronic machines for the formation and solution of the normal equations has also brought forward a preference as to the methods of elimination depending on the type of machine. Some of these factors are discussed below. Until recently the U.S. Coast and Geodetic Survey were using the Gauss–Doolittle Method of elimination with their I.B.M. punch card system. The procedure is practically the same as that used for desk calculators. With the I.B.M. Type 650 Magnetic Drum Data Processing Machine, which was installed in the spring of 1957, they are using a variant of the Cholesky Method. This method requires less storage space than other schemes of elimination. It is also generally agreed that the extraction of a square root for the leading element in each row of the elimination reduces rounding–off errors. It is interesting to note that the U.S.C.G.S. uses the variation of co–ordinate method for the adjustment of extensive triangulation systems.*
>
> *In the past, the solution of large groups of normal equations by iterative procedures, such as the Gauss–Seidel Method, has been suggested to us. Such a procedure could prove to be costly and time–consuming because of possible instability in the set–up. For some systems of equations the successive results obtained for some of the unknowns diverge from the true values, or, at best, fail to converge to the true values. The work required to determine whether a system of equations can be successfully solved by such an iterative method is so great that the advantages of this method become completely negligible.*

For a modern treatment of numerical linear algebra, see [119,181], and [219] where historical notes can be found.

4.9 Nowadays

Nowadays, Cholesky's method is much used in numerical analysis for the solution of systems of linear equations with a symmetric positive definite matrix, in quasi–Newton methods which are used in non–linear optimization for solving the systems of nonlinear equations obtained by differentiation, in Monte Carlo methods for simulating systems with multiple correlated variables, in Kalman filters for choosing a set of so–called sigma points, in optimization [208], etc. But it also has many other applications that we could not describe here. However, let us give some more examples.

Cholesky's method is probably the most commonly used model today in behavior genetic analysis. Behavioral geneticists commonly parameterize a genetic or environmental covariance matrix as the product of a lower diagonal matrix post–multiplied by its transpose, a technique commonly referred to as *fitting a Cholesky* [65]. It also has applications in statistics for multivariate stochastic volatility models [169], the estimation of covariance matrices [238], the likelihood–based analysis of linear state–space models [215], the simultaneous modeling of the Cholesky's decomposition of several covariance matrices [198], Bayesian regression [170], etc. There are also applications in computational physics [214], chemistry [207], geology [28], and biology [249].

When solving a system of linear equations, the method to be used highly depends on its dimension and on the structure of the matrix. At the time of Cholesky and even after World War I, the dimension rarely exceeded 1,000 and, thus, storing everything in the computer memory was not a problem, and direct methods (that is methods giving the exact solution, if no rounding errors occur, after a finite number of arithmetical operations) were used. Nowadays, systems of several millions, or even billions, of equations are considered. Such systems arise in problems involving the solution of partial differential equations and boundary integral equations. Dense systems (that is when the matrix has few zero elements) can be found in airplane wing design, radar cross–section studies, flow around ships and off–shores constructions, diffusion of solid bodies in a liquid, noise reduction, diffusion of light through small particules, and, more generally, electromagnetism [88, 99].

Every computation on a computer is subject to rounding errors since the number of digits of each entry is limited. The accumulation of such errors can lead to a completely false result. This is the notion of *numerical stability* of an algorithm, which was recognized to be crucial a long time ago [232, 239, 244] (see [132] for a more recent treatment). An error analysis of Cholesky's factorization was conducted in [177].

When a system is too big to be stored, direct methods cannot be used, and one has to consider iterative ones (see, for example, [181]). But, hopefully, large systems are usually sparse (very many of its elements are zero) because they are coming out from the discretization of partial differential equations by finite differences or finite elements. When the matrix is sparse, the storage requirement can be reduced. In skyline storage, only the entries from the first nonzero entry to the last nonzero entry in each column are stored. This type of storage is very popular in finite element codes because the skyline structure is preserved by Cholesky's decomposition [38]. When solving partial differential equations by finite elements, the domain can be decomposed into several smaller subdomains, thus replacing the linear system by smaller ones and solving them separately. There are two types of methods according wether the subdomains overlap or not. In the second case, some of these algorithms can be reformulated using a block Cholesky's factorization, an approach which provides a useful framework for the design of domain decomposition algorithms in the symmetric positive definite case [168]. These procedures are related to Schur complements [252]. Cholesky's factorization has also been adapted to parallel computers a long time ago [115]. Some experiments are described in [34]. Let us mention that the dimension of the largest dense system solved by Gauss's method is about 11,870,000 on a machine with over 700,000 cores, and that it took 30 hours. For Cholesky's method, it will be roughly the same. For sparse systems, Cholesky's factorization is used routinely on problems of over a million, and up to around 10 millions of equations.

When the matrix A of the system $Ax = b$ is ill–conditioned, a small error in the coefficients of A or in the components of b can induce a large variation in the exact solution x. Also, in some iterative methods, such as the conjugate gradient algorithm, the speed of convergence depends on the condition number $\kappa(A) = \|A\| \cdot \|A^{-1}\|$ of the matrix (this number is always greater than or equal to 1). Since the identity matrix is well–conditioned (its condition number is 1), the system $Ax = b$ is replaced by $MAx = Mb$ where the matrix M, called a preconditioner, is an approximation of A^{-1}, or by $M^{-1}Ax = M^{-1}b$ where M^{-1} is an approximation of A. Obviously, in both cases, M must be easy to compute. When A is a sparse symmetric positive definite matrix, an incomplete Cholesky's decomposition consists in taking $M = KK^T$ where K is a lower triangular matrix close, in some sense, to L, the lower triangular matrix of the Cholesky decomposition $A = LL^T$. One way of obtaining such a matrix K is to use the exact Cholesky's factorization, and to set to zero any entry of it corresponding to a zero entry in A. Thus, M is as sparse as A. This incomplete factorization is used, for example, in the conjugate gradient algorithm [146]. Most interior point methods for solving linear programming problems need, at each iteration, the solution of a large, sparse, and ill–conditioned system of linear equations with a symmetric positive definite matrix [245]. Such systems can be efficiently solved by convenient modifications of Cholesky's method [246].

This list of applications does not pretend to be exhaustive. Cholesky's method is also related to the singular value decomposition, to multigrid methods for the solution of partial differential equations, and it has applications in several other domains of science.

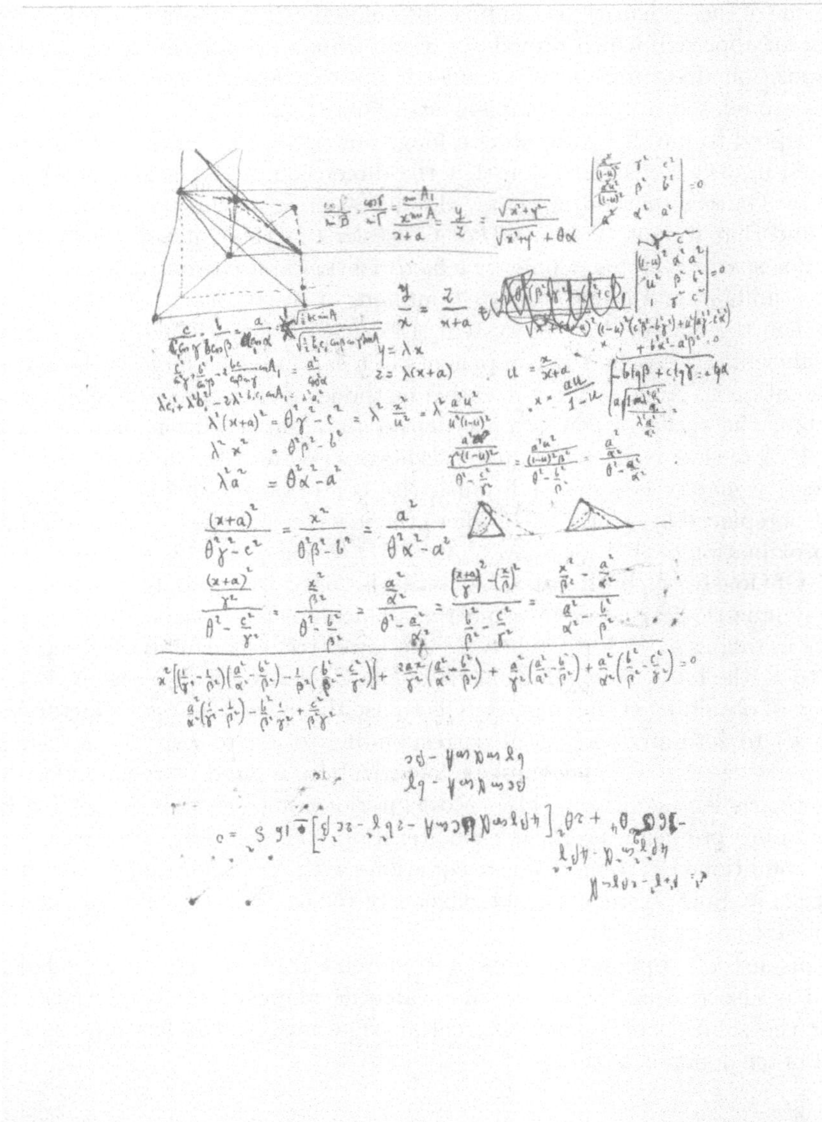

A page of calculations by Cholesky

Chapter 5

Other works of Cholesky

At *École Polytechnique*, many other documents can be found in the *Fonds Cholesky*. They consist of military documents (Section 5.1), documents related to the *École Spéciale du Bâtiment et des Travaux Publics* (Section 5.2), and other manuscripts (Section 5.3). There are also handwritten notes taken by Cholesky during the lectures he followed at *École Polytechnique* when he was a student there, and some printed materials related to them. The most important document is an unpublished hand written course entitled *Cours de Calcul Graphique* of 83 pages on 15.5 × 20 cm sheets. Chapter 7 is devoted to it. It is reproduced in Appendix C. We will detail the two books written by Cholesky in Section 5.4.

5.1 Military documents

There are many scientific documents written by Cholesky at the occasion of his activities during the war. They show that he put his scientific knowledge at the service of the Army. Some of them are handwritten and some others are typed. Their title in French (in *italics*) is followed by their English translation.

5.1.1 Manuscripts

- *Canevas de tir pour situer le plus exactement possible la ligne de combat* (Firing grids to locate as precisely as possible the line of battle) (2 p.).
- *Surveillance des aéronefs* (Surveillance of aircrafts) (1 p.).
- *Protection contre les incursions des aéronefs* (Protection against the incursions of aircrafts) (1 p.).
- *Lieu du point apparent d'émission du claquement pour une pièce de 77 du Grafen Wald pour la Section B, quand on fait varier le plan de tir de la pièce* (Location of the apparent point of emission of the crack for a gun of 77 of

the Grafen Wald for Section B, when the plane of fire of the gun is changed) (1 p. and 3 p. of draft).

- *Tenue à jour du programme de la photo aérienne* (Maintenance of the program of aerial photography) (2 p.).
- *But du groupe de canevas de tir* (Goal of the group of firing grids) (2 p.).
- *Canevas de tir pendant la marche en avant* (Firing grids during the advance) (5 p.).
- *Instructions particulières à l'artillerie pour l'emploi des contre–batteries (complément à l'ordre général No. 12)* (Specific instructions for the use of artillery counter–batteries (supplement to the general order No. 12)).
- *Instructions pour la surveillance des aéronefs* (Instructions for the surveillance of aircrafts) (1 p.).
- *Étude d'artillerie* (Investigation on artillery) (1 p.).
- *Projet de répartition du travail dans le GCTA* [1] (Project of distribution of labor in the GCTA) (3 p.).
- *Sur le combat aérien* (On aerial fight) (10 p.).
- *Organisation du tir de l'artillerie* (Organization of artillery firing) (36 typed pages corresponding to the talks given by Cholesky and Capitaine de Fontanges in May 1915).
- *Programme de l'instruction à donner aux officiers pour coopérer à l'organisation du tir* (Program of the instruction to be given to officers to cooperate in the organization of firing).
- *Appareil de pointage pour mitrailleuse sur avion Nieuport* (Pointing device for a machine gun on the aircraft Nieuport) (3 p.).

Let us discuss this last contribution. In 1914, at the beginning of the war, airplanes became important weapons. Raymond Saulnier (Paris, 27 September 1881 - Chécy, 4 March 1964), a French aeronautical engineer, invented a synchronized device for firing through the propeller, but the system was unreliable. The manufacturer reverted to a less ambitious process combining some synchronization and a shielded part of the propeller located opposite the output of the machine gun. Some bullets went through while others crashed against the fuselage. To insure an unimpeded path for bullets, the aircraft itself had to point at the target. In collaboration with Roland Garros (Saint–Denis de la Réunion, 6 October 1888 Saint–Morel, 5 October 1918), who realized the first crossing of the Mediterranean Sea on 23 September 1913 and was a pilot on a Morane–Saulnier airplane, he perfected the device and made it operational. The gun was located on the wing, in

[1] Groupe de Canevas de Tir d'Armée.

the field of vision of the pilot, and it fired in the axis of the plane between the propeller blades which were protected by a deflector. In the undated work mentioned above, Cholesky describes a pointing device to be installed above or, preferably, under and on the left of the gun (see the figure). He writes:

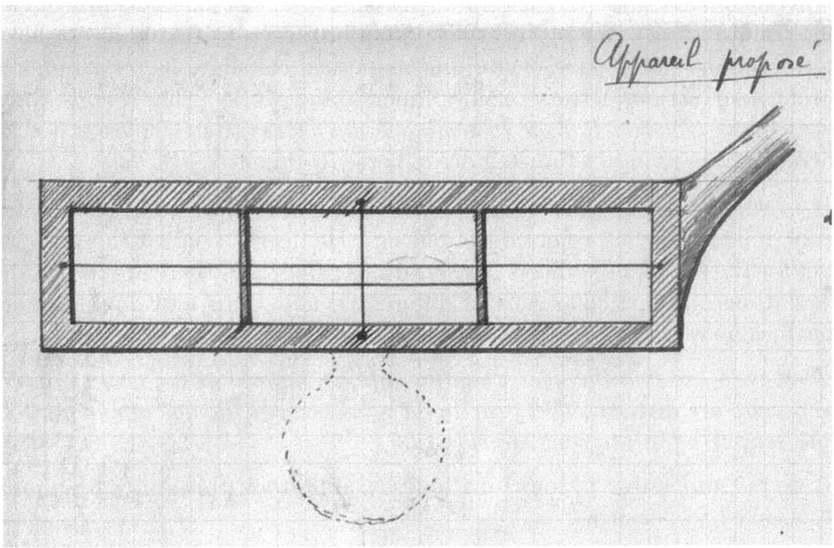

Cholesky's pointing device for a gun

The gun of the fighters (Nieuport type) allows to shoot in the direction of travel in a small angle to the right and a slightly greater angle (approximately 20 degrees) to the left.

It is the pilot who shoots, it must be asked very basic maneuvers, hence there can be no question of using adjustable pointing devices...

The requirements for the pointing device are thus

1)- Line of sight easy to take and precise.

2)- Possibility of having a simple indication of the distance from the plane.

3)- Possibility of taking into account, to a certain extend, the rise and the lateral displacement of the target.

[The pointing device has] *to be fixed at the location of the guidon* [dashed line on the figure], *either above the gun or, preferably, under it and on the left which allows, with the current devices, to lower the line of sight and make it more accessible.*

For what reason did Cholesky work on this device? Was he asked to study the problem? Had Cholesky met Garros?

There are also three manuscripts without titles, one on the work of the geographer officer (5 p.), one on locating potential targets with the use of aircrafts (3 p.), one large double page containing numerical calculations on arrangements of objects, and various letters on instruments and works. The *Fonds Cholesky* possesses many *Tableau R pour le régimage des pièces* that Cholesky had to fill up when he was serving in the 202nd Artillerie Regiment.

It is strange to find also many pages and statistical tables, handwritten by Cholesky, related to the promotion of officers. He deeply studied the proportions to be set for the number of officers in the various ranks, and there are pages covered by calculations, in particular partial differential equations and double integrals. His conclusions were:

That is to say that the ages when the officers arrived at the mean raised seniority are distributed in exactly the same way as the age of entry into the service or in the rank considered.

It seems impossible to know if Cholesky undertook this study by himself or if he was asked to perform it.

5.1.2 Typed documents

- *Étude sur le tir d'artillerie contre les batteries masquées* (Study of artillery firing against masked batteries), sent on 5 November 1914 from Somme–Suippe to the Chef d'Escadron Girard, commanding the 3rd group of the 23th Artillery Regiment, 12 pages.
- *Projet d'organisation de l'observation et du tir d'artillerie sur un front de Corps d'Armée* (Project for the organization of observation and artillery firing on a front of an Army Corps), 24 December 1914.
- *Opérations topographiques effectuées par l'artillerie pour relever les batteries et leurs repères* (Topographic operations performed by the artillery to localise the batteries and their marks).
- *Notes sur le tir avec observateurs latéraux et aériens* (Notes on firing with side and aerial observers), April 1915.
- *Fonctionnement du Service Géographique de l'Armée Roumaine d'octobre 1916 à janvier 1918* (Operation of the Geographical Service of the Romanian Army from October 1916 to January 1918), 83 pages.
- *Étude du développement conforme de la carte de Roumanie* (Study of the conformal development of the map of Romania), 16 pages.

- *Quelques remarques au sujet de l'organisation et du fonctionnement du Service Géographique Roumain* (Some remarks about the organization and the functioning of the Romanian Geographical Service), 19 pages dated 1 February 1918.
- *Étude d'artillerie relative à des opérations devant se développer sur la partie sud du front du D.A.L.* [2] (Study on the artillery operations to be developed on the southern front of D.A.L.), 23 pages. This report was asked to Général Diez, commanding the artillery of the 8th Army Corps having Cholesky as his assistant.
- *Note sur le fonctionnement de la section de repérage par les lueurs No. 61* (Note on the working of the section of glow tracking No. 61), dated 20 August 1916.

The official military documents (instructions, reports, memos, etc.) do not always bear a signature. However, according to the topic and the style, it is almost sure that some of those which are in the *Fonds Cholesky* were written by him. This is the case of the handwritten corrections which are in *Instruction sur l'organisation et les attributions des groupes de canevas de tir des armées* (Instruction on the organization and the allocations of the groups of firing grids of the Army), dated 23 December 1915, and signed by the Maréchal Joseph Jacques Césaire Joffre (Rivesaltes, Pyrénées–Orientales, 12 January 1852 - Paris, 3 January 1931), chief commander of the operations.

5.2 Courses at ESTP

Various manuscripts are related to the work of Cholesky as a professor at the *École Spéciale des Travaux Publics, du Bâtiment et de l'Industrie.*

5.2.1 Complément de Topographie

This *Complément de Topographie* is a handwritten course of 239 pages on sheets of 15.5×20 cm. There is also a printed version in cursive characters with corrections from the hand of Cholesky. This document is certainly a preliminary version of his book *Levés d'Études à la Planchette* already quoted.

5.2.2 Cours de Topographie

The *Fonds Cholesky* has pages 44 to 218 of the manuscript of the book *Cours de Topographie* by Cholesky and published by the École Spéciale des Travaux Publics at an unknown date. Section 5.4 is devoted to it.

[2] Détachement d'Armée de Lorraine.

5.3 Other manuscripts

The *Fonds Cholesky* also contains the following:

- Three pages with the title *Sur la détermination des fractions de secondes de temps* (On the determination of fractions of seconds of time).

- A 15 pages manuscript entitled *Instructions pour l'exécution des nivellements de précision* (Instructions for the execution of precise leveling).

  It contains 11 sections:

I	Terms of leveling outside the main roads of communication
II	Recognition of the route
III	Benchmarks
IV	Fixed and temporary marks
V	Installation of the level
VI	Adjustment of the level
VII	Rods
VIII	Execution of leveling
IX	Keeping of records
X	Computation to be done on the ground
XI	Final computations

- A manuscript of 8 pages with the title *Équation de l'ellipsoïde terrestre rapportée à Ox tangente au parallèle vers l'Est, Oy tangente au méridien vers le Nord, Oz verticale vers le zénith* (Equation of the terrestrial ellipsoid with respect to Ox tangent to the eastward parallel, Oy tangent to the meridian northward, Oz vertical toward the zenith). There is also a typewritten copy with handwritten mathematical formulae.

- A 16 pages manuscript *Étude du développement conique conforme de la carte de Roumanie* (Study of the conic conformal development of the map of Romania). There is also a typewritten copy with handwritten mathematical formulae.

- A manuscript of 3 pages *Instructions sur l'héliotrope–alidade (modèle d'étude 1905)* (Instructions on the heliotrope–alidade (study model of 1905)). This manuscript was written at La Charpenne on 18 August 1905.

- Five pages of description of the holometric alidade.

- Five pages of description of the compass with clinometer.

- Three pages and three plans on the construction of railway lines.

- Three pages entitled *Remarque au sujet du calcul de correction de mire* (Note on the computation of rod corrections).

- Various courses and exercises for students by correspondence of the ESTP on trigonometry, algebra, mechanics, arithmetic, geometry, and even on the strength of materials.
- Two sheets with the title *Compléments de Topographie. Levés d'études à la planchette. 5 séries et 2 exercices pratiques. Tâches à remplir.* It presents the ideas of Cholesky on a course he was required to give, and it might be the preliminary notes for his book *Levés d'Études à la Planchette.*

5.4 Books

Cholesky published two books.

5.4.1 Cours de Topographie

In 1909, Léon Eyrolles, Eugène Joachim Prévot (an engineer of *Ponts et Chaussées*, born in 1866, who worked for the leveling Service from 1882 to 1928 and was its Deputy Director from 1901), and Edouard Quanon (engineer–geometer in chief of *Ville de Paris*) published, at the *Librairie de l'Enseignement Technique Léon Eyrolles*, a book with the title *Cours de Topographie. Livre I: Topométrie.* The second volume, with the title *Cours de Topographie. 2e partie: Topographie Générale*, was written by Cholesky. It contains a description of the method of least squares.

It was not possible to find the date of the first edition of this book. It was first published in cursive writing, and then typed. The second edition dates from 1913, and has 631 pages. The third is from 1920, and has only 394 pages of size 17.0×21.5 cm. The fourth edition is dated 1924. It has 594 pages, contains fourteen chapters, and 18 sheets of photographic plates, some folding, and exactly 100 figures.

This book had, at least, seven editions. The seventh edition of 1937 (and, maybe, an earlier one) was revised by Henri–Albert Noirel (Charleville, 21 May 1870 - 1933), former student and lecturer at *École Polytechnique.* The National Library of France possesses this edition (*Bibliothèque Nationale de France*, cote 4-V-15365 (2)). It is typewritten, has 442 pages, 100 figures and 18 plates or photos of instruments. Being in poor condition (paperback, on brittle and yellow paper, and with loose pages), it is prohibited for consultation.

This book was still relevant nearly thirty years after it first appeared and nineteen years after the death of its author. It is curious that after seven editions it is so difficult to find it in libraries and in antique bookstores. Moreover, it is still mentioned among the books published by Eyrolles on the fourth page of the cover of the 21st edition of the book *Notions de topographie* by Prévot and Quanon in 1945! It took me more than three years of waiting, being listed on many web sites, to get a copy of it (4th edition, 1924). In it, there is a dedication, dated July 1948, on the label of a bottle of champagne: *Remembering of the closing dinner*

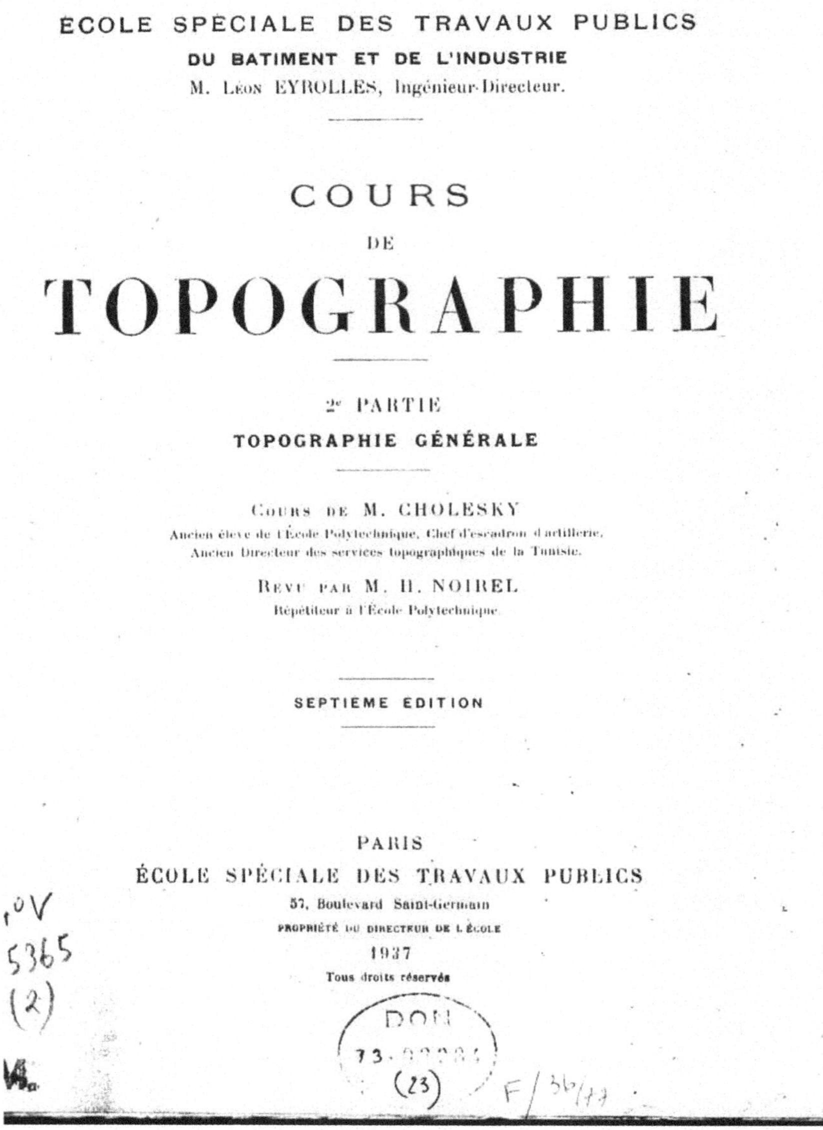
ÉCOLE SPÉCIALE DES TRAVAUX PUBLICS
DU BATIMENT ET DE L'INDUSTRIE
M. LÉON EYROLLES, Ingénieur-Directeur.

COURS
DE
TOPOGRAPHIE

2e PARTIE
TOPOGRAPHIE GÉNÉRALE

COURS DE M. CHOLESKY
Ancien élève de l'École Polytechnique, Chef d'escadron d'artillerie,
Ancien Directeur des services topographiques de la Tunisie.

REVU PAR M. H. NOIREL
Répétiteur à l'École Polytechnique

SEPTIEME EDITION

PARIS
ÉCOLE SPÉCIALE DES TRAVAUX PUBLICS
57, Boulevard Saint-Germain
PROPRIÉTÉ DU DIRECTEUR DE L'ÉCOLE
1937
Tous droits réservés

Cover of *Cours de Topographie*
© Public domain

of two years of study at T.P. [3] Does this mean that the book was still in use at that time?

The table of contents is

In Chapter III, Cholesky gives a complete and clear account of the method of least squares. In a footnote of page 67, he writes:

> *It is necessary to make an important remark which greatly facilitates the application of the above principles; regardless of the relations that bind the quantities calculated to the quantities measured, we can always turn them into linear equations, using as an intermediary sufficiently approximate values of these quantities so that we can develop each of the relations with respect to the increases of each of the quantities contained therein by limiting this development to the first degree term.*

Cholesky does not give any method for solving this system of linear equations. However, in Chapter X, in Section 119 on adjustments, he wrote (p. 358):

> *Whenever you make a calculated triangulation, it is advantageous to also make an adjustment by calculation. We are then brought to write a number of equations representing the geometric relationships between the various elements of the figures of the triangulation and, as there are usually many more unknowns than equations, the uncertainty is lifted by writing that the sum of the squares of the corrections is minimum.*

[3] Travaux Publics.

Topographical instruments from Cholesky's book

We will not enter into the details of this operation which is in the field of geodesy. It generally leads to quite lengthy calculations, but the methods currently in use allow easily to reach the result, limiting as much as possible the chances of making mistakes. We can, therefore, introduce in the same adjustment a large number of measured elements, which is very advantageous, because the consequences deduced from the theory of errors, that is to say from probability theory, are, on average,

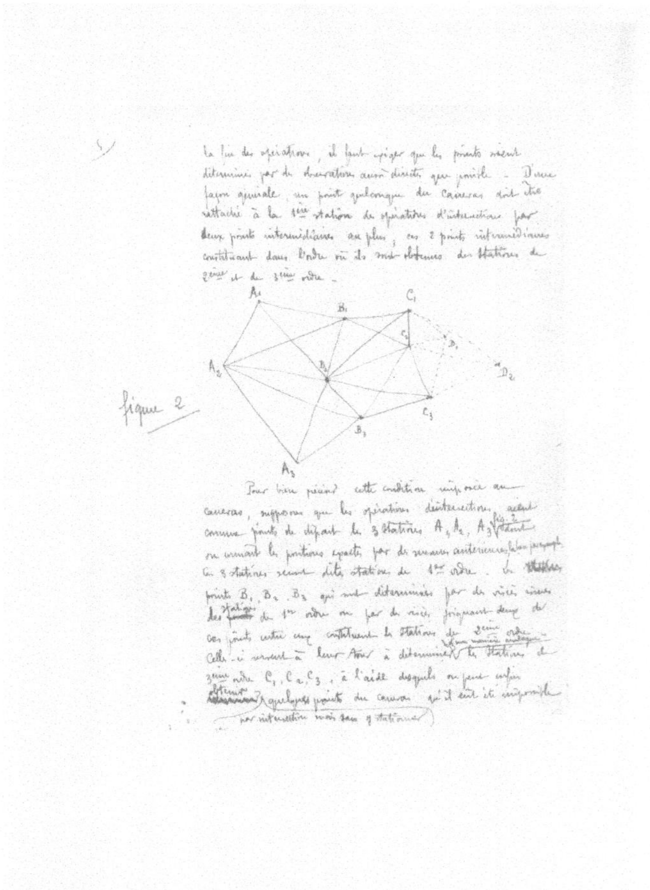

Triangulation from the manuscript of Cholesky's book

closer to the truth when they apply to a larger number of quantities comparable between them.

This book is intended for students who have studied trigonometry, geometry, and differential calculus. The instruments and their use are described with great care. All their settings are clearly explained. The leveling operations are examined in detail. The amount of knowledge needed to write this book is absolutely huge. The presentation of the various topics is given with an unrivaled clarity. The text really gives the impression that the author is directly speaking to the reader. It

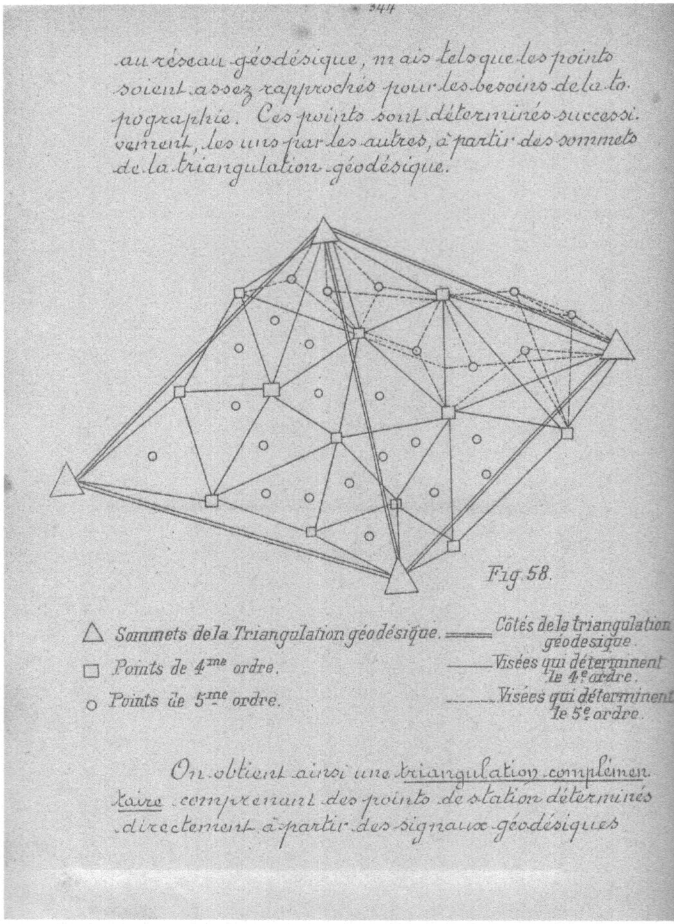

344

au réseau géodésique, mais tels que les points soient assez rapprochés pour les besoins de la topographie. Ces points sont déterminés successivement, les uns par les autres, à partir des sommets de la triangulation géodésique.

Fig. 58.

△ Sommets de la Triangulation géodésique. ═══ Côtés de la triangulation géodesique.

□ Points de 4me ordre. ——— Visées qui déterminent le 4e ordre.

○ Points de 5me ordre. - - - - Visées qui déterminent le 5e ordre.

On obtient ainsi une triangulation complémentaire comprenant des points de station déterminés directement à partir des signaux géodésiques

A page from the 4th edition of Cholesky's book
© Public domain

is obvious that Cholesky knew extremely well all the subjects he touched, and he had a great educational sense.

5.4.2 Levés d'Études à la Planchette

There is a second book by Cholesky whose title is *Levés d'Études à la Planchette.* The National Library of France retains no copies of it. I own its third edition, dated 1923 (five years after the death of its author), with a foreword by H. Lafosse, Deputy Director of the *Service de la Reconstitution Foncière et du Cadastre au*

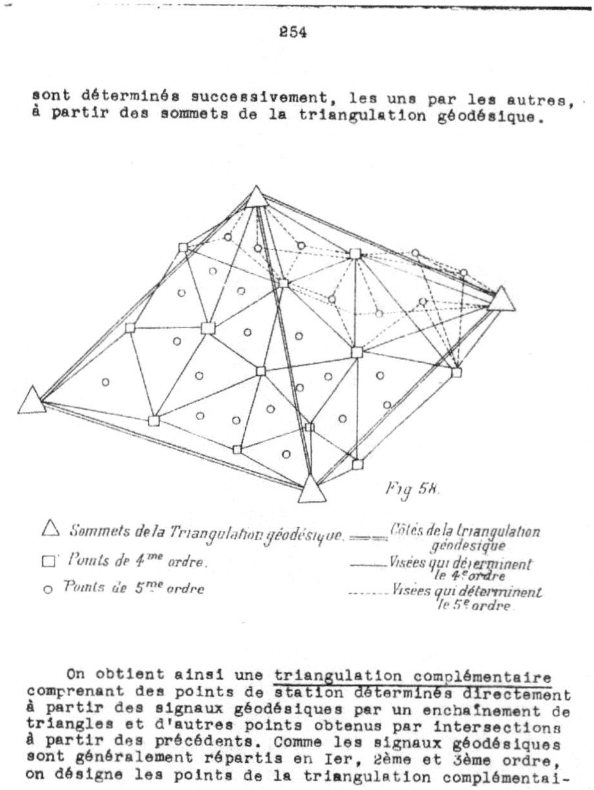

254

sont déterminés successivement, les uns par les autres, à partir des sommets de la triangulation géodésique.

On obtient ainsi une triangulation complémentaire comprenant des points de station déterminés directement à partir des signaux géodésiques par un enchaînement de triangles et d'autres points obtenus par intersections à partir des précédents. Comme les signaux géodésiques sont généralement répartis en Ier, 2ème et 3ème ordre, on désigne les points de la triangulation complémentai-

The same page from the 7th edition of Cholesky's book
© Public domain

Ministère des Régions Libérées (sic). The book has 269 pages, 56 figures, and is followed by two photographic plates with the corresponding captions. A fourth edition, revised by H. Noirel, is dated 1931. I also have it. Its table of contents is

Chapter I	Generalities
Chapter II	Maps with detailed topography
	Comprehensive study of a survey at 1:1,000th
	Performance of the survey
	Marking and logging the points of the grid
	Surveying of the grid

Telescope from the book of Cholesky
© Fonds Cholesky - École Polytechnique

	Installation of the plane table. Construction of the grid leveling
	Surveying of the details
	Leveling of the details
Chapter III	Surveying with a compass clinometer
Chapter IV	Graphical triangulation

As we have already seen, the plane table is certainly the simplest instrument to carry out topographical surveys. It is simply an alidade (sighting system with a peepsight at each end) attached to a board, all mounted on a tripod. At the time Cholesky wrote his book, it was no longer used for small–scale maps and the process was *"generally contempt for the execution of maps with a scale greater than 1:10,000"*, as written in the introduction due to Lafosse who advocated for it and continued:

This abandonment is unjustified: in the hands of a careful operator, the precision obtained, if one uses the telescopic alidade with a stadia leveling rod, is entirely comparable to that of an excellent tachymetric survey, with a very valuable gain on the time spent on operations.

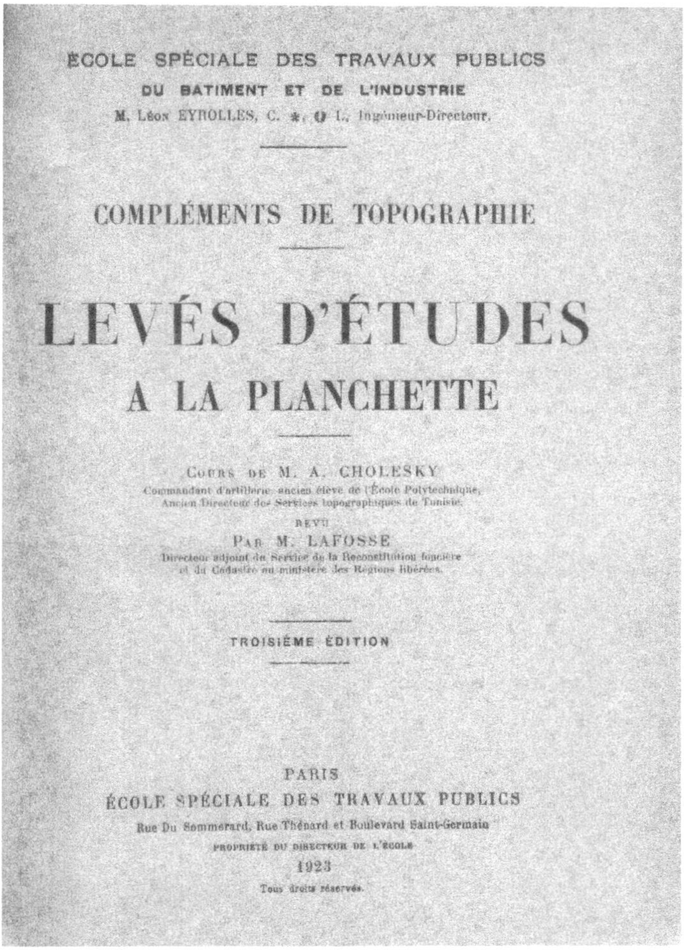

ÉCOLE SPÉCIALE DES TRAVAUX PUBLICS
DU BATIMENT ET DE L'INDUSTRIE
M. Léon EYROLLES, C. ✱, I., Ingénieur-Directeur.

COMPLÉMENTS DE TOPOGRAPHIE

LEVÉS D'ÉTUDES
A LA PLANCHETTE

Cours de M. A. CHOLESKY
Commandant d'artillerie, ancien élève de l'École Polytechnique,
Ancien Directeur des Services topographiques de Tunisie,
REVU
Par M. LAFOSSE
Directeur adjoint du Service de la Reconstitution foncière
et du Cadastre au ministère des Régions libérées.

TROISIÈME ÉDITION

PARIS
ÉCOLE SPÉCIALE DES TRAVAUX PUBLICS
Rue Du Sommerard, Rue Thénard et Boulevard Saint-Germain
PROPRIÉTÉ DU DIRECTEUR DE L'ÉCOLE
1923
Tous droits réservés.

Levés d'Études à la Planchette by Cholesky
© Public domain

This book has the same educational qualities as the preceding one. It only requires some basic knowledge in geometry. At the end, the author gives a list of books to read:

Recherches sur les instruments, les méthodes et le dessin topographique by Colonel A. Laussedat (Gauthiers–Villars, 1898),

Cours de topographie by Lehagre (Gauthiers–Villars),

Cours de topographie and *Instructions diverses* of the École d'Application de l'Artillerie et du Génie de Fontainebleau,

Cours de cosmographie et topographie by Combette and Porchon (Félix Alcan),

Études théoriques et pratiques sur les levés topométriques by Colonel Goulier (Mémorial de l'Officier du Génie),

Notes sur divers instruments topographiques by Colonel Goulier (Mémorial de l'Officier du Génie).

Chapter 6

Léon Eyrolles and his superior school

Since Cholesky was involved in the school founded by Eyrolles, let us now spend some time on this man and his activities.

6.1 The man and his school

Léon Eyrolles was born at 20 rue de la Barrière, in the maternity school of the city of Tulle in the department of Corrèze, on 14 December 1861 at noon. The family originated from Sainte–Fortunade, a small village in the same department. His mother Anna was single. The birth certificate was signed by the midwife and by Antoine Var (a 30 year old farmer), domiciliated at 33 rue du Trech in Tulle. The young Léon was prevented from studying in his youth due to the poverty of his family circumstances, and the need to support his mother. He began his professional career in 1880 as a subalterne employee in the administration of *Ponts et Chaussées* (civil engineering). From 12 November 1881 to 12 Novembre 1882, he was a conditional volunteer in the 108th Artillery Regiment. He ended with the grade of sergent and was proposed as a reserve officer. Then, he returned to his previous position until 16 December 1882 when, having spent all his spare time studying, he became *Conducteur des Ponts et Chaussées*, that is a foreman of civil engineering. He stayed in this office until 1 May 1899.

Foremen who, issued from less prestigious institutions than *École Polytechnique* or *École des Ponts et Chaussées* founded in 1747, had long claimed their suitability for the best of them to be admitted into the corps of civil engineering. Various laws taken from the 1850s had indeed stated that a large fraction of the places of civil engineer should be reserved for foremen recruited by a competitive examination, but they remained practically ineffective due to the resistance of the

Léon Eyrolles
© Fonds Eyrolles - Archives du Val de Marne

members of the corps who wanted to maintain a high recruitment synonymous with social prestige. Thus, the administration was lacking of such employees. However, in 1895, foremen where given the right to sign their own projects and, in 1904, chief foremen having served as engineers for at least 5 years were allowed to use the title of *ingénieur assistant* (assistant engineer). This title was obtained only by about fifty of them. After a long administrative and legal battle, the foremen were given the title of *Ingénieur Civil de l'État* (Civil Engineer of the State) in 1920.

Simultaneously to his work as a foreman, Eyrolles was preparing for the competitive examination to become an engineer. He was offered a position similar to one that existed in the construction of railways, but he soon realized that it neither corresponded to his intellectual aspirations nor to his hopes for his future career. Therefore, he left his region of origin and came to Paris in 1891. Then, in the morning and at night, in addition to his work as an engineer, he began, in a small apartment, rue du Val–de–Grâce and then at 3 impasse Nicole in Paris, to help his colleagues in preparing the examination to become foreman. At that time, the *École des Ponts et Chaussées* was virtually the only institution of higher education in the field of civil engineering, while the epoch was oriented to the idea of "great works" due to the technical revolution (advent of electricity, telephone, birth of metal architecture, reinforced concrete, etc.). It was difficult for students with an ordinary background to succeed in schools of higher technical instruction. But one of his colleagues having to leave Paris and go to one of the provinces, Eyrolles, encouraged by Auguste Kleine (1848 - 1925), director of staff at the *Ministère des Travaux Publics* in 1898 and future director of the *École des Ponts et Chaussées*, decided to help him by sending him written courses and exercises. The colleague had to read them and send his solutions back. Eyrolles then returned them to him with corrections and annotations. With Pierre Jolibois, chairman of the *Société des Conducteurs des Ponts et Chaussées*, he also started to publish a series of high–quality technical books, the *Bibliothèque du Conducteur de Travaux Publics*. This was the starting point, and, in 1892, as the number of his students expanded rapidly, Eyrolles established a school by correspondence called *L'École chez soi* (The school at home). Notice that, in 1890, the American newspaperman Thomas J. Foster (Pottsville, 1 January 1843 - Scranton, 14 October 1936) founded the International Correspondence School in Scranton, Pennsylvania. This school originally provided correspondence courses for coal miners, but later expanded into many other areas. In 1891, Eyrolles began with only one student. The following year, he had 15, 40 in 1893, 101 in 1895, 454 in 1898, and 687 in 1899. In 1893, he entered the *Société des Ingénieurs Civils* as a civil engineer and teacher in applied mathematics [236].

By 1898, since the applications for teaching and his own work as a foreman were too heavy a burden (once, he was found unconscious), Eyrolles resigned as a foreman of civil engineering. He then went on to create the *École Spéciale des Travaux Publics* (ESTP), 61 Boulevard Saint–Germain and 12 rue du Sommerard in the Latin Quarter in Paris, where it still is. The premises were scattered and inconvenient, and the rooms were cramped. The first students and teachers remember the old metal staircase, in the bottom of the garden of rue du Sommerard, which rang under their feet each time they entered the classroom.

In 1903, Eyrolles bought a large piece of land in Cachan, near Paris, to install the necessary workshops for practical tests. Eyrolles was attracted to Cachan by the availability of inexpensive land in the vicinity of the Arcueil–Cachan train station, thus allowing his students to travel between Cachan and the courses they attended in Paris [1]. A *Maison de Famille* (Family house), where students from outside the Paris area were housed, was built in 1903 by the architect Eugène Robinot, 26 avenue du Président–Wilson in Cachan. Each student had his private

[1] On the ESTP, see, in particular: *Technical Education News*, March 1949, pp. 11–13; *Léon Eyrolles (1861–1945)*, anonymous 19 pages booklet with no date, printed after Eyrolles' death (in 1945) with a preface by Raoul Dautry, Ministre de la Reconstruction et de l'Urbanisme, impr. G Dalex, Montrouge, pp. 7–12; Carat, *Cachan a 70 ans*, p. 33.

room, and there was a billard room, meeting rooms for those wanting to work together, a reading room, tennis courts and a football playground. Students organized their duties as they wanted. They had the right to move freely within the school, to smoke, to read, and to openly express their views either orally or in writing. For Eyrolles, each student was a particular case with his own personality. All these conditions were quite innovative at that time. The private house of Léon Eyrolles was built in three steps between 1905 and 1910 near the buildings of his school.

The ESTP in Paris
© Photo Kollar - Fonds Eyrolles - Archives du Val de Marne

On 11 January 1907, Eyrolles gave a talk at the *Société d'Encouragement pour l'Industrie Nationale* on teaching by correspondence. This talk was reproduced in the journal *L'École chez soi*, No. 218, 25 March 1935. Since it relates the genesis of the school and the ideas of its founder, we will give large extracts translated from it:

> *Teaching by correspondence can be defined as follows:* To replace the oral lesson of the professor, with all its explanations and its developments, by a written course and to oblige the student to learn the content of the course by solving assigned exercises that could not be solved

without a thorough mastery of that content and demonstrating an understanding of its main applications.

I hasten to point out that the exercises intervene here as a means and not as a goal. The course is thus the main part, exercises are only an accessory...

Today, I will tell you how I got the idea of creating this teaching in France.

When I belonged to the administrative services of public works, I was able to note that there was no formation for the secondary staff of this administrative services - nor besides for company staff. For years I had gathered notes and prepared courses and, in 1891, I decided to begin a teaching for the training of this staff. First I had 4 students, 3 in Paris and 1 in the country. This one, not being able to come to attend classes, I had to limit myself to send him exercises to solve and to correct them as all professors are doing. But here the solution presented special problems since the courses I was teaching were not existing in bookshops.

It suggested me, of course, to regularly send him, after writing them down, the lessons I was teaching to his fellows and to let him solve the exercises allowing to be sure that he has understood the courses since I was unable to test him differently.

The results obtained by this single student were excellent; he competed favorably with his comrades who seemed however more fortunate than him.

This results encouraged me to extend the method; I had the courses printed by an economical process at a few number of copies, then I started to let me be known; I was helped in this task by my first students who all succeeded at the difficult competitive examinations they were preparing and wrote to their fellows that they owed their success to my teaching...

The number of registered students each year was

year	*students*	*year*	*students*
1892	*14*	*1899*	*687*
1893	*40*	*1900*	*935*
1894	*80*	*1901*	*1,620*
1895	*101*	*1902*	*1,787*
1896	*177*	*1903*	*2,214*
1897	*264*	*1904*	*2,716*
1898	*454*	*1905*	*3,095*

Adding to this last number the students who were already following the courses on 1 January 1905, we come to the total number of 6,094 students who benefited of our teaching in the current year 1905...

The baggage of correspondence education consists in

1. *a written course with all the developments and clarifications which can be given in an oral lesson,*
2. *a series of exercises carefully graded and chosen in order that the student could solve them only if he perfectly knows the course taught to him,*
3. *a timetable indicating to the student the progress of his studies.*

The course is necessarily divided into a certain number of lessons *each of them corresponding to a period of work whose duration varies, at the pleasure of the correspondant, from eight to fifteen days, according to the time at his disposal; at the end of each period, the student has to send his exercises solved which, immediately, are corrected, sent back to him with annotations and useful advices, and with their solutions as given by the professors. The Director adds, if necessary, his personal remarks.*

During the correcting time of a series of exercises, the student goes to the next task which can belong either to the same course or to a different one if the student is following several courses at the same time. Thus, he continues his studies without any loss of time.

The timetable indicates to the student the part of the course he has to study at each period of time, the series of exercises he has to solve; it also recalls him the dates he has to send his exercises. The student should constantly refer to this timetable which can besides be modified if the student asks to increase or to decrease the length of the working period...

As we already saw, the written course has to replace the oral lesson; thus it must be extremely developed in order not to oblige the student to learn the part of the course in other books.

In principle, the student, who is often at a great distance, should never have to ask any explanation, any clarification.

The exercises to be solved must be chosen with great care; it requires that the student cannot copy the course and that, for answering, he is obliged to study it in all details. The teaching is thus quite delicate and it can be entrusted only to experienced teachers, specialists in the topic they have to teach, and having a thorough pedagogy of teaching by correspondence. The training of these professors is quite laborious.

As I told before, the student must constantly be kept in suspense and it must be an obligation *for him to send his exercises at the term of each period. For this purpose, a special Service of the school has the mission to follow him step by step, to remind him his duties, and to constantly provide him the necessary encouragements. Moreover, the student gets very quickly the custom to inform and to apologize if he cannot send his exercises on time...*

There is no doubt that the method of teaching by correspondence can be extended to all branches of human knowledge.

Finally, today I can only indicate the considerable influence, at the social point of view, that can have this teaching method, called today to solve the problem of an almost free integral education, dreamed by generous minds.

By giving to the most modest person having the willingness to work the means of ascending to the highest offices, we can also remove a permanent source of conflict. Perhaps there lies the solution of many problems.

Students at École Spéciale des Travaux Publics
© Public domain

Then, Eyrolles developed the advantages of correspondence courses compared to oral ones. They are based on individual teaching; each professor has a privileged relation with each student; students have the obligation of writing their exercises, which helps them to express their results in a clear and exact way; each student can work where and when it is convenient, without losing time in transportation. Thus the student gains time and money since most books are quite expensive and mostly suited for reference rather than study. Finally, the professor is at the disposal of the student and not the student at the disposal of the professor. Thus he can develop his own initiative. In 1907, 168 different courses of civic engineering were given by 80 professors with the help of 25 administrative employees. These courses were divided into three levels (see the next Section). The students were young people wanting to obtain a professional qualification or adults already working and wanting to improve their position. They came from various countries. Each year, the printing of the courses cost 100,000 Francs. The school received over 35,000 letters and 100,000 exercises, drawings and projects. Postage cost 20,000 Francs. In 1933–1934, the printing costs were about 337,000 Francs and the postage about 279,000. Let us mention that 1 Franc of 1911 corresponds to approximately 3.24 Euros in 2011, and 1 Franc of 1934 is equivalent to 0.68 Euros, according to the site http://www.insee.fr/fr/themes/indicateur.asp?id=29&page=achatfranc.htm.

In the years following this talk, the number of students was

year	students	year	students
1906	2,607	1920	12,458
1907	2,696	1921	11,448
1908	3,178	1922	10,818
1909	3,582	1923	10,149
1910	4,330	1924	10,445
1911	4,851	1926	9,001
1912	5,994	1927	9,532
1913	6,636	1928	10,684
1914	4,451	1929	10,789
1915	2,627	1930	11,486
1916	3,563	1931	11,641
1917	5,314	1932	9,577
1918	6,990	1933	9,577
1919	12,261	1934	6,985

Thus, from 1892 to 1934, a total of 238,165 students followed the courses. The French Government was considering very seriously these correspondence courses since, on 8 May 1914, the Prime Minister, Minister of Foreign Affairs, wrote to the diplomatic agents of the French Republic abroad a letter insisting on the help they have to bring to the opening of branch offices of the École des Travaux Publics:

In these offices, students by correspondence (13,000 with 2,400 abroad) should find autographed courses specially written for teaching, a series of homework to do, diaries giving the order for carrying out these duties

and the date for sending them to the School for correction, instructions on writing assignments.

Students of ESTP
© Photo Kollar - Fonds Eyrolles - Archives du Val de Marne

The school received a gold medal at the International Exhibition of 1907 in Buenos Aires. From 1914 to 1916, Eyrolles administrated an auxiliary hospital in his private house, 23 Avenue du Président Wilson in Arcueil–Cachan. From 1 September 1915 to 2 November 1916, he was director of the Service of Industry in the office of the Under–secretary of State for Artillery and Munitions. He studied the implantation of new factories, the extension of old ones, the intensification of the production of weapons by new processes, and the costs. He also chose the manufacturers able to work for the war effort, and he established their cooperation with the Ministry of War. Then, until 17 January 1917, he was chief adviser engineer at the Ministry of Munitions, and, until 1 January 1918, head of the Mission of recycling schools in war factories. In 1918, he trained designers for the Direction of Inventions and for the Air Force. Eyrolles founded other technical schools, and, in 1918, the publishing company that still bears his name (see the next Section). He himself wrote several books. For details, see [233, 234].

Initially, the ESTP was only a school of civil engineering and, around 1920, it had 130 professors and over 12,000 students, internal, external or correspondents from all countries. By 1921, Eyrolles added schools of mechanical, electrical, and architectural engineering as well as topography. By a decree dated 5 February 1921, the ESTP was recognized by the French Government.

The company *Danger Frères et Fils* was one of the largest firms of surveyors between the two wars. In his book [80], René Danger (19 August 1872 - 10 June 1954), founder of the Order of Surveyors and professor of surveying and of the first course of urbanism given at the ESTP in 1928, wrote:

> *Currently, for all administrative and private situations, surveyor applicants are asked to provide vocational references. So, the diplomas of the ESTP have they already become crucial as manifested by the colonial statutes above.*
>
> *In addition, since a decree of 20 April 1922 issued by the Under–Secretary of State for Technical Education, the School is authorized to issue a diploma of Engineer–Surveyor after the second year of study. This diploma is recognized and guaranteed by the Government.*
>
> *This is also the only official one in this vein and it attributes valuable benefits for the issuance of the diploma of Expert–Surveyor recognized by the Government...*
>
> *The teaching of French Surveyor exists at the ESTP since 1905, following representations made to Mr. Eyrolles by the National Society of Surveyors of France, chaired by Mr. Frère.*
>
> *Correspondence gathered, since that time, many students each year and it is by hundreds that we now count surveyors or their employees who have completed their studies at the School.*
>
> *However, it has been recognized since the war, when the serious problems of rebuilding the devastated land were raised, how insufficient were the resources of staff in the profession.*
>
> *The ESTP has firmly adopted the necessary solution and obtained the assistance of relevant departments: Civil engineering, Rural, Cadastre, General Leveling of France, Geographical Service of the Army, Plan of Paris, with authorized representatives of the French Union of Expert–Surveyors.*
>
> *From the collaboration of these different personalities came out the teaching program of the surveyor at the School, over two years...*
>
> *This is why we must thank Mr. Eyrolles for his judicious foresight and his tenacity when he wanted to create the course of Geometer at the ESTP.*

In the preface of the second edition of this book, Maréchal (Field Marshal) Louis Hubert Gonzalve Lyautey (Nancy, 17 November 1854 - Thorey, 27 July

1934) explained how much, as an officer, he always enjoyed drawing nice contour lines with his plane table.

A service of scientific research was organized at ESTP in 1923. By 1931, the ESTP occupied 7.5 hectares (807,293 sq ft) and included classrooms, libraries, workshops, and laboratories. The industrial revolution ensured a rapid success to the ESTP by generating a need for more engineers than the main traditional schools could furnish. The year of record for the schools culminated with 27,759 students following about 30 different courses. The German occupation reduced the number of students to around 900 during the Second World War but, by 1949, the number again climbed to around 2,000.

A new building for the school was constructed, 57 boulevard Saint–Germain, in 1934. Unfortunately, the birthplace of the French dramatist, poet, and novelist Alfred de Musset (Paris, 11 December 1810 - Paris, 2 May 1857), 33 rue des Noyers, had to be destroyed for that purpose. In 1935, Eyrolles put the laboratories and the workshops of the school at the disposal of Frédéric Joliot–Curie (Paris, 19 March 1900 - Paris, 14 August 1958) and his wife Irène Joliot–Curie (Paris, 12 September 1897 - Paris, 17 March1956) who got the Nobel Prize for Physics the same year. There, they built an electrostatic generator of the Van de Graaf type for forcing the potential barrier of atomic nuclei, and worked in Cachan until 1937.

Léon Eyrolles remained director of the school right up until his death in 1945 at which time his son Marc, also an engineer, succeeded him. The ESTP paved the way for other centers of tertiary education and vocational training in Cachan.

On 23 January 1901, Eyrolles was nominated *Chevalier de la Légion d'Honneur*, promoted to *Officier* on 20 October 1911, and to *Commandeur* in 1920. A banquet was offered to him, on 2 July 1921, at this last occasion, and for the recognition by the Government of the *École Spéciale des Travaux Publics, du Bâtiment et de l'Industrie*. On 15 January 1922, a presentation to him was organized. He also had several other foreign decorations. In the years 1920s, he had a manor in Saint–Malo–de–la–Lande, in the department of Manche. In 1931, he was the chairman of the civil engineering section at the international exhibition in Paris and again in 1937.

Municipal councillor of Cachan in 1924, he was mayor of this city from 17 May 1929 to 21 August 1944. He was certainly the oldest mayor of the department of Seine at the end of his last term. It is within his mandate that the current town hall was built in Cachan, inaugurated in 1935. On his municipal administrative work and his attitude during World War II, see [63]. His wife, Cécile Fleurette Hertz, painter, member of the *Salon d'Automne* (she signed Hertz–Eyrolles), was herself one of the senior citizens of Cachan since she died almost centenary in 1974. They were married on 30 April 1906 and they had three children.

Although the ESTP and Eyrolles played a key role in the Resistance at Cachan, he was accused of collaboration and he was incarcerated on 21 August 1944 by the Liberation Committee of Cachan. Eyrolles' arrest followed a unanimous decision taken at a meeting of the *Légions Françaises Anti–Axe* on

25 September 1943 which condemned Eyrolles to death for intelligence with the enemy. It was his apparent friendship with Pierre Laval (1883 - 1945), head of the French government from 18 April 1942 to 19 August 1944 and executed for high treason, which suggested that he was sympathetic to collaboration. However, on 4 September 1944, the Minister of Interior ordered his release. Eyrolles not only denied any links with the Vichy government or the German occupiers, but he also claimed that he had not had any personal relations with Laval since the mid–1930s. A letter from the Prefect to Eyrolles, dated 10 July 1942, shows that he was even admonished for refusing to cooperate with the Vichy government, and that he was intransigent toward German requests for assistance, even though this attitude had put him in danger since his wife was from Jewish origin and was imprisoned for a time by the Germans. Moreover, Eyrolles also had a son, Marc, who was a prisoner of war (who succeeded his father as the director of the ESTP in 1945). Also, when a student of the school killed a German while on vacation and was subject to a Gestapo inquiry, he was hidden at the ESTP, which gave the German police a false lead on him. Moreover, Eyrolles offered teaching positions to two men who were fired for their opposition to the Vichy government. In 1954, the ESTP received the *Croix de Guerre 1939–1945 avec palmes* to have been, throughout the period of occupation and under the guidance of its leaders, a model of patriotism and a permanent center of resistance.

Léon Eyrolles died in Cachan on 3 December 1945 [63, 85].

6.2 The publisher

Léon Eyrolles founded his publishing company *Librairie de l'Enseignement Technique* in Paris in 1918. Its aim was to offer engineers and technicians a practical and efficient documentation. The catalog, originally devoted to issues of building, public works and civil engineering, has spread to related fields such as photography, telephone, water, and then business and management. In 1931, it already contained 410 volumes.

The library of the *Institut Mémoires de l'Édition Contemporaine* (IMEC) possesses one or two copies of the books published or distributed by the company, and 160 books published by Léon Eyrolles between 1884 and 1920; see http://www.imec-archives.com/fonds_archives_fiche.php?i=EYR.

The Archives of the department of Val de Marne in Créteil possess the private *Fonds École Spéciale de Travaux Publics de Cachan* (reference 122J, 644 boxes from 1877 to 1996). Its contents can be found at the address:
http://archives.cg94.fr/consultation/cdc/ead.html?id=FRAD094_000025.
It contains, in particular, the following books written for the courses on topography

- Cours Élémentaire: *Notion de Topographie (levé des plans et nivellement)*, by Eyrolles, Quanon, Simonneau and Seyer, 3rd edition, 1903, 175 pages.

- *Cours de Topographie. Première partie: Opérations à la surface du sol (levé des plans et nivellement)*, by Eyrolles, Prévot and Quanon, 10th edition, 1906, 383 pages.
- Cours Moyen: *Cours de Topographie. Opérations à la surface du sol, Deuxième partie: Topographie Générale*, by Quanon, 1901, 133 pages.
- Cours Moyen, Section des Contrôleurs des Mines: *Cours de Topographie. Troisième partie: Opérations souterraines*, by Douat, 1901, 40 pages.
- Cours Supérieur: *Compléments de Topographie: Notions sommaires sur les erreurs, Tachéométrie, Méthode du levé du Service du cadastre*, by Prévot, 1903, 128 pages and 2 plates.
- *Cours de Topographie. Première partie: Topométrie*, by Eyrolles, Prévot and Quanon, 12th edition revised and augmented, 1909, 460 pages.
- *Cours de Topographie. Deuxième partie: Topographie Générale*, by Cholesky, 2nd edition, 1913, 631 pages, 1 page of errata, 18 plates and 100 figures.
- *Cours de Topographie. Troisième partie: Opérations souterraines*, by Douat, 3rd edition, 1906, 40 pages and a supplement of 10 pages. The same, 4th edition, 1912, 52 pages.
- *Compléments de Topographie: Tachéométrie*, by Prévot, 2nd edition revised and augmented, 227 pages and 2 plates.
- *Notions de Topographie (levé des plans et nivellement)*, by Eyrolles, Prévot, Simonneau and Caussin–Yvon, 5th edition, revised and augmented, 1909, 184 pages.
- *Notions de Topographie (levé des plans et nivellement)*, by Prévot, Caussin–Yvon and Simonneau, 6th edition, 1911, 184 pages.

This list (which may be incomplete) is enough to show that the books covered the whole topic, that they were regularly updated, and that each of them had many editions. We also see that the courses had three levels: *Cours Élémentaire* (Elementary Course), *Cours Moyen* (Medium Course), and *Cours Supérieur* (Superior Course).

Chapter 7

An unpublished book by Cholesky

The *École Spéciale des Travaux Publics* (ESTP), the innovative establishment discussed extensively in the previous chapter, asked Cholesky to write four correspondence courses: *Topographie générale* (General Topography), *Levés d'études à la planchette* (Advanced Plane Table Surveying), *Calcul graphique des contenances* (Graphical Calculation of Capacities), *Notions de cosmographie et d'astronomie de position* (Notions of Cosmography and Positional Astronomy). The first two courses were completed, taught and later published with success by *Librairie Eyrolles*, to the extent of being reprinted many times: they are analyzed in Chapter 5. We have no trace of the fourth course which, it seems, remained at the planning stage, without being started. As for the course on graphical calculation, we know that Cholesky undertook significant work on it between 1910 and 1914, but was unable to finish it due to the war and his death in 1918, a few months before the Armistice. We are left with an unfinished, unpublished manuscript preserved with various working documents in the archives of the *École Polytechnique*. It seems appropriate to publish this manuscript (see annex C) and to analyze it in detail here, as it displays two characteristics which make it attractive as much for the historian of graphical calculation as for the historian of teaching, and which makes it different from other courses on graphical calculation published at this time: firstly, that this is a course for those who, in general, have a more modest mathematical background than the students of the *École Polytechnique*, the *École des Ponts et Chaussées* (National School of Civil Engineering) and other prestigious engineering schools for whom the majority of known treatises are written, and secondly it is a correspondence course which, by necessity, displays characteristics of particular interest both pedagogically and editorially.

In Section 7.1, we start with a quick overview of the history of graphical calculation before 1914 and the mathematical notions underlying nomography, in order to position Cholesky's course in its scientific and historical context. Narrowing our view, in Section 7.2, we look at the position and the organization of graphical calculation at the ESTP. After carrying out an initial description of Cholesky's manuscript in Section 7.3, we try to reconstruct the process of commissioning and the development of the course in Section 7.4, then conclude with a detailed analysis of its content in Section 7.5.

7.1 Graphical calculation before 1914

7.1.1 What is graphical calculation?

Under the generic term *graphical calculation* you find all the processes using a graph made on a plane support with the help of any type of instrument with the aim of avoiding, completely or in part, recourse to numerical calculation for the resolution of a problem [228].

In numerical calculation, you follow an algorithm which leads to a desired result; alternatively one can read the result directly from a numerical table, that is to say, in a collection of results calculated in advance. In graphical calculation the situation is similar: either you use various instruments to create a geometric construction or you use a previously–made graphical table. The result is a subdivision of graphical calculation into sub–disciplines, which, towards the end of the 19th century were already firmly established under the names of, respectively, *calcul par le trait* and *nomography.*

Calcul par le trait is true geometric calculation, whose foundations were clearly presented by René Descartes (La–Haye–en–Touraine, 31 March 1596 - Stockholm, 11 February 1650), at the beginning of his 1637 *La Géométrie* [84]. Having selected a unit of length, numerical data are represented on the page by line segments. Constructing geometric figures with a ruler and compass, and possibly other mechanical instruments gives rise to new line segments representing the unknown values sought.

Nomography, meanwhile, is the science of graphical tables, also known as *abaques* or *nomograms.* When you have to repeat the same graphical operation many times over (the construction of a value of a given function, the graphical resolution of a certain equation, etc.), it is advantageous to have, once and for all, a graphical table, in the form of marked lines or points, with suitably graduated scales, movable or not, giving the value sought in terms of those parameters by simple reading. The most famous nomogram is probably the slide rule, the graphical equivalent of logarithm tables.

During the second half of the 19th century in particular, the practice of graphical calculation developed considerably. How can we explain such enthusiasm and the fact that it flourished right up to the appearance of the first electronic calculators and personal computers? It can be argued that there are four groups of reasons:

1) Graphical calculation is generally recognized as more intuitive and demanding less mental concentration than numerical calculation. It is quick and efficient, at least when you are happy with the limits of its precision. This is the case for most technical applications and in the domain of engineering, particularly as the calculations are often based on imprecise data coming from experience.

2) Numerical tables are longwinded and tedious to construct and are particularly difficult when there are more than two entries. Graphical tables, on the other hand, allow relations to be represented easily, whatever the number of variables. Furthermore, interpolation is done by sight, without additional effort.

3) When you need more precision, as in astronomy, graphical calculation can serve an auxiliary function to numerical calculation. In methods using successive approximations, graphical calculation is still useful for finding the initial values quickly, or, at the other end of the chain, to determine the final corrections involving only a small number of decimal places. In a more general way, it retains a role in preparation, as a supplement and in verification within a more rigorous calculation.

4) The last advantage, and by no means the least: graphical calculation is available to those even with a lower level of mathematical training, especially technicians who have not benefitted from higher education. This can result in substantial savings, since the use of qualified professional calculators is expensive.

7.1.2 A short overview of the history of graphical calculation

We will now outline the history of graphical calculation.

We know that graphical methods have been developed and used since Antiquity. In particular, in the Middle Ages, Arab mathematicians used such methods in astronomy and in their attempts to solve third–degree equations. With Descartes, the link between arithmetic operations and geometric constructions became clearer, and the possibility arose of systematically representing a relation between two variables graphically, by plotting a curve. 17th and 18th century mathematicians were first and foremost geometricians; many, like Newton, put Cartesian ideas into practice and regularly used graphical methods. However, it wasn't until the end of the 18th century before graphical calculation began to be considered as an autonomous doctrine.

The first graphical tables for use in calculation came from the attempts of the French Revolution to impose a new system of weights and measures. In order to help the common people to adapt to the reform, the law of 18 Germinal, year III prescribed a simplification of the tools used for conversion [105, p. 72]:

> *Instead of tables giving the relations between the old and the new measures as prescribed in the decree of 8 May 1790, graphical scales shall be made to assess these relations without the need for any calculation.*

In response to this request, Louis–Ézéchiel Pouchet (Gruchet–Le–Valasse, 1 June 1748 - Rouen, 30 July 1809), a cotton cloth manufacturer from Rouen, proposed, in 1797, abaques made up of two bundles of marked straight lines, horizontals and verticals, through which a number of marked curves also pass [197]. These abaques are known as *concurrent–line* or *Cartesian abaques.* For example, for the multiplication $\gamma = \alpha\beta$, the method allows multiplication tables to be replaced by a set of equal–sided hyperbolas. On this graphical table (see figure), the multiplication is carried out instantaneously by simple reading. Values that do not appear directly on the table are estimated by sight. Finally, the same table can clearly be used to carry out divisions and the extraction of square roots. Pouchet was the first to use the expression *graphical calculation* and most later authors of treatises on the discipline consider him as its founder [105, 156, 192].

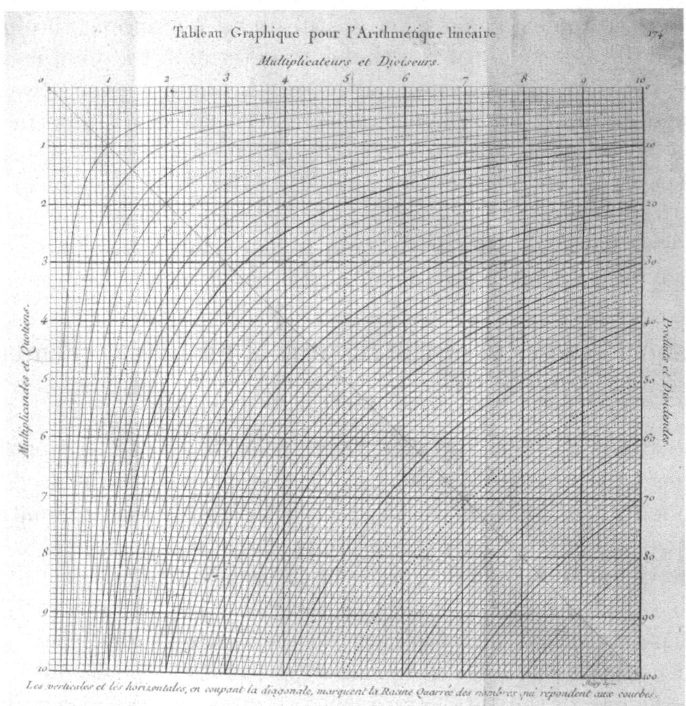

Pouchet's graphical multiplication table [197, p. 174]

Following Pouchet, there were a few isolated examples of the graphical interpretation of double entry tables, but it was, in particular, Léon–Louis Lalanne (Paris, 3 July 1811 - Paris, 12 March 1892), a *Ponts et Chaussées* engineer, who took the decisive step in the theory of abaques. In 1843, he had the idea of using non–regular scales [154]: by replacing the original variables with appropriately selected auxiliary functions of them, he also succeeded in transforming the third bundle of curves into straight lines. By analogy with an optical phenomenon, Lalanne described this transformation as *geometric anamorphosis* [155, p. 13]. By means of Lalanne's transformation an abaque representing an $F(\alpha, \beta, \gamma) = 0$ relation between three variables now takes the form of a set of three bundles of straight lines, and a solution to the equation corresponds to the intersection of the three lines (one for each bundle). In the example of multiplication, after noting that the $\gamma = \alpha\beta$ relation can also be written $\log \gamma = \log \alpha + \log \beta$, you simply have to graduate the axes with the new $x = \log \alpha$ and $y = \log \beta$ variables so that the bundle of Pouchet hyperbolas becomes a bundle of $x + y = \log \gamma$ straight–line equations (see figure).

Lalanne's Cartesian abaques were very widely distributed in the Public Works sector, where his ideas arrived at a very opportune moment. Indeed, the law of 7 June 1842 decided the establishment of a network of major railway lines radiating from Paris and, to execute that decision rapidly, it was deemed necessary to make use of new methods of considering the considerable amount of grading work to be carried out. In 1843 the French government sent all the engineers concerned graphical tables for calculating the volumes for excavation and embankment for the railways and roads.

A little earlier, another French engineer from the *Ponts et Chaussées*, Barthélémy–Édouard Cousinéry (Marseille, 29 October 1790 - 1851), wrote the first complete treatise on graphical calculation, entitled *Le calcul par le trait* [71]. Cousinéry, who falls within the lineage of Jean-Victor Poncelet (Metz, 1 July 1788 - Paris, 23 December 1867), did not limit himself only, as Pouchet did, to elementary operations, but was also interested in graphical methods for quadrature, cubature, the interpolation of functions and the resolution of equations. If Cousinéry could relate to Poncelet, it was due to the many digressions in the latter's works on the use of graphical methods in engineering.

Cousinéry's work passed more or less unnoticed in France, but this was not the case in Germany, especially with Carl Culmann (Bad Bergzabern, 10 July 1821 - Riesbach, 9 December 1881). Culmann was an alumnus of the *École d'Application du Génie et de l'Artillerie* (School of Applied Engineering and Artillery) in Metz where he also learned the graphical methods taught by Poncelet and his disciples. In 1855, Culmann was recruited as a teacher of engineering sciences at the *Institut Polytechnique* in Zurich. There he developed the application of graphical calculation relating to the stability and strength of constructions (beams, bridges, roofs, arches, retaining walls, etc.), thus establishing a new discipline, *graphic statics*, in which he introduced the systematic use of the concepts of the *polygon of forces* and the *funicular polygon*. Culmann's main work *Die graphische Statik* (Graphic Statics), published in 1866 [77] was rapidly assimilated and adopted by a wide circle of engineers and industrial designers, sparing them lengthy and tedious calculations. Culmann, in the lineage of Monge, Poncelet, Cousinéry and Lalanne,

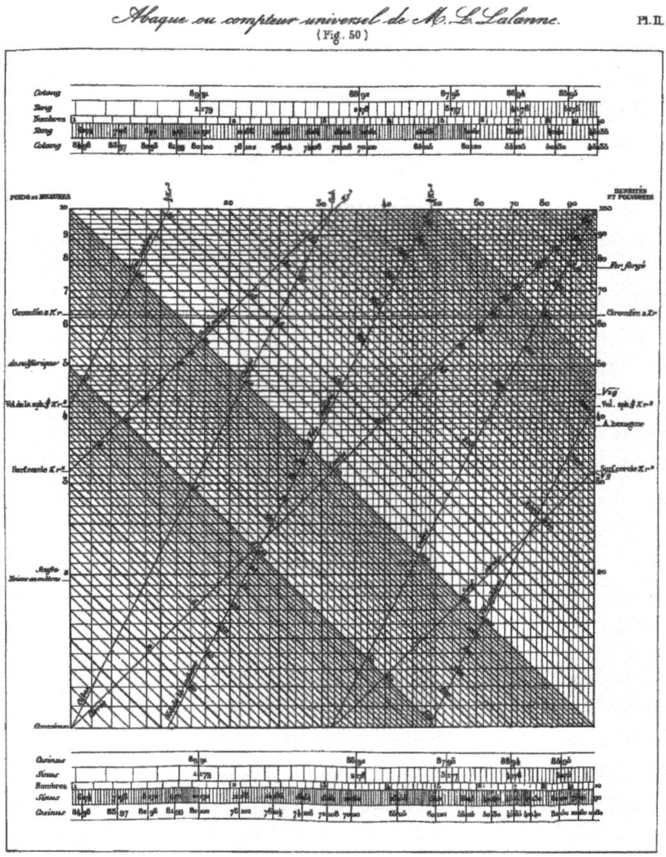

Lalanne's Abaque or Universal Calculator [106, pl. II]

paid tribute to French engineers on the occasion of the publication of a translation of his treatise in Paris.

However, outside Germany it was in Italy, and not in France that Culmann's ideas gained ground. Two key names can be cited here: Cremona and Favaro. First of all Luigi Cremona (Pavia, 7 December 1830 - Roma, 10 Juin 1903) who led research in Pure Geometry at the *Università di Bologna* (University of Bologna) until 1867. In that year he became a Professor at the *Politecnico di Milano* (Polytechnic University of Milan), where his new duties led to an interest in graphical computation and graphic statics He published important works in this field, his principal contribution was to transpose Scottish scientist James Clerk Maxwell's (Edinburgh, 13 June 1831 - Cambridge, 5 November 1879) theorems on reciprocal figures into graphic statics and interpret them in terms of duality in projective space. The practical contribution of this new perspective is that if three forces in equilibrium are shown in the figure by the three sides of a triangle, then they are represented in the reciprocal figure by three intersecting segments.

The second great figure in Italian graphical calculation is Antonio Favaro (Padova, 21 May 1847 - Padova, 30 September 1922) who, from 1870 was assigned to classes in graphic statics at the *Università degli Studi di Padova* (University of Padua). His lectures on graphic statics were translated into French, annotated and expanded by Paul Terrier, a French graduate engineer from the *École Centrale des Arts et Manufactures* (Central School of Arts and Manufactures) [105, 106]. The sub–titles of these two volumes *I. The Geometry of Position, II. Graphical Calculation*, clearly show that Favaro's objective, well beyond graphic statics, was to provide engineers with a complete course in technical drawing, whose two components are descriptive geometry and graphical calculation.

A new extension was given to graphical calculation by Junius Massau (Gosselies, 9 April 1852 - Gent, 10 February 1909), a professor at the *Universeit Gent* (Ghent University) for more than thirty years [229]. His principal ambition was to make *calcul par le trait* independent of graphic statics. In a series of works published from 1878 to 1904, he established a true graphical integral calculation, taken as far as partial differential equations. Leaving aside the use of funicular polygons, he developed techniques of graphical integration equivalent to the quadrature formulae used in numerical analysis (the trapezoidal method, the midpoint method, etc.). Then he systematically applied these methods of approximate integration to the construction of curves, the calculation of areas, statics moments, moments of inertia, grading calculations, the stability of masonry constructions, calculations for straight beams, and the integration of ordinary differential equations [175]. Finally, not satisfied with having renewed the classic subjects, Massau devised, probably for the first time in history, graphical methods for the integration of partial differential equations [176].

The progress of graphical integration is also demonstrated by the invention and the manufacture of apparatus intended specifically to replace *calcul par le trait* with regard to infinitesimal calculus [140]. Planimeters (for measuring the surface area defined by a given curve), integrators (for the measurement of integral values of a function given graphically) and integraphs (which plotted the curve representing the integral) were designed around an integrating wheel linked to a pointer which followed the perimeter of the surface; the wheel's angle of rotation is proportional to the surface area. Two of the best-known of these instruments are the planimeter invented by Jacob Amsler (Stalden, 11 November 1823 - Schaffhausen, 3 January 1912) and the integraph by Abdank–Abakanowicz (Vilkmergè, Lithuania, 6 October 1852 - Parc-Saint-Maur, 29 August 1900). Two more of these devices, Louis-Frédéric Jacob's (Charleville, 29 October 1857 - 1930) cutting-blade integrator and Ernesto Pascal's (Napoli, 7 February 1865 - Napoli, 25 January 1940) integraph were even thought to be of use in the integration of certain differential equations. In 1876, William Thomson (Belfast, 26 June 1824 - Largs, 17 December 1907), better known as Lord Kelvin, adapted the principle of the planimeter to the calculation of Fourier coefficients of a function, thus creating the first harmonic analyzer. Lord Kelvin also considered the modification of integraphs for the resolution of differential equations, launching the ideas that would later lead to the manufacture of the great differential analyzers of the 1930s, 40s and 50s.

Returning to another aspect of graphical calculation, namely, the theory of abaques. Following Lalanne, concurrent–line abaques spread rapidly and to such an extent that they became standard tools for French engineers in the third quarter of the 19th century. In particular, the decade from 1880 to 1890 was especially auspicious and saw many simultaneous theoretical advances.

Firstly, Massau, following in Lalanne's footsteps, introduced a generalized concept of anamorphosis in 1884 [175], with the underlying idea that a change in coordinates of the type $x = f(\alpha, \beta)$, $y = g(\alpha, \beta)$ results in a deformation of the graph which retains the property of the intersection of the three lines, and consequently does not affect the operation of the abaque. From this perspective, Lalanne's geometric anamorphosis corresponds to the particular case of $x = f(\alpha)$, $y = g(\beta)$, in which you are content to perform independent lengthenings and shortenings in the direction of x and of y. Massau sought then to determine which are the functions that you can represent using three bundles of straight lines, without requiring the first two bundles to be parallel to the coordinate axes, which leads to the general notion of *concurrent–straight–line abaques*. The advantage of these abaques came naturally from the fact that they are easy to construct; a straight line is determined by only two points, whereas with curves other than straight lines, the plot of each curve in the abaque requires the prior calculation of a large number of points.

Independently of Massau, particular concurrent–straight–line abaques were devised by Charles Lallemand, an *École des Mines* engineer. We are at a time when a large program of public works was being drawn up in France. Carrying out the work would require a more precise knowledge of the relief of the ground and hence the decision to undertake what geodesists call leveling the entire area: complementary to triangulation which fixes the position of points on the ground in a horizontal projection, the object of leveling is to determine altitudes (see Section 3.2 where the subject has already been addressed in detail). From 1880, Lallemand was assigned the task of creating a National Leveling Bureau for France, which officially came into being in 1884. It is in this context that he invented *hexagonal abaques*, designed as a graphical method of automating the lengthy and tedious calculations necessary for processing the countless measurements made on the ground. In an internal memorandum written for his bureau [156], Lallemand shows that in most standard cases, a three–variable relation can be represented by three bundles of parallel lines making 60° angles between each other, hence the term *hexagonal*. The graphical tables are particularly simple to construct and use insofar as there is no need to really draw the parallel straight lines of a bundle, but only to locate the positions on a bundle perpendicular to it. In addition, it is possible to represent relations involving more than three variables with abaques such as these. For example, the abaque in the following figure, used by the National Leveling Bureau, is made up by the juxtaposition of four basic abaques linked by common scales. If you had opted for a classic concurrent-straight-line abaque, the superimposition of bundles of straight lines corresponding to the four equations would have led to a totally unreadable working drawing. Lallemand's efforts enabled his administration to triple the accuracy of earlier results, while significantly reducing the cost price of the calculations.

At the same time as Massau and Lallemand's publications were taking concurrent-straight-line abaques into a phase of maturity, a new figure intervened and steered research on graphical tables in a different direction. Philibert Maurice d'Ocagne (Paris, 25 March 1862 - Le Havre, 23 September 1938), entered the *École Polytechnique* in 1880, then spent his entire career on the staff of *Ponts et Chaussées*. In particular, he was drafted to the National Leveling Bureau to work as Lallemand's deputy from 1891 to 1901. At the same time, and for 45 years he taught tirelessly at the *École Polytechnique*, at the *École des Ponts et Chaussées* and at the *Sorbonne*. Closely linked to this dual activity as an engineer and teacher, d'Ocagne carried out important research throughout his life, which resulted in over 400 publications. Within this astonishing body of work that addresses many themes, it is essentially the work on abaques that made him famous.

The first contribution d'Ocagne made to the science of abaques was in 1884, when he was only 22 years of age. Making the best use of advances in projective geometry, he defined two new systems of tangential plane coordinates, which he called *parallel* and *axial coordinates* respectively, and used the first of these systems to develop a new method for graphical calculation, to which he managed to give a very general scope in 1890, which led to the publication, in the following year of a summary treatise entitled *Nomographie. Les calculs usuels effectués au moyen des abaques* (Nomography. Standard Calculations Carried Out with Abaques) [187]. In this seminal text you find the first appearance of the term *nomography* (from the Greek *νόμος*, law, to evoke the graphical representation of a law linking several variables, namely an equation).

To introduce his method, d'Ocagne starts from the observation that most of the equations encountered in practice can be represented by an abaque made up of three systems of straight lines, and that three of these straight lines, one taken from each of the systems, correspond when they intersect at a single point. His basic idea was then to construct, by duality, a second figure correlative to the first: by substituting the use of tangential coordinates with point coordinates: each line of the initial abaque is thus transformed into a point, and three concurrent lines are transformed into three aligned points. The three systems of marked straight lines become three marked curves, forming what d'Ocagne called an *alignment abaque*.

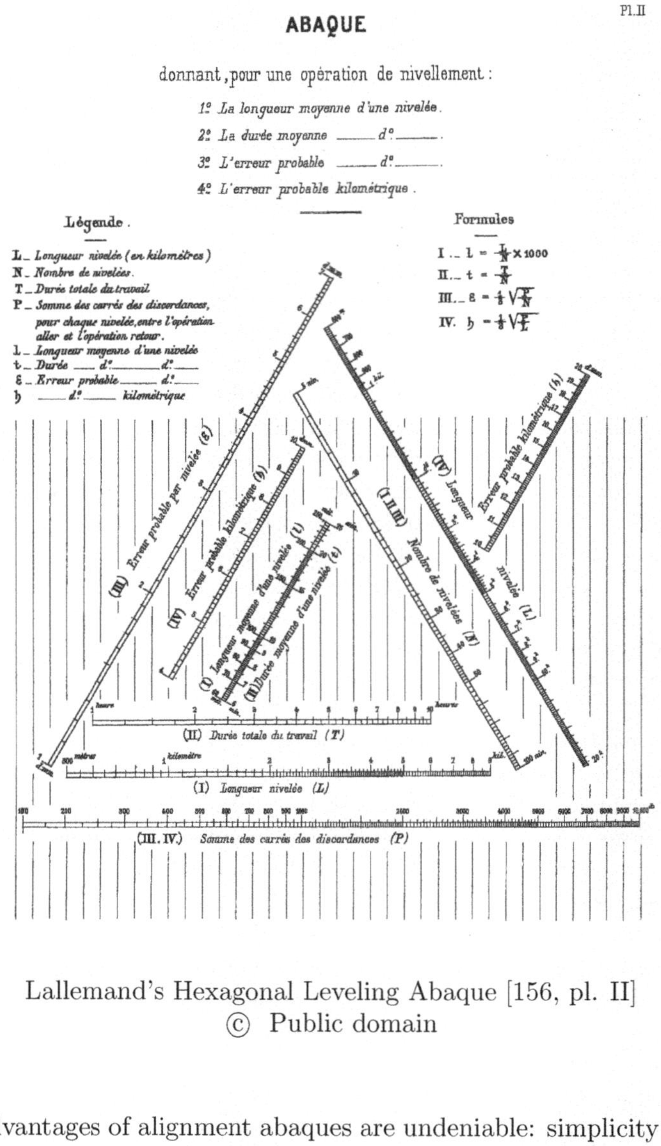

Lallemand's Hexagonal Leveling Abaque [156, pl. II]
© Public domain

The advantages of alignment abaques are undeniable: simplicity of construction, elimination of reading errors due to previously having to follow the lines to read the marks, greater precision of by–sight interpolations. Furthermore, the alignment is maintained by homography, one can easily transform a graphical table of this sort so that its curves are arranged better on the page. Finally, just like Lallemand's hexagonal abaques, the alignment method allows three systems of straight lines to be replaced by three linear scales in a process of a sort of data compression, but d'Ocagne's method includes the general case of any three bundles of straight lines, while Lallemand's method only applies to three bundles of parallel straight lines. Thanks to the space saved on the page, you can, once more, easily juxtapose several three–variable abaques and thus graphically represent a relation of any number of variables, provided that this relation can be broken down into a series of three–variable relations, each having one variable in common with the next. The following figure thus gives an example of a twelve-variable abaque, credited to Commandant of Military Engineering Léopold Bertrand (1852 - 1918), intended to help in dealing with problems relating to the distribution of water.

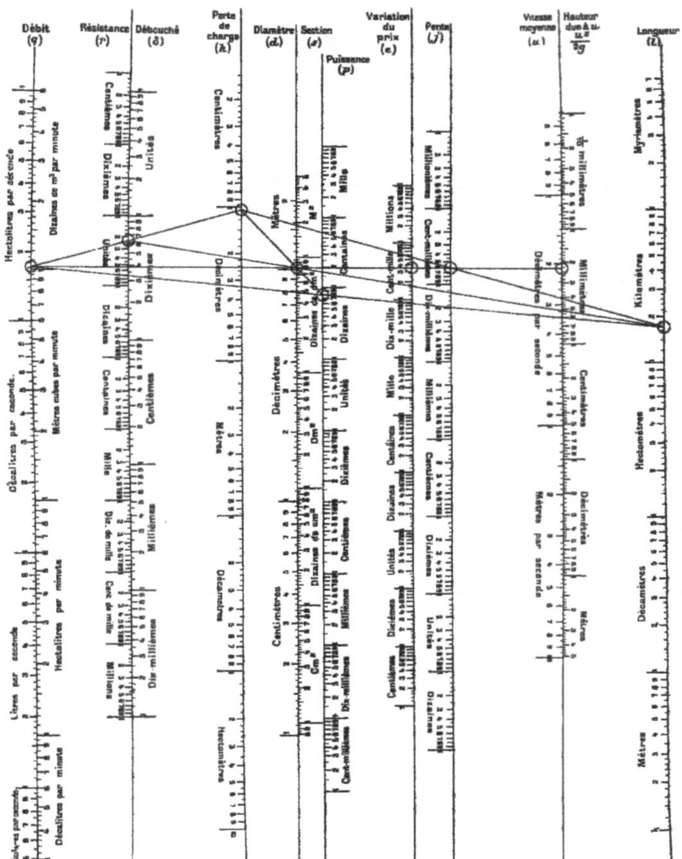

Bertrand's Alignment nomogram for the distribution of water [188, p. 162]

After the publication of his 1891 tract, d'Ocagne extended his theory and the applications of the alignment method until the publication of his much more imposing treatise in 1899, the famous *Traité de nomographie. Théorie des abaques. Applications pratiques* (Treatise on Nomography. Theory of Abaques. Practical

Applications) [188], which became the reference work for the new discipline for many years. A little later in 1902, he introduced the generic term *nomogram* to replace *abaque*, reserving the latter for graphical tables with a checkerboard appearance. From this moment, nomograms were quickly adopted by a large number of engineers who used them in a wide range of applications. At the turn of the 20th century, the doctrine of nomography was expanding rapidly and had already firmly established itself in the landscape of applied sciences [230].

7.1.3 Theoretical principles of nomography

The part of the course on graphical calculation written by Cholesky relates entirely to nomography, so before focusing on it directly, it seems indispensable to return in more depth to the fundamental mathematical notions hidden behind the graphical tables evoked in the previous section. The central problem in nomography is that of the plane representation of three–variable relations $F(\alpha, \beta, \gamma) = 0$. The overall idea of those abaques known as *concurrent–line* abaques, is to demonstrate this relation as the result of the elimination of two auxiliary variables in three equations, each dependent on only one of the three principal variables:

$$F(\alpha, \beta, \gamma) = 0 \iff \exists\ (x, y) \begin{cases} F_1(x, y, \alpha) = 0 \\ F_2(x, y, \beta) = 0 \\ F_3(x, y, \gamma) = 0. \end{cases}$$

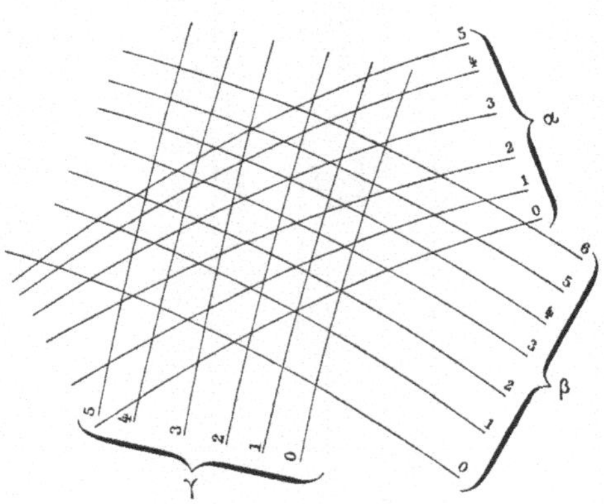

Abaque with three bundles of curves [187, p. 10]
© Public domain

The abaque is then made up of three bundles of marked curves, the respective equations are: $F_1(x, y, \alpha) = 0$, $F_2(x, y, \beta) = 0$ and $F_3(x, y, \gamma) = 0$, inscribed on a graph having Cartesian coordinates x and y (see figure). For each α parameter value, the first equation determines a curve which can be identified on the graph by marking this α value next to it. It is the same for the two other bundles.

On the abaque, a (α, β, γ) solution of the equation simply corresponds to the single point where the three curves, one for each family associated respectively to the marks α, β and γ, intersect. If you give values to any two of the variables, for example α and β, you find the corresponding γ value directly by reading the mark of the curve from the third family that passes through the point of intersection of the curve of the first two families whose marks are α and β, (in the figure, for example, you read $\gamma = 2$ then $\alpha = 3$ and $\beta = 4$). For values between those marked on the abaque you proceed by visual interpolation.

The most common approach, which applies to any three–variable relation, consists in simply taking $x = \alpha$ and $y = \beta$ for the first two equations. In this case, the curves parameterized by α are parallel to the ordinate axis and the curves parameterized by β are parallel to the abscissa axis (see figure). In practice, all you need to do is construct the curves of the equation $F(x, y, \gamma) = 0$ on a sheet of graph paper. You end up with the topographic representation of a surface by its contour lines. This is what Pouchet did for multiplication, as discussed above.

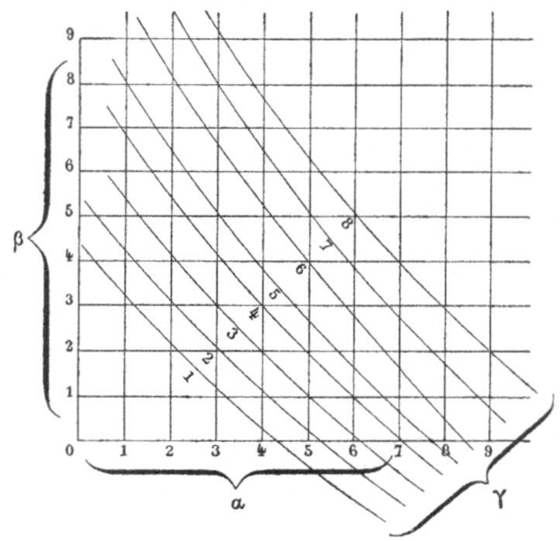

Abaque with two bundles of straight lines and one bundle of curves [187, p. 12]

In Lalanne's geometric anamorphosis method, by marking non–regular graduations on the axes, that is to say, taking $x = \varphi(\alpha)$ and $y = \psi(\beta)$ as the first two equations, where φ and ψ are suitably chosen functions, it is, in certain cases, possible to ensure that the curves of the third family are also straight lines. Thus Lalanne succeeded in reducing Pouchet's hyperbolas to straight lines. To examine, more generally, the conditions permitting an abaque in which three clusters of curves are reduced to straight lines, even if no longer using lines parallel to the coordinate axes, Massau wrote that the equations of the three clusters of curves are straight line equations and obtained the following characterization:

$$\exists\ (x,y)\ \begin{cases} f_1(\alpha)\,x + g_1(\alpha)\,y + h_1(\alpha) = 0 \\ f_2(\beta)\,x + g_2(\beta)\,y + h_2(\beta) = 0 \\ f_3(\gamma)\,x + g_3(\gamma)\,y + h_3(\gamma) = 0 \end{cases} \iff \begin{vmatrix} f_1(\alpha) & g_1(\alpha) & h_1(\alpha) \\ f_2(\beta) & g_2(\beta) & h_2(\beta) \\ f_3(\gamma) & g_3(\gamma) & h_3(\gamma) \end{vmatrix} = 0.$$

Thus, it is when the initial $F(\alpha, \beta, \gamma) = 0$ equation can be put in the form of such a determinant, known thereafter as the Massau determinant, that it will be possible to represent it by a concurrent–straight–line abaque (see figure). This problem, considered as the fundamental problem in nomography has been the subject of much theoretical research up to the present day: most particularly we can mention the work of the Swedish mathematician Thomas Hakon Gronwall (Dylta Bruk, Sweden, 16 January 1877 - New York, 9 May 1932), the American Oliver Dimon Kellogg (Linwood, Pennsylvania, 10 July 1878 - Greenville, Maine, 27 August 1932) and the Pole Mieczysław Jan Warmus (Dobrowlany, 1 June 1918 - Australia, 20 September 2007).

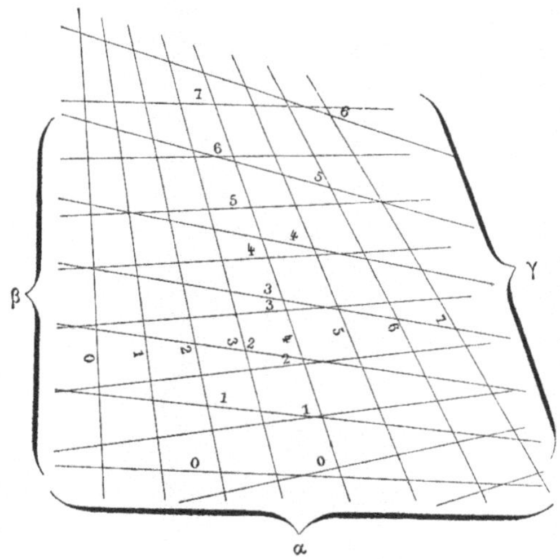

Abaque with three bundles of straight lines [187, p. 14]

The breakthrough d'Ocagne made in 1884 by exploiting advances in projective geometry, notably the principle of duality, to transform concurrent–straight–line abaques into alignment abaques, means finally to consider that if the nullity of the Massau determinant expresses the intersection of three straight lines, this nullity equally expresses the alignment of three points, that is to say, the points of parameters α, β and γ taken respectively from the curves parameterized:

$$\begin{cases} x = \dfrac{f_1(\alpha)}{h_1(\alpha)} \\ y = \dfrac{g_1(\alpha)}{h_1(\alpha)} \end{cases}, \quad \begin{cases} x = \dfrac{f_2(\beta)}{h_2(\beta)} \\ y = \dfrac{g_2(\beta)}{h_2(\beta)} \end{cases} \quad \text{et} \quad \begin{cases} x = \dfrac{f_3(\gamma)}{h_3(\gamma)} \\ y = \dfrac{g_3(\gamma)}{h_3(\gamma)} \end{cases}.$$

Thus, the three systems of marked straight lines become three marked curves. An alignment nomogram is particularly simple to use (see figure): if, for example, it is the values of α and β which are given, you inscribe a straight line passing through the points marked α and β on the first two curves, and this straight line meets the third curve at a point whose position is the γ value sought. In practice, so as not to spoil the abaque, the auxiliary straight line is not actually inscribed on the paper: a thin rectilinear marking on a transparent backing, or a thin wire is used to join the points.

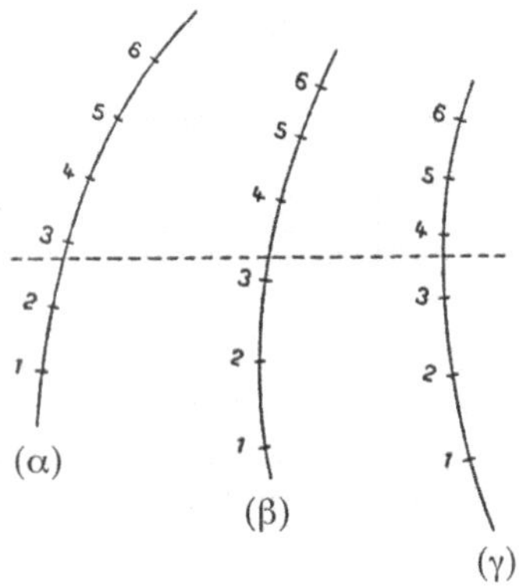

Alignment nomogram [190, p. 178]
© Public domain

The case of Lallemand's hexagonal abaques is a little in the margins of the general mathematical development of the theory of abaques. In fact, they are still concurrent–straight–line abaques, but whose presentation was improved by the addition of the two ideas for simplification already mentioned above. Let us revisit and explain these two ideas. The first consists in noting that, given an abaque with three clusters of parallel straight lines, to identify these straight lines it is enough to cut each cluster by any straight line and mark the position of each straight line next to its point of intersection with the corresponding transversal. You can then erase the three clusters of straight lines, only keeping the three graduated transversal lines, known as *linear scales* (see figure). To join these scales one with the other, you use a transparency, called an *indicator*, on which three axes are inscribed, parallel respectively to the directions of the three erased bundles. Then simply move the transparency on the abaque parallel to itself. Instead of spending time on the numerous lines of a Cartesian abaque, the user can thus concentrate only on those straight lines useful to his calculation.

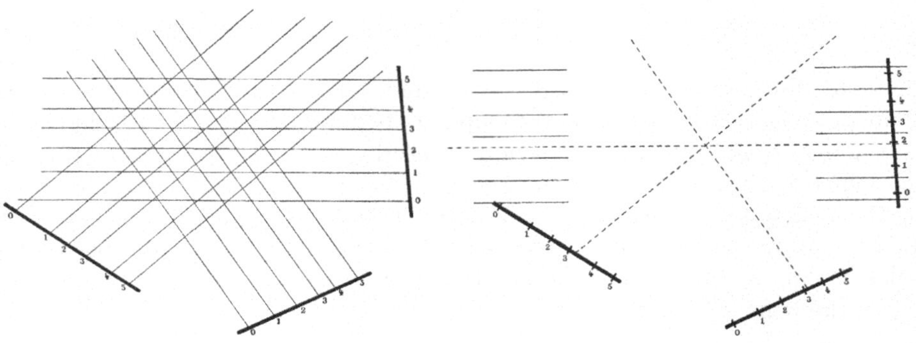

Abaque with three bundles of parallel straight lines [187, p. 31]

The second idea for simplification is contained in a rule for graphical addition that Lallemand stated thus [156, p. 9]:

> *The sum of the projections of a segment of a straight line on two axes having an angle of 120° between them, is equal in magnitude and in sign, to the projection of the same segment onto the internal bisector of the angle between these axes.*

The proof of this result is elementary (see figure): if you call α the angle vector $\overrightarrow{AB}$ makes with the internal bisector AZ of the axes AX and AY is an angle of 120°, everything reduces to equality

$$\cos(\alpha + 60^\circ) + \cos(\alpha - 60^\circ) = 2\cos\alpha\cos 60^\circ = \cos\alpha.$$

If you use the same unit to graduate linear scales AX, AY and AZ regularly, and if the axes of the indicator are placed perpendicular to the three scales, they will intercept the scales at the points whose abscissae verify the relation $x + y = z$. Given that firstly the scales, and secondly the axes are parallel to the diagonals of a regular hexagon, abaques which exploit the rule of graphical addition are said to be hexagonal. By combining these ideas with Lalanne's geometric anamorphosis, you can construct non–regular scales on the $x = f(\alpha)$, $y = g(\beta)$ and $z = h(\gamma)$ axes, and thus represent all $f(\alpha) + g(\beta) = h(\gamma)$ three–variable equations on a hexagonal abaque.

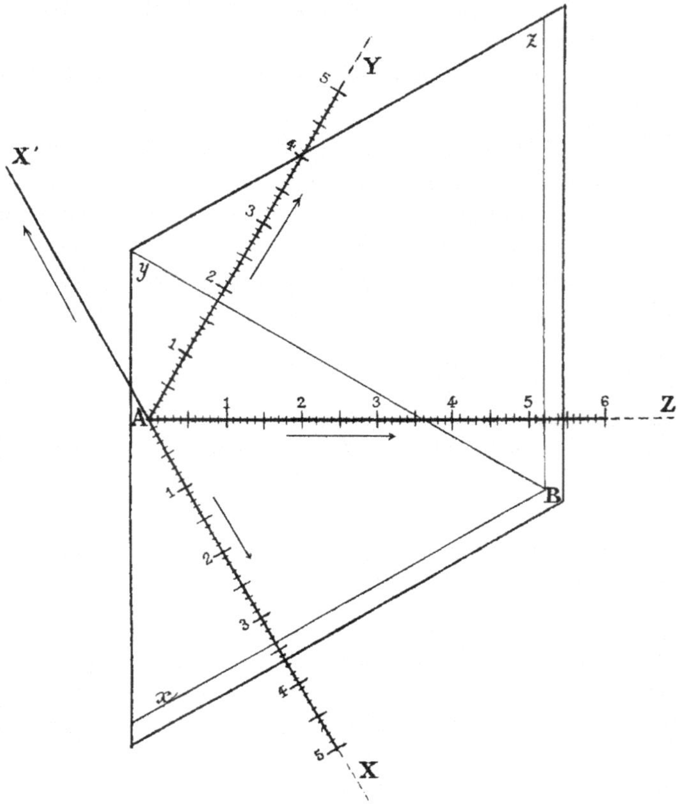

Lallemand's Principle of graphical addition [156, p. 16]

7.2 Graphical calculation at the ESTP

Towards the end of the 19th century, graphical calculation held a significant and growing position in the training of engineers and in their work. In 1907 d'Ocagne, looking only at nomography, listed the applications of the new discipline under 14 headings [191]:

1. Physics (general)
2. Electricity
3. Strength of Materials
4. Hydraulics
5. Shipbuilding
6. Machines
7. Nautical Calculations
8. Geodesy
9. Topography
10. Artillery
11. Aviation
12. Insurance
13. Research on empirical laws
14. Graphical calculation (general)

With the exception of Insurance, this list covers all the traditional sectors in which civil and military engineers work. To meet this new need, courses specializing in the various aspects of graphical calculation were created at the schools of engineering, polytechnic institutes and universities of the time, in Europe and almost everywhere around the world.

The *École Spéciale des Travaux Publics*, seeking, as we have seen in the previous chapter, to provide an education that was both practical and directly operational, could not ignore this important development. In line with the other establishments, it progressively integrated graphical calculation into its courses. In an ESTP publication from 1923 [199] listing on its cover sheets *Cours et Instructions remis aux Auditeurs et Correspondants* (Courses and Instruction provided for Auditors and Correspondents), you find courses entitled:

Graphical Calculation and Nomography
Graphic Statics and Strength of Materials Applied to Metallic Constructions
Graphic Statics
Graphical Calculation of Capacities
Graphical Calculation of Surfaces

The general courses in graphical calculation, nomography and graphic statics were designed principally for future engineers and civil engineering project managers and particularly for those who had chosen *Ponts et Chaussées*, *Mines* and *Travaux Publics*. The courses on graphical calculation of capacities (or surfaces), for their part, were designed specifically for surveyor–topographers. This was a particularly relevant association as the two are, indeed, responsible for the calculation of capacities for the needs of land surveying and Land Registry; in this context, the word capacity is used for surface area: you speak of measuring the capacity of a parcel of land, an everyday and indispensable operation, especially for establishing legal title to a property and the value of the land, allowing the calculation of property tax and division of land in a succession.

To illustrate this question in the context of our study, we will quote Article 91 of the Property Tax in force in Tunisia at the time of Cholesky's sojourn in that country [7, p. 244]:

> *The calculation of surface areas shall be performed in the offices of the head of the topographical department, under the direct supervision of the head of the department.*
>
> *The surface areas shall be calculated twice.*
>
> *If the calculations are performed using graphical methods, as far as is possible, two different methods shall be used.*
>
> *When the surface areas are determined by the use of a planimeter, two different instruments shall be used for the calculations.*

This directive clearly evokes what was expected of a surveyor–topographer and, consequently, what should figure in a course on graphical calculation designed for that profession. For an exhaustive overview of the various measuring techniques that a land surveyor was required to know and use, we can go back, for example, to a course written in 1923 for the ESTP by Joseph Marie Denis Prédhumeau (Poitiers, 10 March 1876 - 16 April 1934), a *Ponts et Chaussées* engineer [199]. Prédhumeau was one of Cholesky's successors at the ESTP; furthermore, his course will serve as a reference to obtain an idea of the parts that Cholesky did not have the time to write on his own course.

We can identify three more or less equally important parts in Prédhumeau's teaching, which correspond to three major types of graphical techniques. Chapter I deals with methods for calculating areas by means of constructions made with a ruler and compass, from basic rectilinear areas (rectangles, parallelograms, triangles, trapeziums) through to curved surfaces that are assessed by breaking them down into a number of smaller basic areas. This part finishes with some of the rudiments of graphical integration and graphic statics; academic theories that make up, in some ways, the culmination of *calcul par le trait*. Chapters II and IV briefly present the various devices that could be used to facilitate, automate and mechanize graphical constructions, from the least expensive to the most costly: strips of paper, *roulettes de Dupuis* (Dupuis wheels), curvimeters, *équerres Beauvais* (Beauvais set squares), *procédé Willotte* (Willotte method), balances, slide rules, planimeters and integraphs. Lastly, Chapter III is devoted to nomographic procedures.

To meet the growing needs of Land Surveyors, particularly in town planning, in 1905 the *Société Nationale des Géomètres de France, d'Algérie et de Tunisie* (National Society of French, Algerian and Tunisian Land Surveyors SNGFAT) concluded an agreement with Léon Eyrolles for the organization of a three–year correspondence course to train surveyor–topographers [234, p. 123].

For further information on how the training for these professionals was organized at the ESTP, you can refer to a booklet published in 1913 by the school for its potential public [5]. This invaluable document, entitled *L'Enseignement par Correspondance* (Learning by Correspondence) tells us much about how teaching and the teaching staff were organized at the very time Cholesky was preparing his course on graphical calculation.

In 1913 the ESTP was organized into five large autonomous specialties: Public Works – Building – Mechanics and Electricity – Mines and Metallurgy – Topography. For topography, the school offered three levels of training. For the first level *Aide–Géomètre* (Assistant Surveyor), there was a course on Graphical Drawing and Notions of Topography, but nothing yet directly linked to graphical calculation. At the second level, that of *Géomètre* (Surveyor), there were courses on Topography, Numerical Calculation of Capacities and Graphical Calculation of Capacities. The training lasted 12 months, and in order to obtain their diploma the students had to sit a large number of examinations spread over a 10 day period. Among these examinations there was one on Surface–area Calculation. Finally, for the third level, that of *Ingénieur–Géomètre* (Surveying Engineer) and *Ingénieur–Topographe* (Topographical Engineer), you find the same courses as the second level, but studied in greater depth, with more series of exercises, and a class on *Levés d'Études à la Planchette* (Advanced Plane Table Surveying). This time, the training lasted 18 months and the examinations were spread over 12 days.

We see therefore that the three courses that Cholesky was responsible for writing – Topography, Advanced Plane Table Surveying and Graphical Calculation of Capacities – were designed for the same public and formed a totally coherent, direct correspondence with the practice of his profession. In the 1913 ESTP brochure, Cholesky is presented as an engineer, Alumnus of the *École Polytechnique*, and director of topographical services for Tunisia, in an impressive list of 164 specialist teachers.

How did Cholesky come to be recruited to teach at the ESTP? We don't know, but one clue may come from the presence in the teaching staff of Frédéric Espitallier (Givrand, 25 May 1849 - 1923), a Lieutenant–colonel in the Military Engineers, former teacher of courses on construction at the *École d'Application de l'Artillerie et du Génie* at Fontainebleau. Espitallier probably taught Cholesky at Fontainbleau, and moreover, he found himself in Tunisia at the same time as him, as can be witnessed in an article published in 1902 in *Le Génie Civil* (Civil Engineering) on the Works at the Port of Bizerte and the arsenal at Sidi–Abdallah. Furthermore, Espitallier was responsible for setting up the colonial teaching provided by the ESTP and was on the committee of the Society of French Civil Engineers along with Eyrolles, the principal and Léandre Étève, the deputy principal. Espitallier was therefore an influential figure at the ESTP and, as it was almost certain that he knew Cholesky directly, it is entirely possible that it was he who was at the origin of his recruitment as a lecturer.

7.3 The unpublished manuscript

Contrary to the Topography and Advanced Plane Table Surveying courses, the course on graphical calculation remained, as we have said above, in draft form. Only the beginning remains, as an 83 page manuscript, written on 15.5 x 20 cm pages.

It is a well–polished manuscript, containing few crossings out, clearly a final version destined to be sent to the ESTP for composition and impression. Proof can be seen in the presence of numerous indications in the margins specifying the position of figures in the text and the scale to use for their construction.

In the Cholesky's Archive at the *École Polytechnique*, this manuscript is not one document in isolation. It must be considered as making up part of a collection of sources which shed light on the context and the process of its production, and

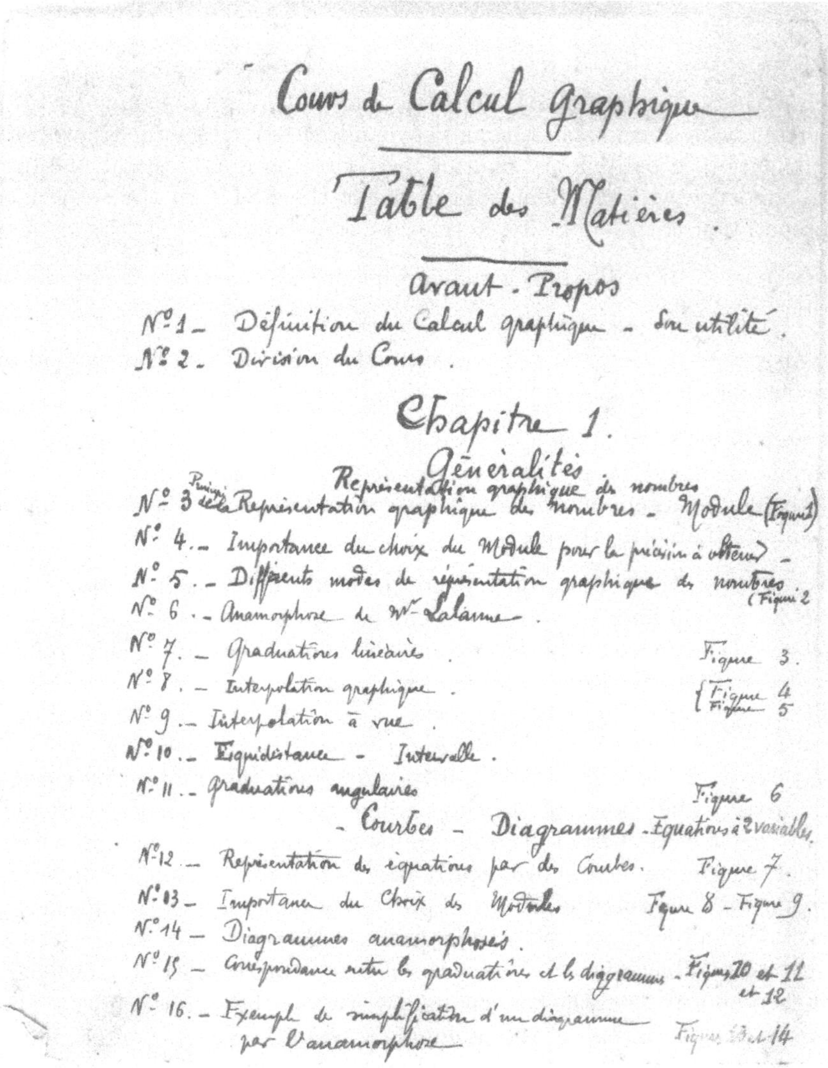

Cours de Calcul graphique

Table des Matières.

Avant-Propos

N° 1 – Définition du Calcul graphique – Son utilité.

N° 2 – Division du Cours.

Chapitre 1.

Généralités.

N° 3 – Représentation graphique des nombres – Module (Figure 1)

N° 4 – Importance du choix du Module pour la précision à obtenir –

N° 5 – Différents modes de représentation graphique des nombres (Figure 2)

N° 6 – Anamorphose de M. Lalanne.

N° 7 – Graduations linéaires. Figure 3.

N° 8 – Interpolation graphique. Figure 4, Figure 5

N° 9 – Interpolation à vue.

N° 10 – Équidistance – Intervalle.

N° 11 – Graduations angulaires. Figure 6

Courbes – Diagrammes – Équations à 2 variables

N° 12 – Représentation des équations par des Courbes. Figure 7.

N° 13 – Importance du Choix des Modules. Figure 8 – Figure 9.

N° 14 – Diagrammes anamorphosés.

N° 15 – Correspondance entre les graduations et les diagrammes – Figures 10 et 11 et 12

N° 16 – Exemple de simplification d'un diagramme par l'anamorphose. Figures 13 et 14

First page of the manuscript *Cours de Calcul graphique*

which bear witness to Cholesky's abiding interest in graphical calculation. This collection is made up of the following elements:

- Carton n° 1, dossier A1: a 31 page handwritten exercise book entitled *Planimètre Polaire de Amsler* (Amsler's Polar Planimeter); a handwritten page containing a bibliography of graphical calculation with library references.

- CARTON N° 2, DOSSIER B1: the first pages of Charles Lallemand's 1885 dissertation on Hexagonal Abaques; a handwritten bibliography of graphical calculation; a preliminary plan of the course on graphical calculation; correspondence with the deputy principal of the ESTP on the subject of the preparation of the course on graphical calculation.
- CARTON N° 2, DOSSIER B3: manuscript of the course on graphical calculation.
- CARTON N° 3, DOSSIER C3: correspondence with René Danger and others regarding various instruments used in graphical calculation.
- CARTON N° 3, DOSSIER C5: two abaques used during the mission in Romania, namely *Abaque de correspondance des grades, millièmes et degrés* (Abaque corresponding grades, degrees and thousandths) and an *Abaque Lamotte* (Lamotte Abaque).

We will return to these documents in detail in the next section. For the time being, we will limit ourselves to a comparison between the preliminary plan Cholesky had drawn up for his course and the plan of the part of this course that was indeed written (see table).

Firstly we note that Cholesky decided to start his course with some general considerations on Cartesian abaques compatible with the representation of two or three–variable relations, linked to some basic information on curves and surface areas. The choice of starting with nomographic methods already sets Cholesky's course apart from the other courses in graphical calculation, which like Prédhumeau's, mentioned above, nearly always addressed nomography after *calcul par le trait*, following the historical order. By his choice of an approach using constructions and lectures on simple graphics, Cholesky's aim is clearly to adapt his course to a public who, to start with, had little mathematical knowledge.

It then appears that the two chapters written in the manuscript develop only in fact, the first paragraph of the preliminary plan. The course on graphical calculation in our possession is unfortunately very incomplete. The missing chapters would have dealt with *calcul par le trait* (the graphical resolution of equations and graphical integration), the various calculating instruments that a surveyor–topographer could make use of, and finally, nomography, dealt with this time at a higher level than in the first two chapters. Cholesky had especially reserved the study of alignment nomograms for the end of the course. While these are certainly easy to use, their elaboration requires far more complex mathematical knowledge than that needed for Cartesian abaques.

Furthermore we note the large amount of space accorded to the description of the numerous calculating instruments. The approach is comparable with that followed for the Topography course, in which 6 chapters out of 14 were also devoted

Preliminary Plan	*Manuscript plan*
Graphical calculation Generalities. Linear graduations. Curves. Surface areas. Abaques Paper distortion. Errors. Approximation. Consequences: choice of variables and arguments Graphical resolution of equations. Graphical integration Calculation instruments: Rulers. Curvimeters. Protractors. Planimeters. Vernier scales. Calculating machines. Proportional compasses. Genaille rods Abaques. Nomograms Graphical compensation	**Course on graphical calculation** Chapter I. Generalities Graphical representation of numbers Curves. Diagrams. Two-variable equations. Surface areas. Three-variable equations Chapter II. Abaques Two-variable abaques General theory of abaques

to the study of instruments, in this instance those used to measure distances, angles, perspectives and elevations (see Section 5.4.1).

Due to the originality of this unpublished manuscript, and its significant contribution to Cholesky's teaching activities, that we have judged it useful to publish it in Appendix C. In the interests of authenticity, we have, with the kind permission of the Archives of the *École Polytechnique*, only reproduced 37 of the original figures.

7.4 Genesis of the course on graphical calculation

We will now turn to the process of the initial request for the course on graphical calculation and its elaboration, that the accompanying documents in the archives, to a large extent, enable us to piece together.

7.4.1 The commission from the ESTP

A series of letters from the ESTP's deputy principal, Léandre Étève to Cholesky show that the course had a long and painful gestation. A first letter (sent to 4 rue Louis Besquet in Vincennes), dated 27 April 1910, gives details of the commission, and provides the course's prospective author with documents that should inspire him:

> *I am sending you the following documents in preparation for the course on the graphical calculation of capacities:*
>
> *The* Mathématique *(Mathematics) by Laisant*
>
> *An edition of* Annales des Travaux Publics de Belgique *(The Annals of Public Works in Belgium) (containing a study on abaques)*
>
> *Mr. Prince's course on graphical calculation*
>
> *Mr. Vasnier's course on analytical geometry*
>
> *The first part of our topography course.*
>
> *Mr. Danger had given us the titles of works that may be of interest to you on this subject, but for Mr. d'Ocagne's works on graphical calculation, I think it best that you settle for acquainting yourself with Mr. Prince's course which already includes the most interesting parts of Mr. d'Ocagne's works.*

In this last sentence the deputy principal is warning Cholesky against the temptation of using works of too high a level. He is also, at the same time, reminding him that the ESTP addresses a very different public to that of the *École Polytechnique* and the *École des Ponts et Chaussées*, for whom d'Ocagne had written his great treatises.

It seems that, despite his military obligations, the ESTP was applying considerable pressure on Cholesky. In June 1910, he had to participate in a seminar to agree the Surveyor's training program. On 25 July of the same year, he received the following letter from the deputy principal, showing the latter's impatience:

> *I trust that you are currently working on the course on graphical calculation of capacities that you promised to write for us.*
>
> *This course has already been promised for a long time and many students have enrolled and have been waiting for two years, it is a matter of particular urgency. I trust that you will do everything you can to accelerate its production.*

In January and March 1912, while he was in Algiers, Cholesky received letters demanding that he submits his topography course. In February of the same year he was asked to contribute to *L'Ingénieur Constructeur*, a journal published by the students and alumni of the ESTP. By 19 March, the deputy principal's impatience was growing:

> *I kindly request that you send us the manuscript of the course on general topography and, at the same time, indicate how far you have advanced on the course on graphical calculation of capacities you are responsible*

for writing. I trust you will give a (favorable: added by hand) response to this question post haste.

Finally, in the last letter we have, dated 26 January 1914 and sent to Tunis, the deputy principal, doubtlessly beginning to despair of ever receiving the course as promised, reiterates his request, but without great conviction:

I would be grateful if you could do everything necessary to finish, as quickly as possible, the course on graphical calculation of capacities which you have said is already at a very advanced stage.

It is easy to see that from 1914, Cholesky, occupied by his missions in the field and then sent to the front soon after the start of the war, didn't have enough time to finish writing his courses. However, as we will see, the archives show that between 1910 and 1914, he carried out regular documentary research; studying graphical calculation and preparing the teaching for which he had been recruited.

7.4.2 Cholesky's preliminary research

Noting, first of all, that through his training and work as a military engineer in the field, Cholesky was already well aware of the various aspects of graphical calculation. We must remember that he entered the *École Polytechnique* in 1895. And that from 1893 to 1912 d'Ocagne was, fittingly, a tutor in astronomy and geodesy at the *École Polytechnique*. To illustrate his teaching, d'Ocagne, at that time, constructed nomograms relating to astronomy: The general nomogram for spherical trigonometry (1894), The nomogram for the approximate resolution of Kepler's equation (1894), The nomogram for diurnal and semidiurnal tide equations (1896), etc. Furthermore, from 1894, d'Ocagne gave lectures on nomography to the students at the school, which resulted in the publication in 1898 of *Une leçon sur les abaques* (A lesson on abaques) in the *Journal de l'École Polytechnique*. Cholesky, who probably had the opportunity of benefitting from d'Ocagne's teaching, could thus have been introduced to the latest advances in graphical calculation during his training.

Subsequently, during his field missions in Crete, Algeria and Tunisia, Cholesky worked regularly under the direction of Commandant Nicolas–Arthur Lallemand, head of the geodesy section of the Army geographic department (see Sections 1.2.2, 1.2.3, 8.3 and 9.1.2). Knowing that Nicolas–Arthur was the brother of Charles Lallemand, director of the National Leveling Bureau for France from 1884 to 1927, and was the creator of the theory of hexagonal abaques, and knowing that the Army geographic department regularly worked with the National Leveling Bureau, especially when producing a new 1:50,000 topographical map of France, we can imagine that Cholesky benefitted from the new graphical methods recommended by Charles Lallemand, which is supported by the presence of part of Lallemand's dissertation on the hexagonal abaques in Cholesky's papers. Similarly, abaques found among his military papers and coming from his mission in

Romania demonstrate in the same way that he actually used graphical calculation in his duties as a topographer and Artillery Officer.

Cholesky was thus the right man for the job, and it was completely natural for the ESTP to entrust to him the courses on topography and graphical calculation. Now we will follow our engineer's efforts to amass specialist documentation on graphical calculation.

On 24 August 1910, at Cholesky's request, the deputy principal sent him a copy of Maurice d'Ocagne's *Instruction sur l'usage de la règle à calcul* (Instruction on the Use of a Slide Rule). On 9 November the same year, he sent several types of arithmetical aids: multiplying rods, Genaille rods, finance rods. The Napier bones are missing from the collection, which are, he says, out of stock at Belin.

Furthermore, following the letter from the deputy principal mentioning René Danger's name, it appears that Cholesky contacted this surveying engineer, living at that time at 6 rue d'Angoulême in Paris to ask him for details. In 1901 Danger had published, through Éditions Dunod a work entitled *Surfaces et divisions de surfaces, calculs trigonométriques suivis d'une table des carrés des nombres de 1 á 10,000* (Surface Areas and Divisions of Surface Areas, Trigonometric Calculations Followed by a Table of Squares from 1 to 10,000). He was also the author of numerous courses for the ESTP (for René Danger's relationship with Eyrolles and the ESTP see [234]). We find, in particular, the existence of a *Cours de calculs numériques et graphiques du topomètre* (Course on Numerical and Graphical Calculation for Field Surveyors) which was in its 4th edition in 1927, which seems to suggest that Danger took over from Cholesky immediately after the war. All in all, everything points to his being an expert in the field of calculating surface areas and that it was natural to turn to him for information.

In the archives, one handwritten page contains Danger's complete bibliography, which Cholesky had obtained, either via the deputy principal or directly from Danger himself. The writing appears to be Danger's, when compared to that in letters written by him and conserved elsewhere. The page is annotated, possibly by the deputy principal, as there are crosses marked against the works that he recommended to Cholesky in his letter of 27 April 1910, but it is equally possible that Cholesky could have himself marked the books that had been recommended to him to use first. Besides the works selected by the deputy principal, Danger's bibliography contains useful books for the topography course, which do not concern us here, and the following additional books on graphical calculation (you find the same titles in another bibliography prepared in Cholesky's hand):

> *Le Calcul simplifié par les procédés mécaniques et graphiques* (Calculation Simplified by Mechanical and Graphical Methods) M. d'Ocagne Gauthier–Villars 1905

> *Nomographie, les calculs usuels effectués au moyen des abaques* (Nomography, Standard Calculations Carried Out Using an Abaque) M. d'Ocagne Gauthier–Villars 1891

Leçons de Topométrie (Lessons on Topometry) M. d'Ocagne Gauthier–Villars 1904

Calcul graphique et nomographie (Graphical Calculation and Nomography) M. d'Ocagne Octave Doin place de l'Odéon

In a letter dated 12 October 1910, Danger wrote to Cholesky to say that he could not provide any information on Genaille rods in addition to that contained in d'Ocagne's *Le calcul simplifié* (Calculation Simplified). However, he did pass on information on the Chenevier triangle, available from Cabasson, a company at 29 rue Joubert, as well as on two types of set squares for calculating surface areas to be found at a supplier called Morin, 11 rue Dulong. He also mentioned a wheel calculator in the form of a sine curve and a book in English presenting all the types of slide rules known at that time (this was very probably *A History of the Logarithmic Slide Rules and Allied Instruments* by Florian Cajori and published in 1909).

In another letter written in the same month, Danger advised Cholesky to pay a visit to the *Conservatoire des Arts et Métiers* (Conservatory of Arts and Trades), where he would be able to see instruments that would interest him: the Beghin slide rule, the Chevenier triangle and graphical calculation tables, the Berthelemy mechanical abaque for calculating coordinates, Genaille rods, the Lalanne Universal Arithmometer and reduction quadrants. Danger also mentions some abaques that did not exist but would be useful, for calculating the surface areas of a triangle for example. Could it be that Cholesky had asked him to write down some ideas for examples, exercise and topics for his forthcoming courses?

It appears that Cholesky followed Danger's advice, as another handwritten page conserved in the archives contains a list of bibliographical references relating to graphical calculation, and more generally to calculating instruments, accompanied by references that correspond with those in the Library of the *Conservatoire des Arts et Métiers*, as we have been able to verify in the library's catalogue:

501.A
Arithmoplanimètre (Lalanne Arithmoplanimeter) Lalanne Pu 23
Rapporteur de précision (Precision Protractor) Cousinery Pu 6
Planimètre de Beuvière (Beuvière Planimeter) Gal Morin Y 70
Inst^{on}/Règle à calcul à 2 réglettes (Instruction on Two–Sliding–Strip Slide Rules) Van Hyfte Pu 23

512.A
Traité du triangle arithmétique (Treatise on the Arithmetical Triangle) Pascal Py 6y
Machines algébriques (Algebraic Machines) L. Torrès Pu 27

513.A
La statique graphique et ses applications aux constructions (Graphic Statics and Its Application to Constructions) Lévy D 14
Statique graphique des mécanismes (Graphic Statics of Mechanisms) Schmitz et Gustav D 12
Éléments de statique graphique appliqués aux constructions (Elements of Graphic Statics Applied to Constructions) Müller-Breslau et Seyrig La 42 *atlas* La 34
Statique graphique appliquée aux constructions (Graphic Statics Applied to Construction) La 43 *atlas* La 37

The 31 page handwritten exercise book on the Amsler's Polar Planimeter, that is in the first carton in the archive, comes, perhaps, from research Cholesky undertook at the *Conservatoire des Arts et Métiers* where several examples of this instrument were kept. Danger also passed on various letters taken from his own correspondence, starting with a letter dated 1 December 1904 from *Arts et Manufactures* engineer Jules Jean Désiré Pillet (1842 - 1912), in which Pillet speaks about the reduction quadrants used in the Merchant Navy, stressing that they, and proportional compasses were increasingly abandoned in favor of more complex graphical tables, and abaques for the complicated formulae in navigational problems.

Two other letters, this time from Cabasson, dated 19 and 22 September, give details on various instruments: the Bertrand mechanical abacus, built by Berthelemy to carry out rectangular coordinate calculations, the Lallemand 50 cm long slide rule with divisions in hundredths, Lasselannes set squares and profiles, the Lalanne abaque or counter, the Chevenier calculation triangle and the Prévot abaque.

Independent of this correspondence with Danger, a few other isolated letters complete our impression of the depth of Cholesky's research work. From December 1911 to January 1912, one of his colleagues, Henri Marliny, acted as an intermediary when he is forwarded a wooden device and cardboard tables from a certain Mr. Bec, a writer in some government department or other. An undated letter from I. Goldschmidt, manufacturer, 12 rue Chabrol in Paris describes a wheel calculator with a logarithmic sine scale for calculating triangulations, which allowed all the trigonometric calculations to be carried out. According to Goldschmidt, this wheel, adopted by the Land Registry department, is called on to offer the greatest service to all experts: engineers, geodesists, topographers, land surveyors, navigators and astronomers.

7.4.3 Sources of inspiration for the course

Once all the documents and instruments discussed earlier were assembled, who finally could have inspired Cholesky in the design and the writing of the first two chapters? Let us examine then, the principal works mentioned in the bibliography above – works that Cholesky had in all likelihood consulted – to look for any possible similarities with the manuscript of his course.

La Mathématique (Mathematics) by Charles–Ange Laisant (Indre, 1 November 1841 - Asnières–sur–Seine, 5 May 1920), the first edition of which appeared in 1898 [153], is a general reflection on mathematics, addressed to all who practice or use the discipline from day to day (engineers, teachers, students, etc.). Laisant particularly supported the idea that the science of calculation and geometry support each other, and in particular that *all operations in calculation can be shown as a geometric construction* [153, p. 168]. Laisant then outlines, over just a few pages, a story completely relevant to the history of graphical calculation, in the middle of considerations on topography, geodesy, graphic statics and nomography. As with the Analytic Geometry course given at the ESTP by Pierre–Louis–Maurice Vasnier (Paris, 14 February 1873 - Verdun–sur–Meuse, 16 January 1916), a course that was already in its 5th edition in 1913 [237], it did indeed contain paragraphs entitled *Évaluation des aires planes* (Evaluation of Plane Areas) – *Méthodes d'approximation pour l'évaluation des aires* (Approximation Methods for the Evaluation of Areas), but their development was very succinct. These first two works, undoubtedly too rudimentary for the composition of a specialist course in graphical calculation of capacities, certainly had nothing to teach a confirmed mathematician like Cholesky. At most they could help define the spirit in which such a course for future engineers should be designed.

It is rather in d'Ocagne's works that Cholesky could, if necessary, have legitimately enquired about the latest developments in nomography. 1891's short *Nomographie* [187], the first summary work by d'Ocagne, is a very clear and quite elementary presentation of the principles underlying the various types of abaque, presented in the chronological order of their apparition. *Leçons sur la topométrie et la cubature des terrasses* (Lessons on Topometry and the Cubature of Terraces 1904) [189], produced by the *École des Ponts et Chaussées*, contains an annex presenting an outline of the principles of nomography, as, according to d'Ocagne, this theory applied to all branches of engineering, and not only to the problems of excavation and embankment found in civil engineering. In twenty dense pages [189, pp. 201-220] you find a general and rather abstract treatment of the nomographic representation of three-variable equations: Cartesian abaques, concurrent–straight–line abaques, hexagonal abaques, alignment nomograms. *Le calcul simplifié* (Calculation Simplified), in its second edition in 1905 [191], is an enhanced version of a series of conferences given at the *Conservatoire des Arts et Métiers*. The work looks at all calculating instruments from calculating machines and numerical tables through to graphical construction and graphical tables, but, this is a book for a general readership, and could not be expected to provide suitable subject matter destined for future engineers. Finally, 1908's *Calcul graphique et nomographie* (Graphical Calculation and Nomography) [192], that comes from an open course given at the Sorbonne in 1907, is by far the most complete and the most technical work contained in Cholesky's bibliography, and clearly the best candidate for what could have served him as a reference document. The edition of the *Annales des Travaux Publics de Belgique* to which the deputy principal made reference is most probably number 67 from 1910, which contains a study by G. Schoofs, a captain in the Military Engineers, on transversal abaques. This article praises the alignment nomograms popularized by d'Ocagne, which surpass Pouchet and Lalanne's old concurrent–scale abaques by far, but it adds nothing more compared to the previously cited works.

To finish, it is necessary to closely examine the course written by Louis–Camille Prince (1867 - 1939) *Calcul graphique et nomographie* (Graphical Calculation and Nomography). This course is doubly important: firstly it was a course taught at the ESTP in 1909, just before Cholesky came on the scene, and secondly, it is the work which was recommended as a priority by the deputy principal of the school. The course begins with Prince's preliminary considerations on tangential coordinates, parallel coordinates, correlative figures and the principle of duality. With a classic graph similar to Prédhumeau's mentioned above, the book continues with the graphical resolution of equations and graphical integration, then addresses nomography at the end. It is, in fact, a high–level course, inspired directly by d'Ocagne's research and designed for candidates at the *École des Ponts et Chaussées*. It is not suitable for the training of surveyor–topographers. There was no way Cholesky could use its content as it was.

Finally, it becomes clear that the first chapters of Cholesky's course bear no direct resemblance to anything that can be found in the works which he had in front of him and that we have reviewed. In the order in which the subjects were addressed, in the level of the mathematics adopted, and in the language used, it is indisputable that he produced a very personal course, suited to the particular context of the ESTP: students lacking extensive schooling in mathematics, and the constraints of a correspondence course.

7.5 Analysis of the course on graphical calculation

Now we turn to the detailed analysis of the course content. To allow the interested reader to refer to the original text in Appendix C, as we go along we will mention, in brackets, the numbers of the paragraphs referred to in our comments.

7.5.1 Subject and organization of the course

The manuscript starts (No. 1) by giving a definition of graphical calculation and developing some arguments in favor of its use in the day to day calculations carried out by an engineer. This introduction merits being quoted in its entirety:

> Graphical calculation is intended to replace numerical calculation with a drawing, a sort of working drawing, whose construction permits one to pass directly from the data to the results. The working drawing used is generally known by the name Abaque.

> *The execution of such a working drawing is often complicated and the greater the precision required, the greater the care that must be taken in its execution. Furthermore, it is not in the general interest to seek the resolution of an isolated case graphically, for which the construction of the graph would very often require spending much more time than for numerical calculation.*
>
> *On the contrary, methods of graphical calculation become very advantageous when it is necessary to reproduce a calculation very frequently, in the application of a formula in which only the data vary. The drawing can be prepared in such a way so as to provide, through simple operations, results corresponding to all the systems of values of the variables. The establishment of the working drawing may, in this case, be very long, but it does result in much noticeable time-saving, if it is used each time that one carries out a much repeated operation, you save an appreciable proportion of the time required for the numerical calculation that has been eliminated.*
>
> Graphical calculation therefore applies itself primarily to calculations that must be repeated very frequently.
>
> *There is no point in emphasizing the interest presented by the reduction in the time spent on calculations, this reduction translates itself into an economy of time and therefore of money, it is for this reason that one has sought to reduce the time spent carrying out numerical calculations in advance for the* numerical tables *constituting the exact equivalent of the working drawings used in graphical calculation. The advantage of the latter is that generally they are easier and quicker to establish, and their use is less tiresome for the person doing the calculation.*

After this clear introduction, paragraph No. 2 simply tells us that *The course will include ... chapters*, with the number of chapters left blank. At this point of writing, Cholesky therefore didn't know how many chapters there would need to develop the forward plan he had prepared (see Section 7.3).

We should note that in his introduction, Cholesky uses the term *abaque* for all working drawings and all geometric constructions that could be used in calculation. This very broad notion of abaque would serve him as a constant theme in the organization of his course. Thus, in the first chapter where he focusses on the standard graphical representations of numbers and of two and three–variable equations, he is already using the concepts and vocabulary of abaques, which will allow him, in the second chapter, to present the theory of abaques, this time understood in the more specialized sense of the term, as a generalization of the graphical representation of the relations between any number of variables. This perspective gives his course great unity, in the measure that the first chapter offers students the opportunity to discover or revise the basic analytical geometry of curves and surface areas, while gently preparing them for the study of true nomography

7.5.2 The graphical representation of numbers

Paragraphs Nos. 3 to 11 deal with the various modes for the graphical representation of numbers in detail. All the methods envisaged come back to the choice of a unit of length M, called the *module*, a curve C having an origin and a direction of travel, and a function f. A number n is then represented by a length $Mf(n)$, measured along the curve C from the origin in the chosen direction. If you draw a line to the extremity of length $Mf(n)$ for equidistant values of n, and if you mark each number n opposite the corresponding line, you obtain a *linear graduation* (see Figure 3). In the simplest case, where $f(n) = n$, this is a regular graduation, whose lines are equidistant. Otherwise, you find, in Lalanne's terms, an anamorphic graduation and this graduation represents function f completely. With regards to curve C, the most common situations are those where a straight line or a circle is graduated, the latter being in fact equivalent to the use of an *angular graduation* in which the numbers are not represented by lengths, but by angles.

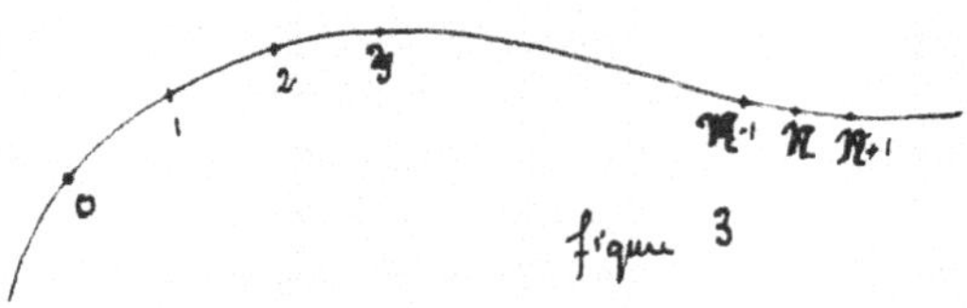

In this part, Cholesky presents at length the practical considerations in the choice of module and the spacing between the graduation marks for the greatest precision, as well as the techniques of graphical interpolation and by–sight interpolation when it comes to reading numbers between the graduating marks. Here he devotes 13 pages to the development of these areas which, in most treatises are regarded as self–evident or make up no more than one or two introductory pages. Naturally, it is a characteristic of a correspondence course that forces the author to explain the basic savoir–faire in the finest detail, while for a traditional teacher, this would instead be the subject of supplementary oral transmission without it appearing explicitly in the text book.

7.5.3 Representation of two–variable equations by curves

The second section of the first chapter (Nos. 12 to 19) addresses the question of the representation by curves of two–variable equations. Cholesky starts by reminding us (No. 12) that, once you have chosen two axes of coordinates Ox and Oy, all $F(x,y) = 0$ relations can be represented by a plane curve. He refers to this curve as the $F(x,y) = 0$ *equation diagram*, or y *diagram* in considering y as a function of x implicitly defined by the equation $F(x,y) = 0$. The word *diagram* is quite surprising here: to the best of our knowledge, it does not figure in d'Ocagne's vocabulary nor was it used by other authors of the period; it seems that it was Cholesky's personal choice. Moreover, as much as the linear graduations seen in the previous section and in the two–variable equation diagrams are *abaques* in the broader sense he gave to the word, it could very well be used to designate all these figures.

Next Cholesky turns the reader's attention to the importance of the choice of the modules (No. 13) for the best use of the available space on the page and to guarantee the greatest precision. The example of the equation $y^2 - \dfrac{x}{100} = 0$ is striking: if you take the same unit for x and for y, the resulting diagram does not allow you to obtain precise y values (see Figure 8), however, with a unit 100 times greater on the y axis, which comes down to drawing the curve with equal units representative of the new equation $Y^2 - 100X = 0$, the second diagram (see Figure 9) provides much greater precision.

By generalizing the transformation that has been made in this example on the graduation of the y axis, you can envisage employing any of the graduations on each of the two axes, which leads to the notion of a *doubly anamorphic diagram* (No. 14). The graduations are to be chosen in the aim of simplifying the drawing of the curve (No. 16), as Lalanne's classic example shows, transforming the hyperbola equation $xy = 10$ (see Figure 13) into a straight line equation $X + Y = 1$ (see Figure 14) thanks to a logarithmic graduation on each of the two axes.

In the same vein, several paragraphs explore the relations between the two fundamental notions of anamorphic graduation and equation diagrams, which are two equivalent ways of representing a $y = f(x)$ function graphically (No. 15): given a regular graduation on the x axis, data from an anamorphic graduation on the y axis allow a diagram to be constructed, and vice versa (see Figure 10). As a result you are able, on one hand, to construct a doubly anamorphic diagram, using auxiliary curves for the construction of each of the anamorphic graduations of the axes (No. 17) and, on the other, you can carry out successive anamorphoses graphically, conveying the notion of function of a function (No. 18).

The last two paragraphs of this section (Nos. 19 and 20) address the graphical representation of two equations in the form:

$$\begin{cases} f(x,y) = 0 \\ f_1(x,z) = 0. \end{cases}$$

A first method (No. 19) consists in representing the equation $f(x,y) = 0$ as a diagram C (see Figure 17), then graduating curve C by marking the corresponding value of z given by the equation $f_1(x,z) = 0$ next to each point K on the x abscissa.

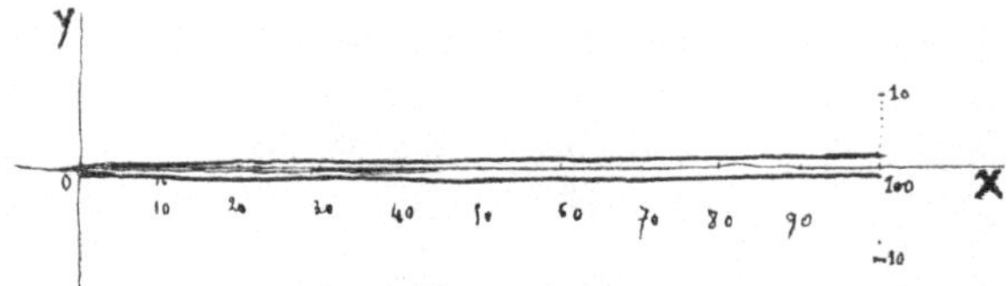

Figure 8

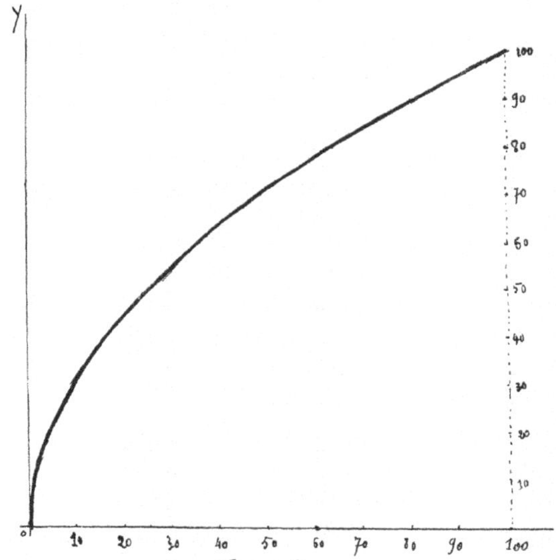

Figure 9.

This abaque, obtained by combining a diagram and a linear graduation is called a *graduated diagram.* You can see that it allows the corresponding x and y values to be read directly, in other words, producing the graphical elimination of the unknown x in two equations.

Another representation of the system of equations (No. 20) is obtained with the aide of two diagrams having a common axis corresponding to the unknown x to be eliminated (see Figure 18). The two other axes, with the values y and z, can be chosen to be opposed, as in the figure, or identified. In that case, they are called *conjugated diagrams.*

As Cholesky remarks later (No. 26), graduated diagrams and conjugated diagrams can be used to represent a three–variable relation, but a very particular relation in that it is split into two relations with only two variables linked by a common variable. This is the graphical representation of a curve in space, produced either with a projection marked on a single plane, or with two projections on two intersecting planes.

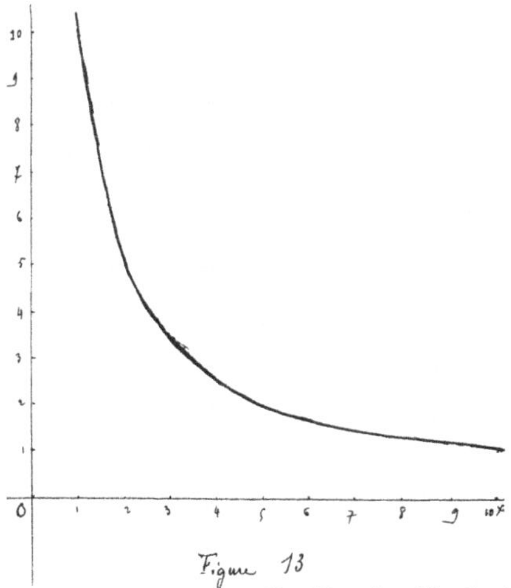

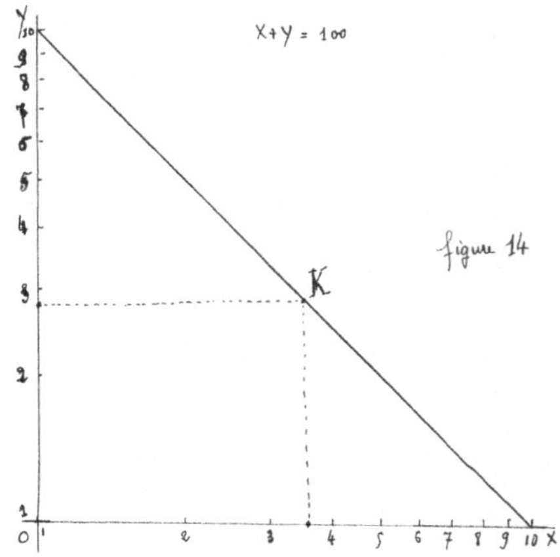

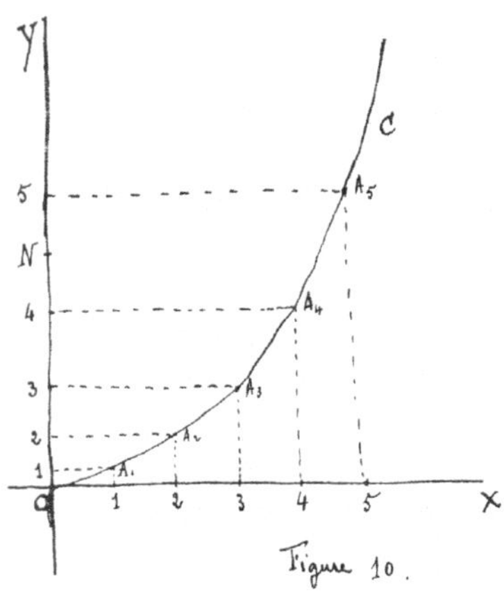

In summary, all this section on curves is elegantly constructed from the first notion of linear graduation. A plane curve, or diagram, appears from the connection of two linear graduations. A curve in space, be it represented by a single graduated diagram or two diagrams having a common graduation, comes down to, in both cases, the connection between three linear graduations. In addition, the final study of curves in space naturally provides a good transition between two–variable relations and three-variable relations, between curves and surfaces.

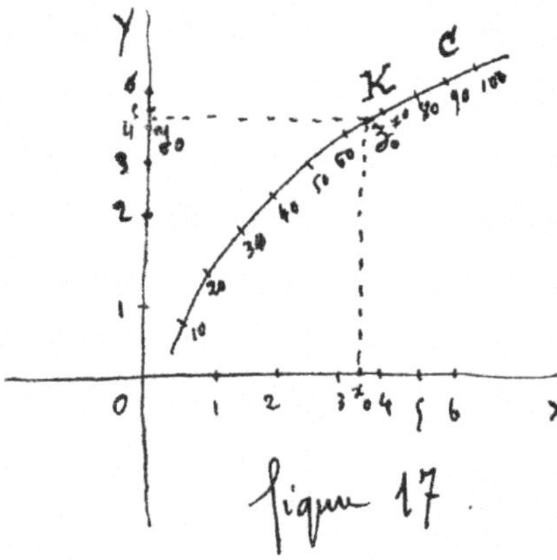

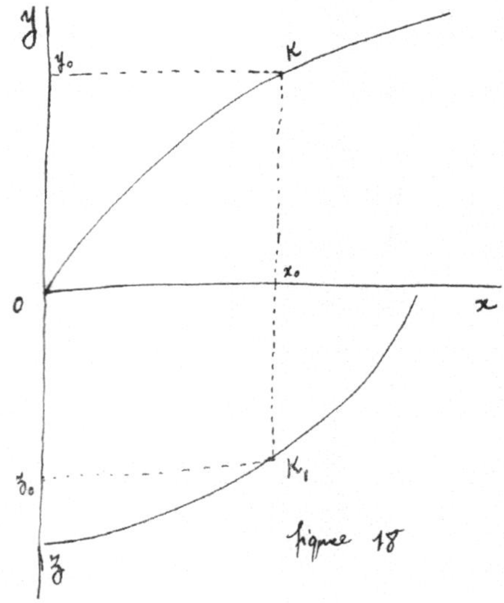

7.5.4 Representation of three–variable equations by surfaces

The final section of the first chapter (Nos. 21 to 27), devoted to the representation of surfaces and to three–variable equations, is organized in a similar way to the part on curves. A three–variable equation $f(x, y, z) = 0$ is represented geometrically by a surface S. The effective construction of a surface in space being impracticable in concrete terms, the simplest method for this representation is the dimensional graph, similar to topographical charts in which the lie of the ground is obtained with contour lines (No. 21). In other words, the surface is represented by a set of sections corresponding to the constant z ordinate values, each value of z being written next to the curve which corresponds to it (see Figure 19).

As he does for curves, Cholesky then launches into a long practical development on the choice of the modules, the optimal spacing between the marked curves and the methods of graphical interpolation to apply to this new type of abaque. With regard to the modules, he reminds the reader that in cases where one chooses regular graduations, and the same module for three axes, this module is none other than *what in topography is called a scale on a chart.* In the general case, choosing one, two or three irregular graduations results in a single, double or triple anamorphosis (No. 23). A simple example shows the advantage of the anamorphosis: that of the sphere of equation $x^2 + y^2 + z^2 - R^2 = 0$, that, when you adopt the graduations defined by $X = x^2$, $Y = y^2$ and $Z = z^2$, becomes a graph of the equation $X + Y + Z - R^2 = 0$. We compare the resulting dimensional graphs in Figures 20 and 21.

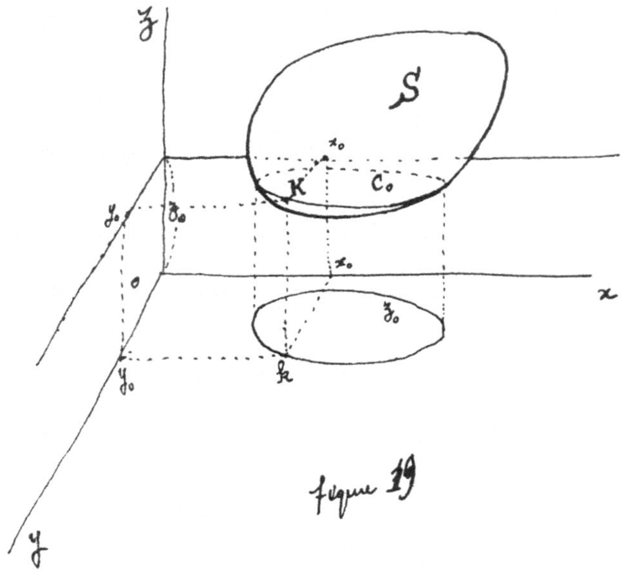

Finally, as is the case for curves, the section on surfaces finishes with a paragraph (No. 27) devoted to graphical elimination. This time it is a question of the graphical representation of a system of equations of the form:

$$\begin{cases} f_1(x, y, z) = 0 \\ f_2(x, y, t) = 0. \end{cases}$$

To do this, you simply construct two dimensional graphs related to the same coordinate axes Ox and Oy. You then obtain two families of curves, one marked z and the other t. Two curves, one from each family, pass through a point on the graph at the coordinate (x, y), which allows x and y to be associated graphically to the values z and t which correspond to them by the system of equations. With this graphical liaison, it is possible, for example, to find the relation between y, z and t directly, by eliminating x.

Cholesky concludes his first chapter by writing: *It is the generalization of the principle of elimination that leads to the notion of abaques.* Ultimately, his intention in this first chapter was to make the student surveyors and the student topographers work on the notions of plane curves and topographical surfaces, doubtlessly notions already known and practiced by this public, at least at a basic level, to make them feel, without saying so explicitly, that these standard geometric objects were already graphical tables that allowed them to calculate with numbers and solve equations. It would certainly have been interesting to have the work

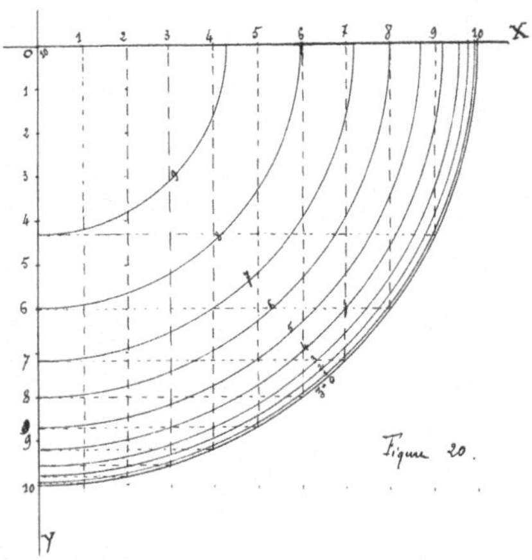

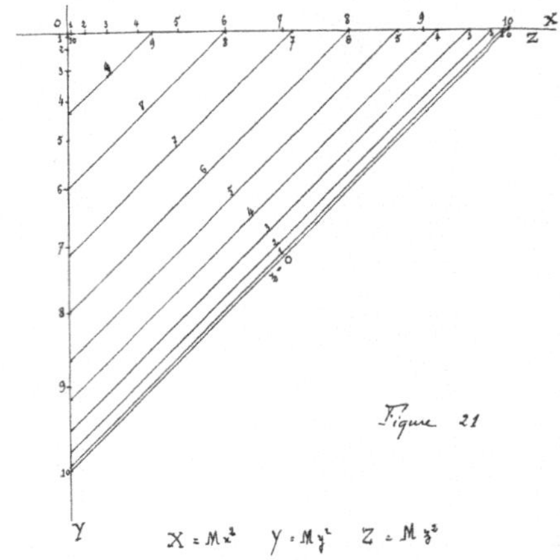

sheets associated with this first chapter to see how, by which examples, by which geometric constructions and by the application of which calculations, Cholesky envisaged for it to be assimilated by the students at the ESTP.

7.5.5 A personal theory of abaques

Chapter II, entitled *Abaques* is divided into two sections. The first is devoted to two–variable abaques (Nos. 28 to 36), the second to a general theory of abaques (Nos. 37 to 38) which appears incomplete.

Cholesky starts by noting (No. 28) that a relation of the form $f(x, y, N) = 0$ can be represented by a family of curves on a graph related to two coordinate axes Ox and Oy, this family being parameterized by the values N. A value of N corresponds to each point on the graph and a particular curve on the graph corresponds to each value of N. The set of curves drawn thus on the graph therefore constitutes a new graphical representation of the set of numbers, which receives the name *planar graduation* by analogy with the linear graduations seen in the first chapter. In a linear graduation the numbers being represented by a set of points on a line; in a planar graduation, they are represented by a set of curves on a graph (see Figure 24).

Linear, and planar graduations are very closely related (No. 31). Firstly, a linear graduation is a particular case of planar graduation: it is the borderline case where each curve of the family is reduced to a point. Secondly, when the curves of a family are superposable, given by the continuous displacement of an invariable curve on the graph, it is impossible to replace, at least in theory, a

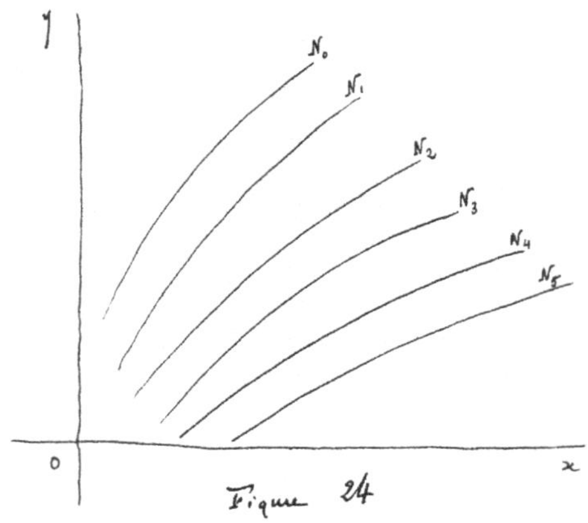

planar graduation with a linear graduation. For example, if you are dealing with a family of parallel straight lines, it is sufficient to cut this family with a transversal and to replace each straight line of the family with its point of intersection with the transversal (see Figure 25). You can then find any straight line of the family using a transparency on which you mark a single straight line making the desired angle with the transversal and that you then slide along the transversal.

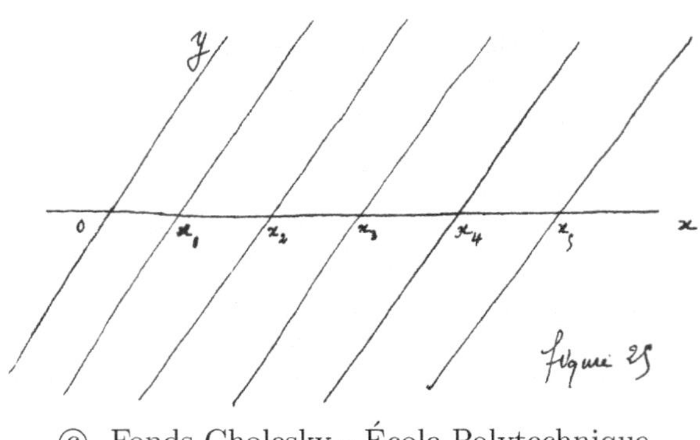

Another example is provided by a family of circles with the same radius (see Figure 27), that can be replaced by a single circle orthogonal to the family (see Figure 28). There again, a single circle drawn on a transparency with a tangent passing through the origin is sufficient to find any circle of the family.

Replacing a planar graduation with a linear graduation is of considerable practical importance. In effect, a linear graduation is generally long and com-

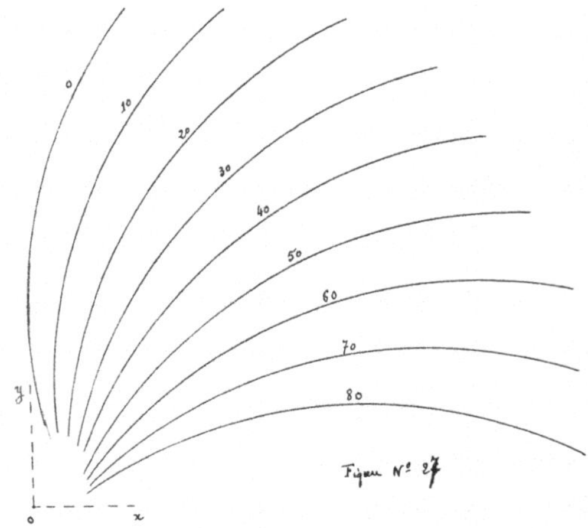

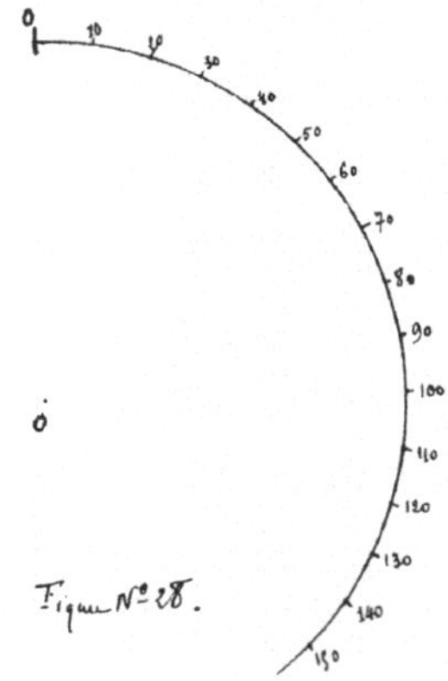

plicated to construct, as you have to draw a large number of curves sufficiently close together. On the other hand, a linear graduation only requires one curve to be drawn on the page and another on the transparency, with possibly a third auxiliary curve to guide the movement of the transparency. In addition, interpolation is easier on a linear graduation. Replacing a planar graduation with a linear graduation is also a method of compressing data which allows much space to be saved on the paper, which offers the possibility of easily placing several linear graduations on the same abaque, without the superposition of several planar graduations which can leave a drawing totally illegible.

This problem occupied the founders of nomography greatly. Lallemand solved it in 1885 using hexagonal abaques (see Sections 7.1.2 and 7.1.3): noting, thanks to his method of graphical addition, that most standard three–variable relations could be represented by an abaque with three bundles of parallel straight lines making 60° angles between each other, he replaced each cluster of parallel straight lines (a planar graduation in Cholesky's terms) with a single straight line perpendicular to the cluster (a planar graduation in Cholesky's terms), and assured the liaison between the three linear graduation thus obtained by the use of a transparency. Cholesky, incidentally, seems to draw heavily from Lallemand in this section, which is not surprising as we know that he had seen the first pages of Lallemand's dissertation (see Sections 7.3 and 7.4.2). D'Ocagne, for his part, found a totally different solution in the more general case of a planar graduation formed by a family of straight lines: using a projective transformation, he transformed the family of marked straight lines into a family of marked points, also known as a linear graduation, and it is in this way that concurrent-straight-line abaques are transformed into alignment nomograms, the liaison between three linear graduations being obtained this time again by a transparency (see Sections 7.1.2 and 7.1.3).

The general method of using a planar graduation to represent numbers leads Cholesky to a personal theory of abaques. He begins paragraph No. 35 with a theory of two–variable abaques. In order to represent an $F(N_1, N_2) = 0$ relation, you need two planar graduations representing the numbers N_1 and N_2, defined by the equations $f_1(x, y, N_1) = 0$ and $f_2(x, y, N_2) = 0$ respectively, and the curve of the equation $\Phi(x, y, N_1, N_2) = 0$, called an *indicator*, depending, or not, on N_1 and N_2. There must be a geometric relation between the indicator and the elements of the double graduation allowing each value given by N_1 to be associated to the value corresponding to N_2 satisfying the equation $F(N_1, N_2) = 0$.

For example, in Figure 32, the geometric relation between the indicator and the curves representing N_1 and N_2 is a tangential relation: being given a N_1 value, you consider the tangent common to the indicator and the curve representing N_1, and you record the curve of the second graduation that is tangential to this same straight line, which gives the N_2 value sought.

In most cases, the indicator is very simple (a point, a straight line or a circle) and frequently fixed (dependent neither on N_1, not N_2), the case of a movable indicator is only practicable if it maintains an invariable form, which allows the use of a transparency. As for geometric relations, it is more a relation of the alignment of three points or the concurrence of three lines, easier to work with than a tangential relation! Cholesky detailed a series of example to illustrate the various cases (No. 36).

The second chapter finishes with a general theory of abaques with any number of variables. (No. 37), only reviews and extends what has already been presented and illustrated at length for two–variable abaques. To represent a relation between p variables, you construct an abaque made up of p planar graduations and an indicator, with a geometric relation allowing the value of the p–th variable to be determined knowing those of the $p-1$ first. For example, in Figure 37, the p planar graduations are reduced to linear graduations, and the indicator is a variable curve determined by $p - 1$ points from the first $p - 1$ graduations; where it intersects with the p–th graduation determines the value of the variable to calculate.

This last, very abstract paragraph contains no examples, which seems to show that chapter II remains unfinished.

7.5.6 Final comments

At the end of our reading of the manuscript, it appears that Cholesky's course on graphical calculation is the fruit of a very personal reflection. It resembles none of the other graphical calculation courses that we know. Towards the end of the 19th century d'Ocagne was the first to teach nomography, initially at the *École des Ponts et Chaussées*, then at the *École Polytechnique*, and finally at the *Sorbonne*. He established the vocabulary and the concepts of this new discipline for a long

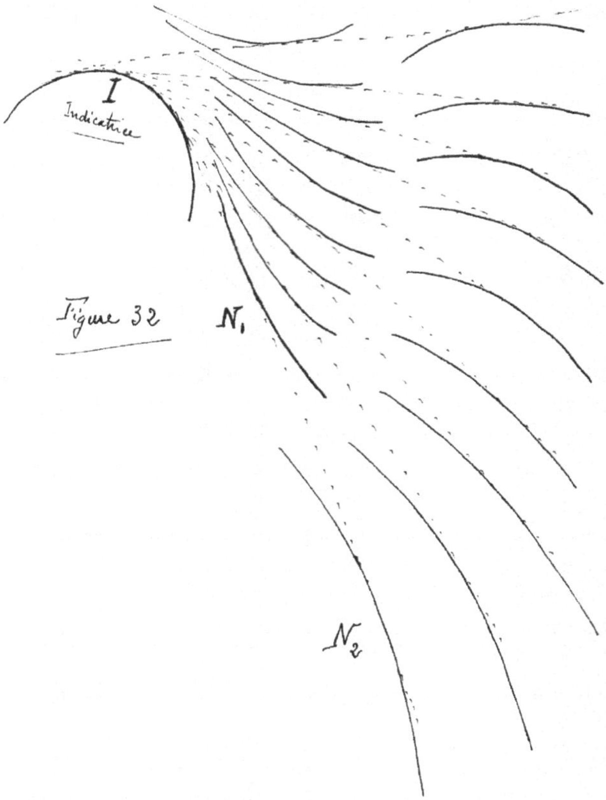

while through his numerous books and publications; most later treatises, be they written in France or elsewhere, have followed his perspective enduringly. It is all the more remarkable as Cholesky strays from the usual path by introducing his own vocabulary (*linear graduation*, *planar graduation*, *diagrams*, *graduated diagrams*, *conjugated diagrams*, etc.) and by outlining an atypical theory of abaques, that neither follows the historical order established by tradition, nor the mathematical order of from the general to the particular that is found in other books.

Cholesky's very detailed, progressive course, is based on numerous examples and much practical advice, proceeding by successive extensions and generalizations, even to the point of repetition, seems well suited to teaching surveyor–topographers by correspondence. The only reservations that can be expressed are about the final pages, which seem excessively abstract when bearing in mind the public he is addressing. You have the impression that towards the end of the manuscript, Cholesky was somewhat carried away by his own personal interest in

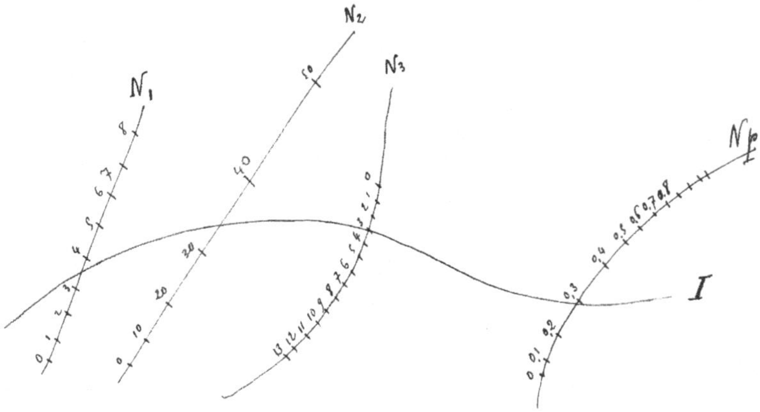

Figure 37

mathematics and his desire to do original work. It is also perhaps that he had embarked on a course that was relatively unfamiliar to him and he could not advance in its composition as quickly as he would have liked. Note, nevertheless, the great effort undertaken to produce the general abstract theory while relying on the minimum of prior mathematical knowledge. Indeed, Cholesky never called on projective geometry, nor determinants, and hardly ever on differential calculus, while these are invariably the tools of many treatises on nomography. There is an incontestable pedagogical effort to explain the principal ideas underlying the theory of abaques to a non–mathematician public.

The final impression this incomplete manuscript leaves confirms the image of Cholesky we have built up throughout the book, namely an "engineer–scholar", having received a high level of education and capable of showing initiative and originality in all the tasks confronting him. From this perspective, the teacher is no less interesting than the soldier, the engineer, the mathematician, the topographer or the artillery officer.

Chapter 8

Ernest Benoît

As mentioned above, the method of Cholesky for solving systems of linear equations was first published in 1924 (that is 6 years after the death of his author) by a Commandant Benoît [42]. Who was he? Even the initial of his first name was unknown, and Benoît is a quite common French family name. In his paper, it is only mentioned that he was *"de l'Artillerie coloniale, ancien officier géodésien au Service géographique de l'Armée et au Service géographique de l'Indo–Chine, Membre du Comité national français de Géodésie et de Géophysique"* (Of the Colonial Artillery, former Geodesist Officer in the Army Geographical Service and in the Geographical Service of Indo–China, Member of the French National Committee of Geodesy and Geophysics). After many searches in the name catalogs of the Army and on internet, and sending messages to various people and to this French national committee, it was still impossible to locate him. However, after a long search on internet, it was finally discovered that the initial of his first name was an E. Then, an idea arises: a geodesist officer could have been a student at *École Polytechnique*. Thus, looking at the site of its former students, it was found that only one of them had a first name beginning with an E: Ernest, born in 1873, deceased in 1956, and who was a student at *École Polytechnique* in 1893. The dates were corresponding to those of Cholesky. Thus, going to the archives of the Army, and looking at his folder (5 cm thick) revealed that he was our man [1].

8.1 Biography of Benoît

Ernest Benoît was born in Morez, in the Jura massif, on 15 July 1873. He was the son of Charles Auguste Benoît, a house–painter, living at 11 rue du Blé in Chalons–sur–Saône, born on 26 April 1827, and of Louise Adelaïde Romand. They had nine children. After getting his bachelor degree (*baccalauréat*) in Lyon in July 1890, Ernest succeeded to the entrance examination to *École Normale Supérieure* and to *École Polytechnique* where he entered in 1893 (66th out of 239). He was *boursier avec trousseau* (which means that he received a scholarship of 2000 French Francs and 457 Francs for his clothes from the French government), he was 1 m 60 tall (that is 5.25 ft), had brown hair and brown eyes. After his two years as a student at *École Polytechnique*, he entered *École d'Application de l'Artillerie et du Génie*, on 1 October 1895 (Cholesky entered *École Polytechnique* only on 15 October 1895, and in the second school on 1 October 1897).

Ernest Benoît as a student at *École Polytechnique*

On 1 October 1896, Benoît was assigned to the 37th Artillery Regiment where he was lieutenant en second until 30 September 1897. Then, he went to the 4th Artillery Regiment, and then to the 2nd Artillery Regiment of the Navy

(29 September 1898 - 26 January 1899). He was assigned to the Group of Africa and Antilles on 27 January 1899, and he was sent to Senegal until 22 January 1900 and then to Guinea (in war) until 14 August 1900. He worked for the commission for the delimitation of the Portuguese Guinea. On 1 September 1899, he was promoted lieutenant en premier, and to capitaine on 4 July 1900. On 12 September 1900, he was sent to the pyrotechnics school of the Navy in Toulon. On 24 May 1901, he was appointed to the Geographical Service of the Army in Paris and was sent to Tunisia from 21 November 1901 to 27 June 1902. After a stay in Paris, he went again to Tunisia from 16 December 1902 to 2 June 1903. He was working on the map of the region of Gafsa. It was probably during this second stay that he met Cholesky [1].

It would be tedious to enumerate in detail the succession of its many assignments. Let us only say that his following missions sent him to Tonkin (1 August 1903 - 25 October 1905) in war, then to Lorient and Toulon, then to Algeria (13 October 1906 - 30 March 1907, south of Oran), and again to Tonkin (11 October 1908 - 5 July 1911 and 15 February 1913 - 1 August 1914) until the declaration of war in 1914. On 25 January 1913, he married Marguerite Jeanne Marie de Lanessan. He embarked for France on 1 August 1914 and arrived in Marseille on 7 October.

From 23 November 1914 to May 1916, he was part of the group of *canevas de tir* (group of firing grids) of the 10th Army, where, maybe, he again met Cholesky. As explained above, these firing grids were special maps for trying to locate, by sound, the position of the gun batteries of the enemy. He participated *"to the preparation of the attacks of 1915 in Artois, to the establishment of working methods, and to the first management plans for these firing grids"* with the 10th Army, and he took part in the offensive on Verdun in December 1916 with the group of Général Mangin. His name is mentioned in [79].

In the meantime, he translated into French the two volumes *Aerial Flight* by Frederick William Lanchester (London, 23 October 1868 - Birmingham, 8 March 1946) on the problems of powered flight [159, 160]. Lanchester was an English mathematician, engineer, and car builder who made important contributions to gas and petrol engineering, aerodynamics, and contributed to the development of operations research.

In March 1916, Benoît commanded a battery of 75 guns on the front near Soissons, and north of Verdun in December of the same year. One month later, he was given the commandment of a group of 320 millimeter guns on a railway track with which he ended the war. In a report, it is said that he was particularly suited for managing this battery as a geodesist since all the preparation for firing was done by calculation. This remark also obviously applies to Cholesky. Benoît took part in the battles in Champagne (April 1917 with the 4th and 5th Army), in Flanders (end of 1917 with the 1st Army), in Aisne (May 1918 with the 6th Army), and again in Champagne (July 1918 with the 4th Army). Between 28 March and

[1] The dates are not exactly the same in all documents.

30 April 1918, he was in the region of Trigny, a village on the Vesle river, not far from Reims. He was promoted to chef d'escadron on 19 (or 23) April 1918.

On 27 May 1918, Benoît managed to save his equipment near the village of Fismes, which earned him reports of great praise from his superiors, and a citation to the Division. In a report, dated 4 October 1921, Général Antoine Gramat (Brive, 7 October 1866 - Turenne, 27 December 1924), head of the French mission in Greece, estimating that Benoît deserved a promotion, wrote:

> *He* [Benoît] *had the opportunity, rare in the great power heavy artillery, of what is probably one of the finest feats of arms in this war, to save his equipment...*
>
> *During the German advance of 27 May 1918, three groups of heavy artillery on railway were cut off from their escape route by Fismes and their situation was considered desperate enough by the Command to give the order to blow up the material. Commandant Benoît estimated that a possible chance of evacuation remained by engaging his guns on the track to Reims...*
>
> *The Chef d'Escadron Benoît, however, had the good fortune to succeed and overcome, with low losses, the continuous barrage established by the German artillery at the outlet of Reims. Only two trains were directly hit and the most deteriorated coaches could nevertheless continue to travel. Thanks to the commendable initiative of the Chef de Bataillon Benoît, 4 trains with 4 guns of 320 mm, one gun of 305 and 6 of 194, saved from the destruction of the neighboring groups, could thus escape and the operation, despite its considerable risks, costs only 3 killed and 5 wounded.*
>
> *This action, entirely inspired and controlled by the Chef d'Escadron Benoît, as no other superior officer took part in it, removed from the spreadsheet of the German victory over 11 guns of large caliber, which were again available against the offensive of the following 15 July in Champagne...*
>
> *From the above statement, it follows that the Chef d'Escadron Benoît has fulfilled his part of war in all circumstances in a very successful way in a position which best suited to his abilities as a geodesist and that, in particular, he was able to take in a complete manner the largest and most successful initiative in particularly critical situations.*

Benoît took part in the American offensive near Verdun in September 1918.

From 21 May 1919 to 31 December 1921, Benoît was part of the French military mission in Greece, as Deputy Director of the Hellenic Geographical Service. A letter from the Minister of War, dated 17 August 1920, mentions:

> ... *Commandant Benoît had the happy idea, on the occasion of the establishment of the Directing Plan of Smyrna, to enforce the Hellenic*

Army officers, previously trained in heavy enough German methods of triangulation and survey, to apply our own methods, much simpler and faster.

He showed them, in this instance, that it was possible to obtain any desired accuracy by a considerably lesser work. These happy results have certainly been recorded with satisfaction by the General Director of the Hellenic Geographical Service.

It is the occasion to congratulate Commandant Benoît and to encourage him to continue his efforts in a way that cannot fail to be beneficial to the Hellenic Geographical Service.

Ernest Benoît

From 1922 to 1924, Benoît gave courses on topography, in particular to pilots. In May 1924, he took part in the study group of practical coast artillery in Toulon. On 25 December 1924, he was promoted to lieutenant–colonel. He is found at sea from 28 May to 8 June 1925 to join the French West Africa, where he remained until 10 May 1927. Back to France, and after a three months leave, he took the boat to Cochin–China (8 March - 7 April 1929). His career continued in Indochina and he became colonel on 25 December 1928. He was then the head of the artillery in Cochin–China and Cambodia until 23 October 1931. On 28 January 1933, he was promoted to Commander of the Legion of Honor. Général Gaston Henri Gustave Billotte (Sommeval, Aube, 10 February 1875 - Bailleul, 23 May 1940), commander of the French troops in Indochina, wrote in a report:

> *Senior officer of a very vast scientific and military culture who gave during 35 years to the Army, his heart, his vast intelligence, his life. Everywhere he scored its trace as shown by the certificates of any kind and even an award from the Academy of Sciences and he merits, at retirement, to receive the stars [2]. It would be fair to give him this ultimate reward.*

On 3 October 1937, Ernest Benoît was admitted to assert his right to retirement, and went to Sainte–Marguerite, by La Garde, near Toulon. He died in 1956. He had many decorations from the various countries where he served.

Benoît has always been esteemed by his superiors, but with some shades

Remarkable technician, specialist of high value in geodesy.

Very extensive culture. Sometimes hesitant by excess of conscience and the desire to do better. He sometimes argues accepted solutions, which slows down the execution of works.

Lack of decision. Endowed with a sometimes exaggerated self–esteem and a great nervousness that make relationships difficult.

Superior officer whose thought is difficult to follow; changes his mind with ease, excess of consciousness, no doubt. Général Billotte.

The hesitations reported in his decisions come from the desire to always do better. Général Billotte.

And this one, that every researcher will appreciate

Inventive and inquiring mind, unfortunately too willing, as inventors, to criticize what exists and to accredit all the benefits to his combinations.

Besides the paper on Cholesky's method, Benoît is the author of other works including a new scientific method for the observation of astronomical latitudes, a method for calculating initial corrections, one about the transportation of firing by slide rule, and, at the Bureau des Longitudes, of a work on formulae for the computation of geographical coordinates [43,44]. In April 1909, while in Indochina, he found formulae for the computations of coordinates which were exclusively used. His military file mentions the following works

- Communication to the Academy of Sciences of a method for calculating latitudes, longitudes and azimuths of geodetic points, published in the *Bulletin de l'Association Internationale de Géodésie*,
- *Artillery - Réglage par observation unilatérale* (Artillery - Setting by unilateral observation), published in *Revue d'Artillerie*, volume 101, January–June 1928,

[2] That is to be promoted to the rank of général.

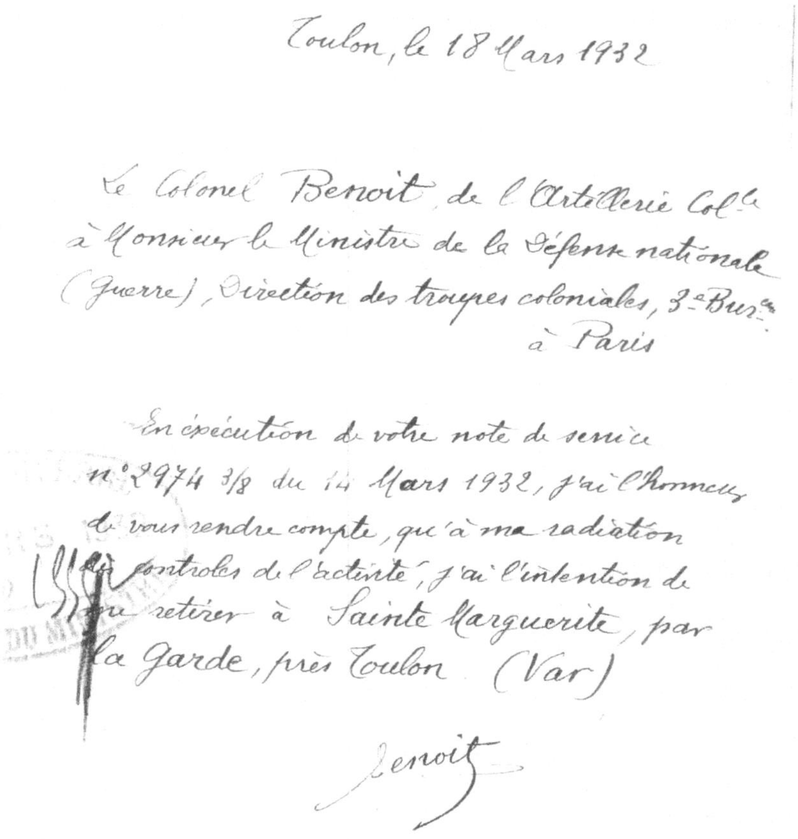
Toulon, le 18 Mars 1932

Le Colonel Benoît de l'Artillerie Col^le
à Monsieur le Ministre de la Défense nationale
(Guerre), Direction des troupes coloniales, 3e Bur.
à Paris

En exécution de votre note de service
n° 2974 3/8 du 14 Mars 1932, j'ai l'honneur
de vous rendre compte, qu'à ma radiation
des contrôles de l'activité, j'ai l'intention de
me retirer à Sainte Marguerite, par
la Garde, près Toulon (Var)

Benoît

Letter of Benoît about his retirement

- *Artillerie de côte - Méthode de réglage des télémètres* (Costal artillery - Method for the adjustment of rangefinders), published in *Mémorial de l'Artillerie Française*,

- *Méthode de calcul des corrections initiales et de transport de tir par règle à calcul* (Method for calculating initial corrections and transport of firing by slide rule), published in February 1934 in *Revue d'Artillerie.*

At its meeting of 22 December 1924, the Academy of Sciences awarded him the Tchihatchef prize (*CRAS*, vol. 179, pp. 1486–1488) intended to reward a naturalist who distinguished himself in the exploration of the Asian continent. Let us give a translation of the report for this prize:

TCHIHATCHEF PRIZE

(Members, MM. Bertin, Lallemand, Fournier, Guignard, Douvillé, Mangin, Gentil; Bourgeois, Ferrié, reviewers.)

The Commission proposes to allocate

A prize of 1,500 francs to Commandant ERNEST BENOÎT, *for his surveying works performed in Indo–China.*

Report of Mr. the Général Bourgeois, on the works of M. BENOÎT.

Commandant ERNEST BENOÎT *made almost his entire career in the Geographical Service of the Army, in the colonial annexes. He took part*

1- In the delineation of Portuguese Guinea during five years, 1901, 1902, 1903, 1907 and 1908.

2 - In 1904, 1905, 1906, 1909, 1910 et 1911, he executed important geodesic works in Indo–China.

This work, very important, have been instrumental in establishing the principal network that was used as the basis for the second order work, the basis of the map, and included two major meridian chains, that of Annam and that of Mekong, related together by six transverses, that is 4,000 km of chains.

The geodetic triangulation work was based on two bases measured under the leadership of Commandant Benoît.

It was during this work that Commandant Benoît established certain new formulae for practical surveying which will be discussed further.

3 - Commandant Benoît also took part in the geodetic work of the Geographic Service, in Algeria and Tunisia.

We said that Commandant Benoît had established very interesting new practical formulae for surveying.

These formulas concern a new method of calculation of the geographic coordinates of the geodetic sheets.

The formulae, named after the topographical engineers, do not give a sufficient accuracy for the principal networks. The formulae named Andraë, very complete, are of a complicated application. Commandant Benoît has fortunately simplified them, while retaining the accuracy required, even for first–order chains. They have been presented at the

Academy in 1908. Commandant Benoît also devised a method for determining astronomical latitudes with a meridian circle, he perfected the method of adjustment of geodetic networks, and finally devised a method for calculating the points obtained by raising a process called inverse intersections.

All these works, some of which are quite remarkable, make Commandant BENOÎT *a most serious candidate for the Tchihatchef prize, and he deserves, in all respects, the proposed distinction.*

At its meeting of 12 December 1932, the *Académie des Inscriptions et Belles–Lettres* of the *Institut de France* awarded Benoît the Delalande–Guérineau prize (*CRAS*, vol. 185, p. 84–85) which was *"to be given every two years to the French traveler in our colonies or other exotic lands who would render more services to the natural history, particularly about the nourishment of man"*:

DELALANDE–GUÉRINEAU PRIZE

(Members: MM. Douvillé, Mangin, Lallemand, Lecomte, Fournier, Bourgeois, Fichot; Georges Perrier, reviewer.)

The Commission proposes to award the prize to M. ERNEST BENOÎT, *Colonel in the colonial artillery, for his theoretical work and his geodetic operations on the ground.*

The Colonel of the colonial artillery E. Benoît, a former student of École Polytechnique, while carrying without cease his military obligations, and in particular by effecting his duty nobly during the Great War, remained for much of his career as an attached surveyor to geographic services (Geographical Service of the Army in Paris, Geographic Service of Indochina in Hanoi, Geographical Service of Hellenic Army in Athens, etc.).

He has personally performed or directed many works on the ground, but his tastes have focused him in particular to the study of theoretical issues to practical applications, eg

In astronomy, the observation and the computation of latitudes.

In geodesy, the resolution of the normal equations coming out from applications of the method of least squares; the calculation of geodetic coordinates on the ellipsoid.

As soon as 1903, as head of the Section of Geodesy of the Geographical Service of Indochina, he had established and passed to the Service, for calculating these coordinates, eminently practical formulae, retaining only the terms of third order, more accurate that those of the geographical engineers, even corrected by Hossard, but of a similar type.

In 1913, he made them known and he demonstrated their use in the centesimal division system of the circumference and for the reference ellipsoid of Clarke (1880).

These formulae allowed to establish the necessary tables for the calculation of the geodetic coordinates of the international reference ellipsoid adopted by the General Assembly of the International Union of Geodesy and Geophysics in Madrid in 1924, and that in both division systems of the circumference, sexagesimal and centigrade.

Throughout all his works, Colonel BENOÎT *brought the resources of a mathematical mind particularly original and fruitful.*

The Academy adopted the proposal from the Commission.

8.2 Benoît's paper

In the paper where he presented Cholesky's method, Benoît writes [42]:

> *The Artillery Commandant Cholesky, of the Geographical Service of the Army, killed during the Great War, imagined during his researches on the adjustment of geodetic networks, a very ingenious method of resolution of the so-called normal equations obtained by application of the method of least squares to linear equations in number fewer than that of unknowns. He concluded a general method for solving linear equations.*
>
> *We will follow, for explaining this method, the same lines than those used by Cholesky to devise it.*
>
> *It is known that the* adjustment *of a geodesic network is intended to obtain from the angular values of observation a system* adjusted *such that all the checks of the diagrams are met, and that the geometrical diagram obtained be distorted as little as possible from the one that would have been given by the values of the observation.*
>
> *These conditions of diagram: closing of the angles of the various triangles, equality of the lengths obtained for the same side regardless of the sequence followed, result in equations called* of condition, *which developed with respect to corrections, can be limited to the 1st order of smallness.*
>
> *One has, in short, to solve a system of p linear equations between the n angular corrections, which became the true unknowns, n being greater than p, otherwise it would be useless to make the observations, the problem being indeterminate. One imposed the additional condition, already mentioned, to deform as little as possible the minimal diagram*

of observation, that is to say, to satisfy the equations with the smallest possible values of the unknowns.

One could, for that, express that the sum of the absolute values *of the unknowns is minimum: but this condition does not lend itself to a convenient algebraic solution and it is the main reason why one prefers to apply the method of* least squares *of Legendre, which also gives, in principle, the most likely corrective system...*

Benoît's paper continues by the following *Abstract*:

Abstract. - *In short, the very complex calculations by ordinary methods, including that of Gauss, and which require as many separate tables as unknowns to be eliminated, resulting in an extreme complication of writing, become, by Cholesky's method and the use of a calculator, relatively easy and much shorter. They are presented in a single table, where the order of setting is easy to recognize and where the operations are always the same. One can, with this method, easily address the resolution of 40 or 50 unknowns, which would have required weeks of hard work by previous methods.*

Finally, it is interesting to quote the last paragraph:

Application of the method of Cholesky to the solution of a system of p linear equations with p unknowns. - *The system being determined, has only a single solution that will be obtained afterwards, also, by looking for the minimal solution provided by application of the method of least squares. The method of calculation of Commandant Cholesky can therefore be adapted to the resolution of any system of linear equations.*

But this adaptation is more theoretical than practical because it leads to calculations that are generally longer than the direct solution of the system by one of the the usual methods: substitutions, eliminations, etc... However, it has two quite important advantages: on one hand, to reduce the writings, and, on the other hand, to provide an homogeneous method, with plain application, presenting continual checks.

Benoît's paper ends by a numerical example of the adjustment of a quadrilateral by the method of Cholesky.

8.3 Benoît's eulogy of Cholesky

Let us give a translation of Commandant Benoît's eulogy on Cholesky [41]. This translation was made in 1975 by Richard W. Cottle, with the assistance of Eduardo Aguado, and presented to *Operations Research 314* on 15 October 1975 — the 100th anniversary of Cholesky's birth:

André–Louis Cholesky, born 15 October 1875, in Montguyon (Charente–Inférieure), entered École Polytechnique at the age of 20 and, upon graduation, went into the artillery branch. Attached to the Geodesic Section of the Geographic Service, in June 1905, he made himself noticed at once by extraordinary intelligence, a great facility for mathematical work, an inquiring spirit, original ideas, sometimes even paradoxical, but always marked by a great dignity of sentiment which he maintained with great conviction.

That was the period when the revision of the French triangulation had just been decided in order to continue the revision of the meridian line of Paris, to be used as the base of a new cadastral triangulation. The problem of the adjustment of the grid preoccupied many officers of the Section, who wished to contribute to fixing, in the sense of speed, convenience and maximal precision, methods which were not yet entirely agreed on. Cholesky approached this problem, bringing in his solutions, as in everything he did, a marked originality. He invented, for the solution of the condition equations in the method of least squares, a very ingenious computational procedure which immediately proved extremely useful, and which most assuredly would have great benefits for all geodesists, if it were published some day.

In the midst of all this, there appeared the question of the map of Crete, then occupied by international troops. After a proposal of Colonel Lubanski, commanding officer of the French troops in Crete, a former geodesist, and, also, a rapid reconnaissance made in March–April 1906 by Lieutenant–colonel Bourgeois, then Chief of the Geodesic Section, it had been decided that the French Geographic Service would undertake the triangulation of the French and British sectors of the island of Crete (departments of San Nicolo and Candie), and the topographical survey of the French sector.

Three officers, of whom Major Lallemand, Chief of the mission, and Lieutenant Cholesky, who had left France in November 1907, proceeded in three months to the fundamental operations: measurement of a base (in the Kavousi plain), determination of an astronomical latitude and azimuth at the southern end. Then, Cholesky, alone, remained to execute the triangulation.

The reconnaissance and the construction of the markers were pursued in mid–winter. If one reflects that Crete, whose breadth is scarcely 40 kilometers, is covered with mountains whose altitude occasionally exceeds 2,400 meters, one appreciates how particularly hard the operations were. At the summit of the Lassithi heights, the water necessary for the detachment was still furnished at the end of May by the melting of enormous drifts of snow. Nevertheless, surmounting all the obstacles,

Cholesky succeeded in finishing the work on 15 June, 1908. Unfortunately, the political circumstances did not ultimately permit the sending of topographers to Crete to execute the surveys.

Afterwards, the geodesic activity of Cholesky underwent an interruption of two years (September 1909 to September 1911), because of the obligation where he found himself carrying out his tour of duty as a battery commander.

Sometime after his return to the Section, he was sent to Algiers to take direction of the precision leveling of Algeria and Tunisia, executed by the Geodesic Section on behalf of the Governor General of Algeria, and that of the Regency of Tunis.

In his new functions, Cholesky brought the ardor he applied to everything. One discerns it from reading the Report on the Operations of the Precision leveling of Algeria and Tunisia during the Campaigns 1910–11, 1911–12, 1912–13.

The primary Tunisian grid chart was finished in the field in the winter 1913–1914. Immediately, all the calculations were reviewed, organized, the grid was checked and adjusted. The publication which will be made will be the result of this effort.

At the same time, for the problem of leveling in Morocco, under somewhat peculiar conditions, Cholesky sought to fit the means to the purpose, and determined the methods to be used.

In May 1913, Cholesky was assigned to the Ministry of Foreign Affairs, and put in charge of a vast service, the Topographical Service of the Regency of Tunis. He hardly had time to show the measure of what he would have been able to do there. In effect, the war broke out shortly thereafter.

During the campaign, he was one of the officers who best understood and developed the role of geodesy and topography in the organization of artillery firing. In October 1916, his technical qualities resulted in his being sent on a mission with the Geographical Service of the Romanian Army. He returned in February 1918, having rendered distinguished service there.

Five months afterward, on 31 August 1918, two months before his comrade Levesque, also the commander of an artillery group, he fell for his country, a sad and irretrievable loss for Geodesy.

Chapter 9

Documents

In the first Section of this Chapter, we will give a translation of the military documents concerning Cholesky: mentions in dispatches, individual campaign leaflet, and a transcription of the marks he received. The second Section is devoted to the everyday notes Cholesky wrote during one of his campaigns. In the third Section, we describe the contents of the *Fonds Cholesky* at *École Polytechnique*. Some trigonometry is recalled in Appendix A. Finally, Appendix B contains the original paper (in French) of Cholesky on his method for solving systems on linear equations, and Appendix C his book (in French) on graphical computations.

9.1 Military documents

We will now give a translation of the military documents concerning Cholesky. They all belong to the Archives of the History Department of the Army [2].

9.1.1 Mentions in dispatches

- *Mention No. 148 to the order of the IVth Army on 21 December 1914 (J.O. [1] of 14 January 1915).*
  The general commanding the Fourth Army cites to the order of the Army for the following: Capitaine André Cholesky: Sent on 8 December to the lieutenant–colonel, director of the attacks of the 83rd Regiment, to indicate the time when the attack could be launched; his mission completed, remained voluntarily with this superior officer to be his assistant, and, under intense fire, did this service until the next day at 9 am, in our trenches and in the enemy trenches which had just seized. Artillery officer exceptionally

[1] J.O. means *Journal Officiel*, where all the laws and decrees of the French Republic are published.

gifted, whose multiple observations greatly contributed last month to give our artillery a complete superiority on the enemy artillery.

- *Mention No. 561 to the order of the 17th Army Corps.*
  The general commanding the artillery quotes to the order of the artillery of the 17th Army Corps Capitaine Cholesky of the 23rd Artillery Regiment for the following reason: has demonstrated activity, initiative, and talent by centralizing, in the artillery of the 17th Army Corps, the service of signals, observation, and firing.
- *Citation No. 344 to the order of the Army of 12 October 1918.*
  Officer of the highest value by his knowledge, his utter disregard of danger, and the high example given to all. He has inspired to his group a irresistible impulse. Killed in his battle station.

9.1.2 Individual campaign leaflet

- **Summary of the records prior to the year 1910**

  Left Fontainebleau [2] with number 4 on 92 in 1898. Looks good, serves with zeal and enthusiasm, a little giddy, rides well. Will become a good officer.
  Detached to the Geographical Service of the Army in Tunisia in 1902, 1903–04–05 temporarily, he is permanently assigned to this Service on 24 January 1905. Very well noted. Appointed capitaine on 25 March 1909, he was assigned to 13th Artillery Regiment by Ministerial decision of 28 August. Good impression from the start. Colonel Grand–Didier.

- **Copy of the records of the leaflet of the staff**

  - **1910**
    April. Robust temperament, keen intelligence, clear and reflective mind. Highly honest nature, caring and strong, Capitaine Cholesky handles with great care the instruction and the administration of his Company. Although he was long detached, quickly became aware of the service of the company. Knows very well his staff. Will be a very good captain.
    Colonel Grand–Didier.
    May. Scores obtained in the regional firing course of La Courtine from 25 April to 4 May 1910: Ability of the officer from the point of view of firing, at the start of the series: well prepared - zealous and constant progresses, very satisfying - Ability at the end of the course for serving as the commander of a company: very suitable for acting as commander of a

[2] Where the *École de l'Artillerie et du Génie* was.

firing company.
Chef d'Escadron Vincent, Director of the military training at École Polytechnique, Director of the course.
October. Capitaine Cholesky commands his section with great zeal, firmness and kindness; he has achieved very good results that can be further improved by monitoring tighter some details. Well versed in all matters of firing, he fires with calm and observes well; he will become, during the next firing lessons, one of the best firers of the regiment. Gave, at the maneuvers in Picardy, entire satisfaction to his group leader.
Colonel Grand–Didier.

- **1911**
  April. Methodical mind, very thoughtful, Capitaine Cholesky is calm, cool, and has a lot of initiative. The battery is tidy, well–educated and well managed. He commands his battery for fire in a brilliant way; he observes surely and takes prompt decisions.
  Colonel Grand–Didier.

  Second capitaine bright and very good at firing, neglecting no detail in both the administration and in educational matters and the maintenance of the staff, Capitaine Cholesky, which has been the subject of a mutation for the Geographical Service, left the regiment with a perfect knowledge, in all respects, of the duties of a capitaine.
  Lieutenant–colonel [unreadable].

- **1912**
  Assigned to the particular General Staff of the artillery to be detached to the management of the Geographical Service by a ministerial decision of 24 September 1912.
  Avril.
  Arrived at the Geographic Service of the Army during the month of October, Capitaine Cholesky has been assigned to the Geodesy Section where he had already been appreciated from 1905 to 1909. Responsible for directing the operations of precise leveling in Algeria and Tunisia, he fulfilled his duties with authority and a remarkable skill. Scientific mind, always alert and at work.
  The Commandant, Head of the Geodesy Section, Lallemand.
  October. Capitaine Cholesky took part in the maneuvers of the Army as an officer cartographer (Staff of Général Marion) he earned high marks; outstanding officer in every respect. Proposed for Chevalier of the Legion of Honor.

The Lieutenant–colonel, Head of the Geodesy Section, Lallemand.

– **1913**
April. Continued to manage the operations of precision leveling in Algeria and Tunisia with a remarkable zeal and competence. Seems to have to get a position in the Direction of Publics Works of the Tunisian Government, as head of the Topographical Service. The Geographic Service and the Army can not but regret his departure, the officers of this value being rare.
The Lieutenant–colonel, Head of the Geodesy Section, Lallemand.

Removed from the rolls of the Geographical Service on 25 May 1913, put out of the list, at the disposal of the Minister of Foreign Affairs (Head of Topographical Service of the Regency of Tunis. Ministerial Decision of 4 June 1913).
Lallemand.

– **1914**
Assigned to a position of capitaine in the 7th group of foot artillery of Bizerte, in case of mobilization. Commanded the 10th battery of the depot since his arrival in the Corps. Assigned to the 16th Field Artillery Regiment (depot of Issoire). Sent by 15 September to his new position. Tel. 119 m of the général commander–in–chief of the troops in North Africa. Received 10 September.
Bizerte, on 13 September 1914. The Commandant of the 7th group, L. Viane [?].

Joined the 23rd Artillery Regiment in Somme–Suippe (Marne) on 23 September, took command of the third group on 27 September, then that of the 9th battery on 18 October with the arrival of Commandant Girard. Very positive mind, very knowledgeable, hard working; showed from the beginning an intense activity in both the organization of the battery and the installation of its bivouacs, and in the completion of fire against aircrafts or the operation of the observatories: fires well, observes well - Entrusted on 4 January 1915 of the making of the plan. Manager of the firing of C.A. [3] *and of the management of the observation of firing. Assigned in late January to the Geographical Service - quits the regiment when*

[3] *Corps d'Armée* that is Army Corps.

he was in excellent position to be promoted or for the Cross [4].
Lieutenant–colonel Malet.

- **1915**
  10 February 1915. Capitaine Cholesky has demonstrated a very comprehensive knowledge of all that concerns the organization of the firing, the observation and the transmission service. He has shown a remarkable activity and organizing talent during January and until his departure.
  The General commanding the artillery of the 17th C.A., Falque.

  Officer remarkably gifted, with a mind always alert, and original ideas, most often fruitful in their application. Was, for the head of the Group of firing grids of the 7th Army, a collaborator fully aware of all the technical issues, ardent in his task. Helped in large part to all the results obtained. Very qualified for the higher rank.
  Général de Boissoudre, Head of the E.M. [5] *7th Army.*

- **1916**
  During the first half of 1916, Capitaine Cholesky continued to demonstrate the same qualities. He now takes command of the group of firing grids of the 7th Army for which he is perfectly qualified. He should however be wary, as the head of a service, of any tend to originality and paradox. He should be left the shortest time as possible in the rank of capitaine, the officers of his value being rare. He will appear everywhere, in a technical service as in the troop, brilliantly and with honor.
  The Commandant, Head of the group of firing grids, Perrier. 26 May 1916.

  Led ably the G.C.T.A. Worker, resourceful, thoughtful, has widely developed the production of the group and has taken successful initiatives. In short, has very well served in the succession of Commandant Perrier.
  The Lieutenant–colonel Hergault, Head of the E.M. 7th Army, on 24 October 1916.

- **1917**
  Technical Director of the Geographic Service in Romania, has taken since his arrival the control of this Service in shaking the torpor, tripling the production and ensuring the continuity in the work despite the transfer of the services as a result

[4] The Cross of the Legion of Honor, which means a promotion to Officer in this order.
[5] *État–Major*, that is Headquarters.

of the operations. Honors the French Geographical Service, active, energetic, man of authority, has found here the full development of his personality. His promotion, already late, should no longer be deferred.
At the Romanian General Head Quarter, on 25 January 1918, Colonel Pétin, Head of the E.M.

- **1918**
  Same excellent marks.
  The Head of the E.M. Pétin.

  Continued until the departure of the French Mission from Romania to lead with authority and competence the geographic services of the Romanian Army. Gained by these functions, all the rights to take up in France a work requiring initiative.
  The Head of the E.M.of the Romanian mission, Colonel V. Pétin.

9.1.3 Transcription of the marks

They are the same as what precedes except what follows:

> *Proposed for the rank of commandant with the following reason:*
> *Very deserving to be immediately appointed to the upper grade. 8 years in the rank of capitaine will not make his promotion an exceptional choice that, however, his personality should deserve. Ardent, energetic, tireless worker, with great authority. He honors the Mission in the functions entrusted to him.*
> *Général Berthelot, Head of the Military Mission in Romania.*

9.2 Notebooks

In these two notebooks, Cholesky gives a vivid relation of his everyday life when he was surveying in the Massif Central, in south–central France, during 1905–1906.

9.2.1 Notebook No. 2

5 August 1905

Durand left at 5:30 am.

I go up to the signal for my pin and to build my leveling plane. I install my two rafters and the two extreme boards. Decidedly, I adopted the square of 3 m. My diagonal equalized, I let nail the rest. I have to do the mason. I made a hole a little bigger for my pin that, besides, enters by simple gravity; as I only have very little plaster, so I adjust small pieces of granite and I poured my plaster. I return to my rod, but the fog around us begins to burst. But we must finish.

6:30 pm. (unreadable name) starts painting, not long, because it's raining now and I am forced to get everyone back to the camp. I prepare the papers for tomorrow. And here is the farmer of [Mount] *Pilat with whom I rule my accounts. He brought the carriage. Tomorrow morning, we could load at the first hour.*

7:30 pm. Here's an opening. I leave the men at the preparation of their soup and mine, and I grab the can of paint. I am back at dark (hands idem) (the rod idem). All is well. Unless the heat that has received a severe blow to. Tomorrow morning at the first hour, alarm for the battle, remembering of the bled [6].

6 August

Pretty good start. Despite the compact fog and the wind rather acute. The camp is quickly up and everything is loaded so that, at 8 am, the carriage can start properly covered will all the tents as it rained several times.

I waited anxiously for this departure with undisguised fear of seeing the carriage turning over before arriving on the road. Fortunately, with a thousand precautions, we get there safely, and leave the car proceed to come back to the rod. Ah, the poor rod, what nice scoops it takes twice before to stand on its feet. Still, in the third round, we make it. It is well centered but the masonry of the basis of the pillar does not allow much to move it. I prefer to measure the eccentricity than trying to place it better. Moreover, it is 11:30 am, I have to send back my people so that they could be in Bourg-Argental before 3:30 pm. I go down to the farm for lunch, hoping that Lunaçon [?] *does not let me wait too long. Besides, I will leave even if he does not arrive, leaving him an order to join. Fortunately, here is he as I start my lunch. After having shown him his benchmarks, I'm leaving.*

Le Bessat: my brake does not apply. I lose a quarter of an hour trying to make it work. Without another result than to take me to the last minute if I want to catch the train. Also, what a speed!

Disappointment, I see my three men who trot on the road about 3 km from Bourg and it is 3:15 pm. My equipment is still on the cart, and I must give up all hope to leave at 4 pm. I fear that it may be one day late. A tire punctured.

[6] Name given to the countryside in North–Africa.

Bourg–Argental. Grumbling of the farmer who was waiting since noon without seeing anything coming. I enter into negotiations with the stationmaster who, sawing my outfit obviously not so much gorgeous, shows some suspicion. Finally, the result is that I delete the transport order, and I take all the equipment with me as luggage. That way, I'm going with everything at 7 pm this evening, while, otherwise, the material would not leave before tomorrow at 2 pm, and would likely take 3 days to come. Many distractions in Bourg where there is a great celebration. I looked dancing a quadrille at the country ball. I almost regretted leaving.

I sleep in two trains until Rives, where I just has the strength to go to the hotel opposite to the station.

7 August

Departure at 5:20 am. Arrival around 6 am at Chabons. I immediately informed the man that must transport the material. One loads while I'm off to Biol. A 1 hour stop. Ascension of Haut–Biol. The host of Ferrand stops me, and not without giving me a glass of white wine, offered to accompany me to La Charpenne. It bothers me a little bit but I finally agree: he may be able to help me get in touch with the owners up there. I made the crossover with Ferrand who was coming down while I was going up. I mark the center of the upper face of the pillar and I conduct a preliminary horizon sweep. I can't do anything. I go back down to drink a glass of wine that might replace, maybe, an absent lunch. Fortunately, I guzzled two eggs in Biol. Finally at 12:30 am, we hear rumors, that's the stuff. It is still necessary to half unload so that the three horses consent to make the last 300 meters. No matter: equipped with the scouting–theodolite, I conduct a little more serious horizon sweep. Bad luck: Granier and Colombier that I see cannot go with the Pilat. It is not well defined so that I remain in doubtful which is annoying.

I put the pins of the theodolite, in order to do something more serious. Nevertheless I still only have adjustments a little hazy, and I only have a vague confidence in the benchmarks direction points I have chosen. I align the knobs of the 3 mirrors and I give complete instructions to [name illegible] *who arrived here this morning a little before me and also to Ferrand that I decided to send to Montellier instead of* [name illegible] *that I want to have with me for Granier and Colombier.*

I'm surprised by the night when preparing my two men for departure. I had to bawl out a little bit everybody to appease a bit stormy discussions. This is probably why I came with great difficulty to get some soup and an omelette around 9 pm. Fortunately I have had no lunch at noon, I am prepared for a bad dinner. Then I have enough to do to leave tomorrow morning for Boussuivre.

8 August

After aligning a mirror on Boussuivre and settled the schedule of the day, I go to Bourgoin. Fixed my tire in the luggage van between Lyon and Tarare.

Lunch at Tarare. Violent digestive: 9 km of slope without drinking, so I make it up extensively in Violay. The sun is pretty strong to go up to Boussuivre. We come there around 3:15 pm. I orient myself on the map and I see soon two flashing in the binoculars.

I set up my mirror giving instructions to [name illegible]. *Horror! my mirror has a collimation which I think is 4 mm. I correct it somehow because I do not see the light of Charpenne, but what will do my two other mirrors?*

I go back to Charpenne a little sorry of my discovery. Will they see my light tomorrow and will they be smart enough, if there is a mistake, to correct it?

I pointed the mirror with -5 instead of -9, slope of Charpenne or Boussuivre.

9 August

Charpenne. I arrived at 3 am after waiting for my bike from 9:15 am to 12 am in Bourgoin. Bad luck. Good luck: I find Durand when returning.

Long [illegible word]. *Short abnormal. Not very fruitful morning. Mail - ordinary. We will work only at 11 am.*

11 am. Shed. The problem becomes exciting. Pilat has beautifully lit in the morning; then, what is the significance of the error of Boussuivre?

Will the two others appear? I trust that Boussuivre will be seen since I went there. It's the very least.

Durand [illegible word] - *2 or 3 minutes barely elapse and I see Boussuivre in the telescope. This is good. We made measurements and immediately we throw a light to Montellier which also responds and wonderfully after 10 minutes. In short, we started at 11:15 am and, at 12:20 am, our three mirrors rival in light.*

A 6:15 pm, the mirrors go less well, the wind may be. I go down to [illegible word]. *All is well.*

10 August

Rest at Chabons.
I intended to go to the Granier.
My two men arrive in the afternoon after the last train that allows me to sleep at Chapareillan. I send them sleep at Grand Lemps. I will join by myself during the night.
There is some good in a bike.

11 August

Few drops of rain in the dark night which leads me with my bicycle to Grand Lemps. My two lascars are at the station. Departure 2:20 am. 1 hour stop in Grenoble. I get acquainted with the city from the station to the Victor Hugo square and the banks of the Isère. Excellent impression. Travel to Chapareillan very attractive despite the fog and the rain.

Chapareillan; very nice mayor, everything seems to go at will; yet I do not go before 9:30 am and I do not know what I will find up there. I just know that the two carpenters were finishing boards and rafters this morning but it was quite a bad weather from Grenoble. Will I find them. It's necessary to climb.

We arrive at 1 pm to the cliff. Lunch, it's greatly time. More especially since it rains in torrents: it is a return of the storm which stopped us around 11 am. If only it does not last.

2 pm. We set off again but it is necessary to carry up the plates and the rafters that we found here. And the weather is very threatening. The experience is also not happy. The rain starts again soon. We try to fight but, drenched in sweat and rain, we can't go very far. My men ask me to return to the shepherd's hut which is a few hundred meters below. I surrender.

It's sweet to find a good fire, even in August in this hospitable neighborhood; but when we came close of it, you must not think to leave it; however, we had no bread but the cool water from the mountain was enough to be convinced that we will get out from there only with a nice weather. The dark comes quickly. I still hope that the night will bring a good weather and, as we have the moon until midnight, I caress the hope to get my rod in the moonlight and to go down with my acetylene lamp that I carried with me.

8 pm. Dark, the rain continued with the same ardor. There is not more than to think about resting. I lay the feet to the fire with a pad of straw under the hips and the head on my bag. I do not know how many times I changed positions during the night. I even slept enough to get some rest.

12 August

At daylight, I open the door of the hut. The weather, without being nice, seems to promise an improvement. I go to the food basket to take the bread that was left yesterday to my men who are still sleeping. For me, I scrunch a little piece of crust that I slipped into my pocket after my yesterday lunch in order to shorten the descent.

5 am. We resume the climb. Arrived first in a fog that does not allow me to judge my advance, I start to clear the pile of stones which I guess being the old signal. I have enough time to recover under 1 m 50 of stones the geodetic point slightly mutilated by the age that I prepare to receive the bronze station mark. Finally, here are my men with their loads of wood; I send back our old guide (who's starting to give us a little heartache) to fetch water which we lack for mixing the cement, and we begin to build the leveling plate. The implementation does not take place without difficulty. The wood is green and heavy and we are only 3 because our old man is totally unable to help. He hardly stands on his legs. Finally, I feel obliged to get on my shoulders a fir that served as a signpost and that I want to make a rest support. Thanks to this new help, we finally draw up the rod. After the final arrangements, we go down at 2 pm.
Ah! what a lunch!
4 pm. New start towards the Colombier. Beautiful journey Chambéry–Aix. Long negotiations with the employees who want to convince me that it's easy to climb from Culoz, but I'm suspicious, I want to go to Virieu le Petit. Sleep at Artemare.

13 August. Sunday

Arrival in carriage at Virieu at 8 am. Very nice weather. I tell myself that I can regain the time lost at Granier, but this hope is of short duration. After a short conversation with the mayor, it is clear to me that there is no hope for today. The heat is not suitable for cattle. I begin the climb at 11 am with the town councillor who, going to [illegible word], *proposed to guide me and put me in touch with people that I can use tomorrow.*
It's hot, but what a plain compared to the Granier!
Everything is settled, I recognize my work site and I'm back at 5 pm.

14 August

New departure at 3 am. We extract the stone for the pillar as soon as we arrived at the top; we also have plenty of time since the material arrives only at 9 am and yet the last part of the journey must be made on the back.
Finally, everything is starting. The pillar first gives me some concerns by its slowness to rise but at last it rises up. The leveling plate goes up slowly but safely. My mirror is finally aligned. I see some points with a bad signal but since I saw a very good signal, I'm sure of what has been done. I'm going down finally. 4:45 pm. At 6 am I'm at the table with the hope of taking the 8 am train at Artemare which through Lyon will take me to Chabons around 3 am.

15 August

I can't even sleep all the morning, the plain light puts me out of the bed. However, I have no desire to go to Charpenne by bicycle. It is too hot. We will go tonight with the carriage.
Travel to Haut Biol. I meet Durand who, despite our exhortations, is returning to the camp after dinner.

16 August

Here I am returning back to Boussuivre. By passing through Lyon, I try to have the mullets for Granier. Without success, besides.
At departure from Lyon, severe thunderstorm. I am concerned about the success of my day. I brilliantly go up the slope of Boussuivre but there is still no sun.
6:30 pm. I come down again from the signal. Here I am forced to sleep in Violay. I only hope that this weather does not last too long.

17 August

It rained all night long and it is still raining. I gloomily look the clouds go, watching with anxiety the arrival of a break; but here is 1 pm; I was wrong to skip my lunch because it's still raining. I wonder if I still have to insist staying here. Yet...

9.2.2 Notebook No. 3

One page of diagrams.

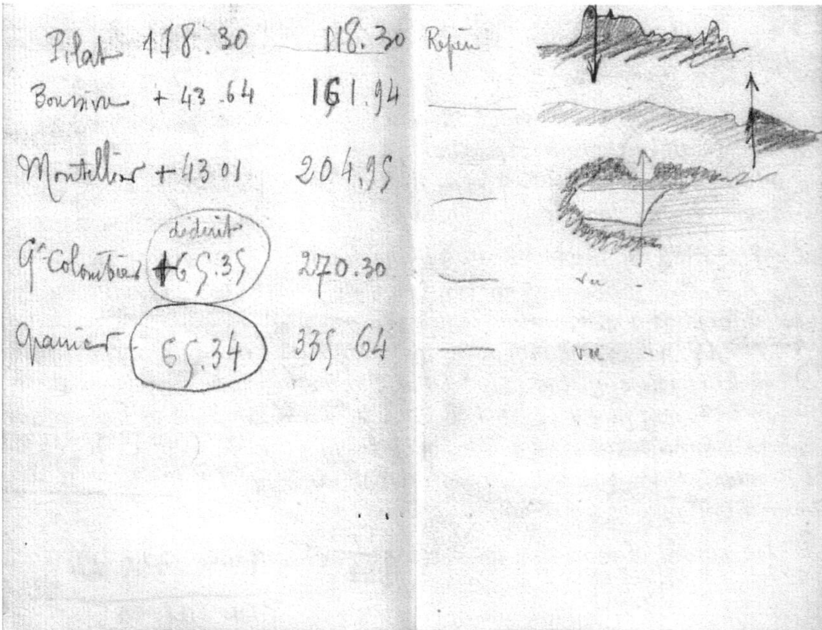

Page 8 from notebook 3

<u>*24 July*</u>

Pierre–sur–Haute.
Measurement of the elements of misalignment. Diagrams and formulae.

<u>*25 July*</u>

Installation of the mirror for Mount Pilat on a auxiliary pillar and of a flat rod in the vertical of the mirror.
Diagrams and formulae.
The hut is much further than that.

<u>*28 July*</u>

Le Montellier (castle).
Arrival to the pillar at 8 am.
The mirror is oriented to the left and too high.
Orientation at 1 cm of the centering sheet.
Pilat was visible, it completely disappeared.

9 am - Lining above the ridge in front at the southeast of [name illegible]
altitude 310.
P=-2 mm.
Height of Pilat above the ridge 14 mm
For the length of the mirror Pelletier 4 mm 55
11 am. I saw two flashes but very fleeting
Nothing more. I check my calculations.

The Pilat is visible. I am looking for the signal according to the map. It seemed to me that my centering sheet was not sufficiently accurate (difficulty of orientation relatively to 3 visible points). The castle of Montellier was not a signal on the map at the 80,000th and we were obliged to carry it on the map at the 200,000th where it did not exist. It seemed to me that I pointed to the left. I see far right of my station mark a summit which, in the binoculars, has the shape of the Crêt de la Perdrix; it seems low but, according to the map, I think it is in fact lower than the [a word illegible].

11:30 am. I move my mirror to the right on the recognized ridge, 4 mm to the right of the primitive mark.
11:40 am. I can see [name illegible]. *It's going well.*
To change my direction without, as far as possible, changing my height, the stiffness of the building bothered me. I had to make by myself a small lever on the back with the blade of my knife to be able to see a station mark on the ridge in front inside the vane.

Mont de Bar - 30 July

Arrival at the pillar at 12:40 am.
The plane table seems a little to the right and too low.
Orientation using the centering sheet. The calculated absolute height seems good, but the height with respect to the ridge of Bellevue seems clearly false. I start again the calculations and found the same result.???? (see the map at the 80,000th). The plane table is nailed on a board designed to increase its distance from the pillar and provided with an eye–glass shade to prevent the sun to hit it.

2:30 pm. The plane table is very nearly fixed in height by oscillating around a point in the middle of the plane table. I did place the plane table much on the left so as to be obliged to force much to bring it in direction and thus ensure a good calibration with another board nailed perpendicularly to the first one on another side of the floor of the woodwork.
3 pm. My light is directed to the Pilat where I have absolutely not yet seen anything.
3:15 pm. That's it, points are made for me. I make Lunaçon (?) leap

to completely set my points.
3:40 pm. From the points, my light might have been almost continuous, and I have seen nothing at the Pilat.
I'm leaving.

Division of 598,47 by 2.

A table of measurements.

A triangulation map on two pages.

One page of computations.

One page of measurements.

One page of drawings.

One page of measurements.

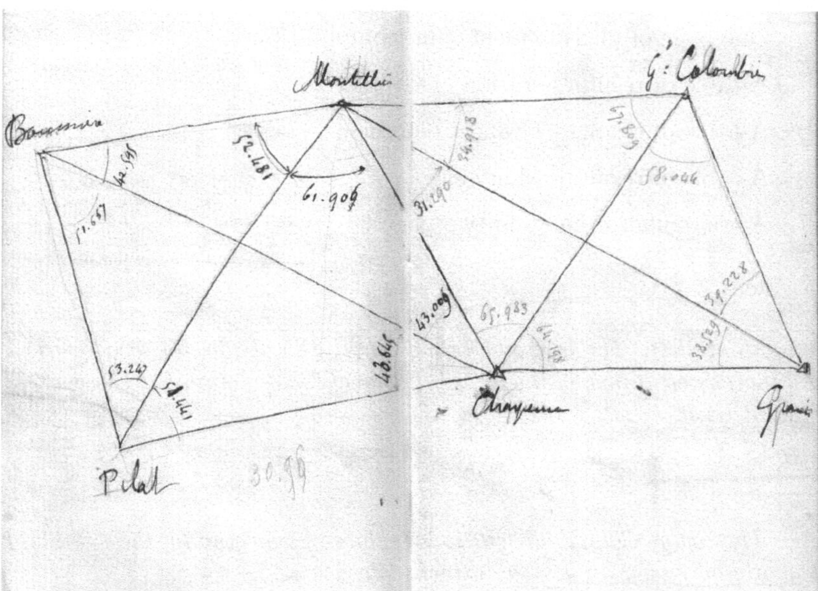

Page from a notebook

<u>*11 and 12 August*</u>

Installation of a rod in Granier and implementation of a marker on the old mark of the I.G. The mark was slightly damaged at the top.

I did it break to get to a solid part. The rod is very much in line with Charpenne and 168 mm farther.

14 August

Construction of the pillar of Colombier and installation of a rod. The mark of I.G. was beautifully preserved. The rod is strictly centered on the station mark.

17 August

Boussuivre, bad weather.
18 August id.
19 August id. in the morning.
In the evening, I rectify the mirror on a signal of the map to the 80,000th, and I join Charpenne. The mirror worked the next day.

One page of measurements and computations.

A table of results - station of Granier.

A table of results - Grand Colombier.

A table of results - Montellier.

A triangulation map on two pages.

27 August

Montellier. The mirror Durand remained outside, the wood worked and there is clearance in the hinge. I correct the direction but I'm leaving without trust.

28 August

The rod of Colombier felt; it is replaced as originally; the wind laid it gently by rising up a stone attached to a wire.
Nothing has changed.

2 September

As a result of the abandonment of the station of Granier, the rod of Colombier must be turned over to Montellier.

$Hc= 0.980+1.000+1.650=3.630$
$Hb= 3.630+1.650=5.280$

The rod is centered on the station mark.

4 and 5 September

Construction of the pillar at Nivigne.
Dimensions on the orientation sheet. The mirror has been aligned according to the map and the slope calculated, I saw no light.

6 September

Recognition of Crêt de la Neige. From the highest peak, one see well, except in a direction that, after an approximate orientation, seems very close to that of Montellier. I orient myself as much as possible and the sight seems then very grazing. I give up this summit. The southernmost peak seems about 2 m lower, but I am sure to see all directions.

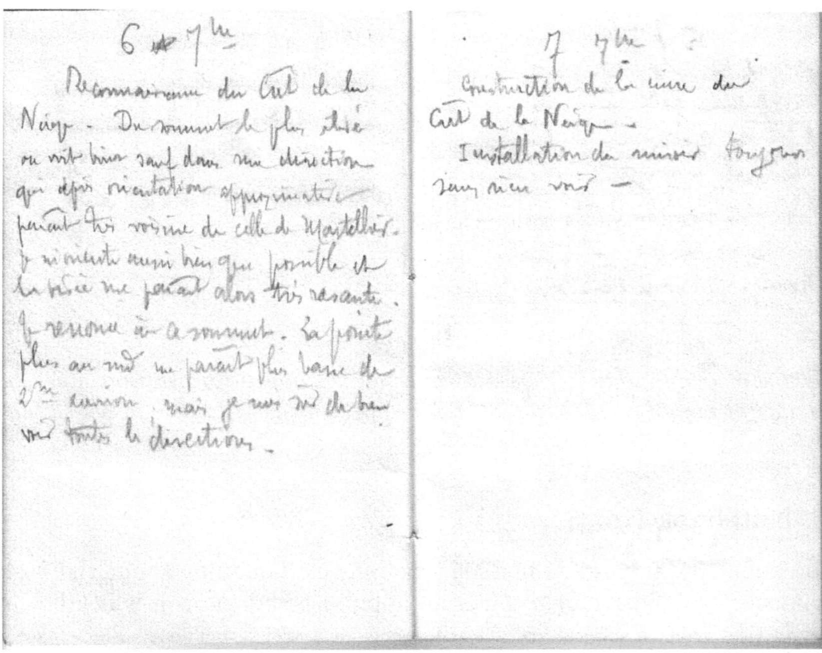
6 7bre

Reconnaissance du Crêt de la Neige. Du sommet le plus élevé on voit bien sauf dans une direction qui après orientation approximative paraît très voisine de celle de Montellier. Je m'oriente aussi bien que possible et la visée me paraît alors très rasante. Je renonce à ce sommet. La pointe plus au sud me paraît plus basse de 2m environ. mais je suis sûr de bien voir toutes les directions.

7 7bre

Construction de la mire du Crêt de la Neige.
Installation du miroir toujours sans rien voir —

Pages of the notebook about Crêt de la Neige

7 September

Construction of the rod of Crêt de la Neige.
Installation of the mirror always without seing anything.

10 September

The rod of Colombier projected only on the black part of the Mont–Blanc. That of Granier is absolutely invisible projecting on my black peak of Belledone.
After conference with the commandant, I will return to Granier.

12 September

My aide–de–camp, who was called back from Crêt de la Neige, does not come. The weather is becoming overclouded. The ascent of Granier seems threatened.

13 September

Arrival at Chapareillan under a torrential rain, I'm muddied from the feet to the head and soaked. The ascension seems more and more far–away.

14 September

Continuous and strong rain all day long.

Granier. 17 and 18 September

Reset the rod.
The shot of Montellier passes 1 centimeter south of the station mark.
Heliostat placed on slab stone whose upper surface is 198 mm above the reference.

9.2.3 Notebook No. 5

The notebook No. 5, written in 1905, contains explanations about the handling of certain surveying instruments. It is an unpublished document which belongs to the *Fonds Cholesky* of *École Polytechnique.*

9.3 Archives

The documents given by Cholesky's family to *École Polytechnique* were sorted by Claudine Billoux, archivist, Michel Gross, Cholesky's grandson, and me. They now form the *Fonds Cholesky* [124].

The work for classifying these records consisted of, first identifying the various documents composing the holdings, to monitor their condition and eliminating

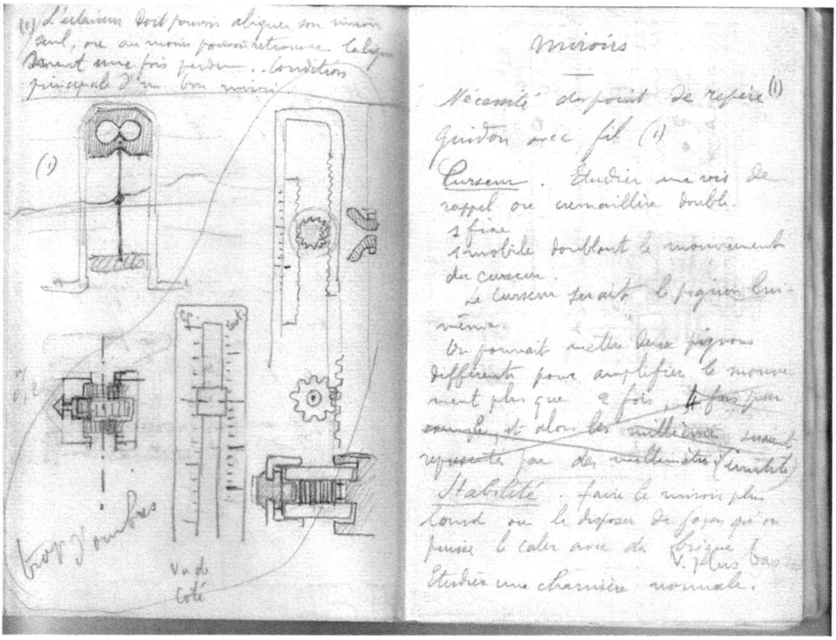

Page 2 from notebook No. 5
© Fonds Cholesky - École Polytechnique

those very few that seemed totally unusable. The second step was to propose an outline for a classification and to list all the documents.

Then, some documents which had been dispersed have to be gathered, such as the manuscripts of the lectures either from *École Polytechnique*, or, the most numerous ones, for courses at *École Spéciale des Travaux Publics.* Then, the papers were grouped into boxes of archives, each box containing a more or less detailed inventory of its contents. In addition, two other tasks were carried out: a decryption by me of the manuscript, dating from 1910, where Cholesky described his method, and the decryption by Michel Gross of three handwritten notebooks.

The documents are displayed in five folders marked A to E.

Folder A

This folder includes courses of *École Polytechnique*, military documents, some written by Cholesky, two books in Cyrillic script, the inventory of the *Fonds Cholesky*, diskettes and one CD's reproducing the notebooks 2, 3 and 5 of the surveying campaign of 1907 (see Folder E below).

Folder B

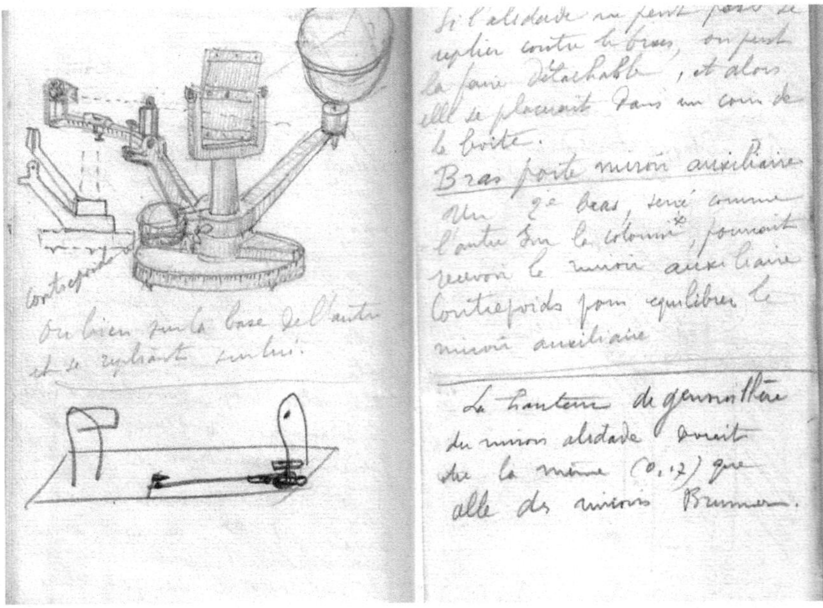

Page 9 of notebook No. 5

Essentially, this folder includes manuscripts, some mimeographed, of courses or for the preparation of scientific works mainly directed at topography and measurements. The manuscript entitled *Sur la résolution numérique des systèmes d'équations linéaires* is in this folder (No. B4). This folder also contains (No. B12) a copy of the notice of I wrote, updated according to the work done on the *Fonds Cholesky.*

Folder C

This folder contains a series of documents concerning the Army and, in particular, the groups of firing grids, studies on the status, the location of batteries, Tunisia and Romania (where Cholesky spent part of his career). Personal manuscripts have been gathered under No. C3. Material not classifiable elsewhere have No. C10.

Folder D

Essentially, these are 70 printed materials which were the subject of a detailed inventory attached to the folder. These 70 documents are of a military nature, some are annotated from the hand of Cholesky. It should be noted that one of them, No. 60 is annotated *Choleski* with an i and not a y. This error is surprising, especially since it seems that the annotation is really from his hand. In addition,

No. 1 is annotated in the name of Michel Mesnier: Jeanne, the sister of Henriette Brunet, the wife of Cholesky, had a son with that name, but it is unclear whether this book does indeed refer to him, and how it arrived in the *Fonds Cholesky.*

This folder also contains maps and atlases, photographs of surveying materials and five printed documents out of the inventory, three in Romanian, two blank records on fire and observation, and a *Guide for the moral education of the soldier* in French.

Folder E

This folder contains, essentially, seven notebooks numbered 1 to 7. Two of these notebooks, Nos. 2 and 3, have been fully restituted and are classified as No. E3 (their translation was given in the previous Section). The three notebooks 2, 3 and 5 have been reproduced on a CD that is in the folder under No. A0.

Appendix A

On trigonometry

Let us recall some formulae about the trigonometry of triangles which are commonly used, in particular by topographers.

Let a, b and c be the sides of a plane triangle, and let A, B and C be the angles respectively opposite to each of them. The following fundamental relations are satisfied

$$
\begin{aligned}
&A + B + C = \pi \\
&a/\sin A = b/\sin B = c/\sin C \\
&a^2 = b^2 + c^2 - 2bc\cos A \\
&(a+b)/(a-b) = \tan((A+B)/2)/\tan((A-B)/2).
\end{aligned}
$$

Resolving a triangle consists, from three of these elements, to calculate the three others. The length of a side must always be among the three elements which are known since, otherwise, an indeterminacy remains. We have

Data	Calculation of the other elements
a, A, B	$C = \pi - A - B, \quad b = a\sin B/\sin A, \quad c = a\sin C/\sin A$
a, b, C	$\tan(A-B)/2 = (a-b)/(a+b)\cot C/2, \quad (A+B)/2 = (\pi - C)/2$ Having obtained $A+B$ and $A-B$, we deduce A and B $c = a\sin C/\sin A$
a, b, A	$\sin B = b\sin A/a$ If $a \geq b$, $B < \pi/2$ can take only one value. If $a < b$, three cases are possibles: 1 - if $b\sin A < a$, B can take 2 values ($B_2 = \pi - B_1$), 2 - if $b\sin A = a$, $B = \pi/2$, 3 - if $b\sin A > a$, the triangle cannot exist. $C = \pi - A - B, \quad c = a\sin C/\sin A$
a, b, c	$r^2 = (p-a)(p-b)(p-c)/p$ avec $p = (a+b+c)/2$ $\tan A/2 = r/(p-a), \quad \tan B/2 = r/(p-b), \quad \tan C/2 = r/(p-c)$

The perpendicular height h on the side a (issued from A) is

$$h = b \sin C = c \sin B.$$

In geodesy, angles are not measured in degrees, as in trigonometry, or in radians, as in mathematics, but in *grades*, also called *gons*. A circle is divided into 400 gons instead of 360 degrees.

In a spherical triangle, that is a triangle on a sphere, the sum of the angles is always greater than π. There exist special relations for resolving such triangles. In topography, they only have to be used when mapping a large portion of territory.

Appendix B

Sur la résolution numérique des systèmes d'équations linéaires

Sur la résolution numérique des systèmes d'équations linéaires
A. Cholesky

La solution des problèmes dépendant de données expérimentales, qui peuvent dans certains cas être soumises à des conditions, et auxquelles on applique la méthode des moindres carrés, est toujours subordonnée au calcul numérique des racines d'un système d'équations linéaires. C'est le cas de la recherche des lois physiques ; c'est aussi le cas de la compensation des réseaux géodésiques. Il est donc intéressant de rechercher un moyen sûr et aussi simple que possible d'effectuer la résolution numérique d'un système d'équations linéaires.

Le procédé que nous allons indiquer s'applique aux systèmes d'équations symétriques auxquels conduit la méthode des moindres carrés ; mais nous remarquerons tout d'abord que la résolution d'un système de n équations linéaires à n inconnues peut très facilement se ramener à la résolution d'un système de n équations linéaires symétriques à n inconnues.

Considérons en effet le système suivant :

$$\text{I}\left\{\begin{array}{lcr} \alpha_1^1\gamma_1 + \alpha_2^1\gamma_2 + \alpha_3^1\gamma_3 + & \cdots & +\alpha_n^1\gamma_n + C_1 = 0 \\ \alpha_1^2\gamma_1 + \alpha_2^2\gamma_2 + \alpha_3^2\gamma_3 + & \cdots & +\alpha_n^2\gamma_n + C_2 = 0 \\ \dots\dots\dots\dots & \dots & \dots\dots\dots\dots \\ \alpha_1^n\gamma_1 + \alpha_2^n\gamma_2 + & \cdots & +\alpha_n^n\gamma_n + C_n = 0. \end{array}\right.$$

Effectuons la transformation linéaire représentée par le système :

$$\text{II}\left\{\begin{array}{l} \gamma_1 = \alpha_1^1\lambda_1 + \alpha_1^2\lambda_2 + \cdots + \alpha_1^n\lambda_n \\ \gamma_2 = \alpha_2^1\lambda_1 + \alpha_2^2\lambda_2 + \cdots + \alpha_2^n\lambda_n \\ \dots\dots\dots\dots\dots\dots\dots\dots \\ \gamma_n = \alpha_n^1\lambda_1 + \alpha_n^2\lambda_2 + \cdots + \alpha_n^n\lambda_n. \end{array}\right.$$

Le système d'équations I donnant les n inconnues γ se trouve remplacé par le système III donnant les n inconnues λ permettant, à l'aide de II, de calculer les valeurs des γ.

$$\text{III}\left\{\begin{array}{l} A_1^1\lambda_1 + A_2^1\lambda_2 + \cdots + A_n^1\lambda_n + C_1 = 0 \\ A_1^2\lambda_1 + A_2^2\lambda_2 + \cdots + A_n^2\lambda_n + C_2 = 0 \\ \cdots\cdots\cdots\cdots\cdots\cdots\cdots\cdots\cdots\cdots \\ A_1^n\lambda_1 + A_2^n\lambda_2 + \cdots + A_n^n\lambda_n + C_n = 0. \end{array}\right.$$

On a d'une façon générale

$$\text{IV)} \qquad \begin{array}{l} A_p^p = \sum_{k=1}^{k=n} (\alpha_k^p)^2 \\ A_p^q = \sum_{k=1}^{k=n} \alpha_k^p \alpha_k^q. \end{array}$$

Le coefficient A_p^q est obtenu en faisant le produit des coefficients des lignes p et q du système I qui se trouvent dans la même colonne et en faisant la somme des produits ainsi obtenus dans les n colonnes, ce qu'on peut exprimer symboliquement en disant que A_p^q est le produit de la ligne p par la ligne q.

L'ordre des facteurs pouvant être inversé dans chaque produit, on voit immédiatement que

$$A_p^q = A_q^p$$

dans le déterminant du système III les termes symétriques par rapport à la diagonale sont égaux, autrement dit le système d'équations aux λ est symétrique.

Proposons-nous donc de résoudre un système d'équations de la forme III.

Remarquons d'après ce qui précède que le système d'équations II si l'on y supposait les γ connus serait un système d'équations aux λ équivalent au système III. On aurait donc un moyen de résoudre le système III si l'on pouvait trouver un système I permettant de calculer facilement les γ.

C'est ce qui arrive si dans le système I

la première équation contient seulement γ_1

la 2ème γ_1 et γ_2

la 3ème γ_1, γ_2 et γ_3

ainsi de suite. On peut en effet calculer ainsi tous les γ successivement à partir de γ_1.

Le problème est donc ramené à la recherche du système

$$\text{V)}\left\{\begin{array}{ll} \alpha_1^1\gamma_1 & +C_1 = 0 \\ \alpha_1^2\gamma_1 + \alpha_2^2\gamma_2 & +C_2 = 0 \\ \alpha_1^3\gamma_1 + \alpha_2^3\gamma_2 + \alpha_3^3\gamma_3 & +C_3 = 0 \\ \cdots\cdots\cdots\cdots\cdots\cdots & \cdots\cdots \\ \alpha_1^n\gamma_1 + \alpha_2^n\gamma_2 + \alpha_3^n\gamma_3 + \cdots + \alpha_n^n\gamma_n & +C_n = 0. \end{array}\right.$$

Ce système étant en effet trouvé le problème devient très facile, puisque le système II est remplacé par le système VI qui permet de calculer les λ de proche en proche à partir de λ_n.

$$
\text{VI)}\left\{\begin{array}{llllll}
\alpha_1^1\lambda_1+ & \alpha_1^2\lambda_2+ & \cdots & \cdots & +\alpha_1^n\lambda_n-\gamma_1=0 \\
 & \alpha_2^2\lambda_2+ & \alpha_2^3\lambda_3 & \cdots & +\alpha_2^n\lambda_n-\gamma_2=0 \\
 & & \alpha_3^3\lambda_3 & \cdots & +\alpha_3^n\lambda_n-\gamma_3=0 \\
 & & \cdots & \cdots & \cdots\cdots\cdots\cdots\cdots \\
 & & & & \alpha_n^n\lambda_n-\gamma_n=0.
\end{array}\right.
$$

Nous calculerons facilement les coefficients α en partant des coefficients A du système III, en appliquant les relations générales IV) au système V. On voit ainsi qu'on peut calculer ligne par ligne tous les coefficients du système VI

$1^{\text{ère}}$ *ligne*

$$
\left\{\begin{array}{lll}
A_1^1=(\alpha_1^1)^2 & \text{d'où} & \alpha_1^1=\sqrt{A_1^1} \\
A_2^1=\alpha_1^1\alpha_1^2 & & \alpha_1^2=\dfrac{A_2^1}{\alpha_1^1} \\
\cdots\cdots\cdots\cdots & & \\
A_p^1=\alpha_1^1\alpha_1^p & & \alpha_1^p=\dfrac{A_p^1}{\alpha_1^1} \\
A_n^1=\alpha_1^1\alpha_1^n & & \alpha_1^n=\dfrac{A_n^1}{\alpha_1^1}
\end{array}\right.
$$

$2^{\text{ème}}$ *ligne*
les α_1 sont déjà connus par le calcul de la $1^{\text{ère}}$ ligne

$$
\left\{\begin{array}{ll}
A_2^2=(\alpha_1^2)^2+(\alpha_2^2)^2 & \alpha_2^2=\sqrt{A_2^2-(\alpha_1^2)^2} \\
A_3^2=\alpha_1^2\alpha_1^3+\alpha_2^2\alpha_2^3 & \alpha_2^3=\dfrac{A_3^2-\alpha_1^2\alpha_1^3}{\alpha_2^2} \\
A_p^2=\alpha_1^2\alpha_1^p+\alpha_2^2\alpha_2^p & \alpha_2^p=\dfrac{A_p^2-\alpha_1^2\alpha_1^p}{\alpha_2^2}
\end{array}\right.
$$

$p^{\text{ème}}$ *ligne*
Tous les α dont l'indice inférieur est plus petit que p sont connus par le calcul des lignes précédentes.

$$
\left\{\begin{array}{l}
A_p^p=(\alpha_1^p)^2+(\alpha_2^p)^2+(\alpha_3^p)^2+\cdots+(\alpha_{p-1}^p)^2+(\alpha_p^p)^2 \\
\qquad \alpha_p^p=\sqrt{A_p^p-(\alpha_1^p)^2-(\alpha_2^p)^2-(\alpha_3^p)^2-\cdots-(\alpha_{p-1}^p)^2} \\
A_p^q=\alpha_1^p\alpha_1^q+\alpha_2^p\alpha_2^q+\alpha_3^p\alpha_3^q+\cdots+\alpha_{p-1}^p\alpha_{p-1}^q+\alpha_p^p\alpha_p^q \\
\qquad \alpha_p^q=\dfrac{A_p^q-\alpha_1^p\alpha_1^q-\alpha_2^p\alpha_2^q-\cdots-\alpha_{p-1}^p\alpha_{q-1}^q}{\alpha_p^p} \\
q>p
\end{array}\right.
$$

Quant au calcul des γ il s'effectue facilement à l'aide des équations V. On obtient

$$\begin{aligned}(-\gamma_1) &= \frac{C_1}{\alpha_1^1}\\ (-\gamma_2) &= \frac{C_2 - \alpha_1^2(-\gamma_1)}{\alpha_2^2}\\ \cdots \quad & \cdots \quad \cdots\cdots\cdots\cdots\\ (-\gamma_p) &= \frac{C_p - \alpha_1^p(-\gamma_1) - \alpha_2^p(-\gamma_2) - \cdots}{\alpha_p^p}\end{aligned}$$

ce qui montre que les coefficients $(-\gamma)$ qui figurent dans le tableau des équations VI) se calculent par rapport aux termes constants C du tableau III exactement de la même façon que les coefficients α par rapport aux A.

Les calculs peuvent être disposés d'une façon commode en un seul tableau. Les équations données étant symétriques, il suffit d'écrire dans le tableau les coefficients d'un seul côté de la diagonale, en-dessus par exemple.

Les équations transformées du système VI peuvent alors être disposées en dessous de la diagonale symétriquement placées par rapport aux équations données, chaque nouvelle équation occupant une colonne en dessous de la diagonale.

Les écritures se bornent à la transcription des coefficients α ; en effet le calcul d'un coefficient α de la forme $\frac{\sum mn}{K}$ est effectué sur la machine à calculer à l'aide d'une série de multiplications qui s'ajoutent automatiquement et algébriquement sur la machine, la somme algébrique étant immédiatement divisée par K. On utilise ainsi l'opération la plus compliquée que puisse faire la machine à calculer, par suite on obtient le rendement maximum de cette machine, tout en se mettant autant que possible à l'abri des erreurs [*rayé et remplacé par "fautes"*] fréquentes dans les transcriptions de chiffres lus.

On trouve, tant dans la disposition des calculs, que dans l'emploi de la machine à calculer, des simplifications pour l'application des formules ou des indications précieuses. Donnons un seul exemple : les machines à calculer du genre "Dactyle" enregistrent le quotient en chiffres blancs ou rouges suivant que la division est faite en tournant dans le sens de l'addition ou de la soustraction, il s'ensuit que tout coefficient α étant le résultat d'une division sur la machine, son signe est indiqué par la couleur des chiffres qui le composent ; les erreurs de signe sont ainsi facilement évitées.

Il paraît inutile d'insister sur le reste de la résolution, c'est à dire sur le calcul des λ à l'aide du système VI). On voit en effet immédiatement comment on peut remonter successivement de λ_n à λ_{n-1}, puis à λ_{n-2} et ainsi de suite jusqu'à λ_1.

On peut mettre en évidence les avantages de cette méthode de résolution des systèmes linéaires, au point de vue de l'approximation avec laquelle les résultats sont obtenus.

Toute méthode de résolution doit nécessairement conduire à un système d'équations du genre du système VI permettant d'obtenir directement une des inconnues et de déterminer successivement toutes les autres. Supposons qu'on ait été conduit au système :

$$\text{VII}\begin{cases} \delta_1^1\lambda_1 + \ \delta_1^2\lambda_2 + \ \cdots \ +\delta_1^n\lambda_n - \varepsilon_1 = 0 \\ \qquad\quad \delta_2^2\lambda_2 + \ \cdots \ +\delta_2^n\lambda_n - \varepsilon_2 = 0 \\ \qquad\qquad\qquad \cdots \ \cdots\cdots\cdots\cdots \\ \qquad\qquad\qquad\qquad \delta_n^n\lambda_n - \varepsilon_n = 0 \end{cases}$$

on peut faire correspondre à ce système un second système fournissant les valeurs des ε, soit :

$$\text{VIII}\begin{cases} \beta_1^1\varepsilon_1 & +C_1 = 0 \\ \beta_1^2\varepsilon_1 + \beta_2^2\varepsilon_2 & +C_2 = 0 \\ \cdots\cdots\cdots\cdots\cdots\cdots & \cdots\cdots \\ \beta_1^n\varepsilon_1 + \beta_2^n\varepsilon_2 + \cdots + \beta_n^n\varepsilon_n & +C_n = 0. \end{cases}$$

Les formules IV sont alors remplacées par les suivantes :

$$A_p^p = \sum_{k=1}^{k=p}(\beta_k^p\delta_k^p)$$

$$A_p^q = \sum_{k=1}^{k=p}(\beta_k^p\delta_k^q).$$

D'où l'on peut conclure que le système unique des coefficients α que l'on a employé précédemment est remplacé dans tous les autres modes de résolution par un double système de coefficients β et δ tels que l'on a toujours $\beta_p^q\delta_p^q = (\alpha_p^q)^2$.

Or les calculs s'effectuent nécessairement avec une précision limitée et on est amené pour éviter des erreurs [*rayé et remplacé par "fautes"*] et rendre le calcul aussi simple que possible, à calculer tous les nombres employés avec un nombre de décimales fixe. Il en résulte que les nombres α, β, δ sont affectés d'une erreur η dépendant des décimales négligées et indépendante de la grandeur du nombre calculé.

L'emploi de la quantité $(\alpha_p^q)^2$ dans les calculs correspond à l'introduction d'une erreur $2\alpha_p^q\eta$.

L'emploi de la quantité égale $(\beta_p^q\delta_p^q)$ correspond à l'introduction de l'erreur $(\beta_p^q + \delta_p^q)\eta$.

On sait que le produit $\beta_p^q\delta_p^q$ étant constant la somme de ses deux facteurs atteint son minimum lorsqu'ils sont égaux. L'erreur la plus faible que l'on puisse introduire est donc

$$2\alpha_p^q\eta.$$

Il en résulte que le mode de résolution des systèmes linéaires qui vient d'être exposé apparaît comme celui qui fournit la meilleure approximation des calculs.

Cette propriété qui permet de réduire au strict minimum le nombre de décimales à employer dans les calculs, compense largement le petit inconvénient d'employer la racine carrée dans la résolution d'équations linéaires.

D'autant plus que la racine carrée peut être obtenue facilement et rapidement à l'aide de la machine à calculer par le procédé suivant complètement différent des procédés généralement indiqués par les fabriquants de machines à calculer.

Soit à extraire la racine carrée d'un nombre N.

Supposons qu'on connaisse un nombre n voisin de la racine r cherchée.

Soit pour fixer les idées

$$\begin{aligned} r &= n+\varepsilon \\ N &= r^2 = (n+\varepsilon)^2 = n^2 + 2n\varepsilon + \varepsilon^2. \end{aligned}$$

Si ε^2 est d'un ordre inférieur à la dernière décimale que l'on veut calculer, on a le droit d'écrire

$$N = n^2 + 2n\varepsilon = n(n+2\varepsilon)$$

c'est à dire qu'en divisant N par n on a pour quotient

$$n + 2\varepsilon.$$

2ε représente l'excès de ce quotient sur le diviseur n et l'on obtient r en ajoutant à n la moitié de cet excès.

Pratiquement, il est avantageux d'avoir à sa disposition une table de carrés qui donne à première vue la racine carrée d'un nombre quelconque avec 3 chiffres significatifs exacts.

$\dfrac{\varepsilon}{n}$ est alors inférieur à $\dfrac{1}{10^2}$

$\dfrac{\varepsilon^2}{n^2}$ $\qquad\qquad\dfrac{1}{10^4}$.

La première division donne la racine carrée avec 5 chiffres significatifs.

$\dfrac{\varepsilon^4}{n^4}$ est inférieur à $\dfrac{1}{10^8}$

donc la 2ème division faite avec la racine à 5 chiffres exacts donnerait 9 chiffres significatifs exacts et ainsi de suite.

On peut énoncer une règle simple en supposant que le nombre comporte une virgule placée à la droite du premier chiffre de gauche. Dans ces conditions chaque division double le nombre de décimales de la racine.

À l'aide de ce procédé, un opérateur exercé obtient en quelques secondes la racine carrée d'un nombre de 5 chiffres avec le même nombre de chiffres exacts.

La méthode de résolution des systèmes linéaires qui vient d'être exposée a été complétée par l'adaptation du système de vérification indiqué par Gauss sous le nom de preuve par sommes.

La vérification est obtenue de la façon suivante : On juxtapose au terme constant C_p de l'équation de rang p un terme V_p donné par la relation $-V_p = A_1^p + A_2^p + \cdots + A_n^p + C_p$.

Dans ces conditions la somme des nombres inscrits dans la ligne p du système des équations est nulle. Si l'on traite V_p dans la résolution de la même façon que C_p, cette relation linéaire se maintiendra et sera encore vraie pour les coefficients α.

De plus il sera encore possible de vérifier le calcul des λ à partir du système des équations VI), car si l'on remplace dans les équations III les termes constants C par les termes de vérification V, l'opération équivaut à changer λ en $(1-\lambda)$; le calcul des inconnues fait avec les V donne donc des valeurs λ' telles que $\lambda_p + \lambda'_p = 1$.

On peut en opérant comme il vient d'être dit réussir à coup sûr et en peu de temps la résolution de systèmes d'équations très complexes.

La résolution d'un système de 10 équations à 10 inconnues peut être faite avec 5 chiffres exacts, en 4 à 5 heures, y compris la vérification des équations et le calcul des résidus.

On a résolu par cette méthode plusieurs systèmes dépassant 30 équations, et en particulier un système de 56 équations. Ce dernier cas fait partie d'un calcul de compensation des altitudes des chaînes primordiales de la triangulation de l'Algérie. En raison de l'importance des calculs et pour éviter l'encombrement, on a dû adopter une disposition spéciale, mais les calculs ont été conduits exactement comme il vient d'être dit.

Vincennes le 2 décembre 1910

Signature

Appendix C

Cours de Calcul Graphique

COURS DE CALCUL GRAPHIQUE

—

TABLE DES MATIÈRES

—

N° 1. – Définition du Calcul graphique. Son utilité. – *Le Calcul graphique a pour objet de remplacer les calculs numériques par un dessin, une sorte d'épure, dont la construction permet de passer directement des données au résultat. L'épure employée est généralement désignée sous le nom d'Abaque.*

L'exécution d'une telle épure est souvent compliquée et demande d'autant plus de soin que l'on désire une précision plus grande. Aussi n'a-t-on pas en général intérêt à chercher à résoudre graphiquement un cas isolé, pour lequel la construction du graphique demanderait le plus souvent beaucoup plus de temps que le calcul numérique.

Au contraire, les procédés du Calcul graphique deviennent très avantageux lorsqu'il s'agit d'un calcul qui se reproduit très fréquemment, de l'application d'une formule dans laquelle les données seules varient. Le dessin peut alors être disposé de façon à fournir à l'aide d'opération simples les résultats correspondant à tous les systèmes de valeurs de variables. L'établissement de l'épure peut dans ce cas être très long, il n'en résulte pas moins une économie de temps très sensible, si chaque fois qu'on s'en sert pour une opération très fréquente, on gagne une partie notable de la durée du calcul numérique qui se trouve ainsi supprimé.

Le Calcul graphique s'applique donc principalement à des calculs qui doivent être répétés très fréquemment.

Il est inutile d'insister sur l'intérêt que présente toute réduction dans la durée des calculs, cette réduction se traduisant toujours par une économie de temps et par suite d'argent ; c'est pour cette raison qu'on a cherché à réduire la durée des calculs numériques en calculant à l'avance des *Tables numériques* constituant absolument l'équivalent des épures employées dans le calcul graphique. L'avantage de ces dernières et que généralement elles sont plus faciles et moins longues à établir ; de plus leur emploi est moins pénible pour le calculateur.

N° 2. – Division du Cours. – Le cours comprendra chapitres :

Chapitre I. Généralités

N° 3. – Représentation graphique des nombres. Module. – Un nombre est toujours le résultat de la comparaison d'une grandeur à une autre grandeur prise pour unité.

Le nombre est indépendant de la grandeur qui a servi à l'obtenir, aussi peut-on lui faire correspondre une autre grandeur convenablement choisie.

On peut même dire qu'à un nombre quelconque on peut faire correspondre une grandeur quelconque à la condition de choisir d'une façon convenable l'unité qui sert à mesurer cette grandeur représentative : cette unité est désignée en général sous le nom de *Module.*

Considérons par exemple une droite indéfinie $x'x$ et convenons que l'unité correspond à la longueur AB prise sur cette droite et dans le sens de x' vers x.

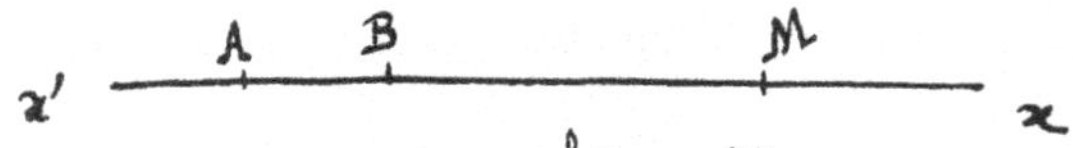

figure N: 1

Un nombre quelconque N pourra être représenté par une longueur AM prise sur $x'x$ et qui contient autant de fois AB que le nombre N contient de fois l'unité.

$$AM = 4\,AB \text{ représente le nombre 4.}$$

D'une façon générale, en représentant le module AB par M, le module N sera représenté par la longueur $L = MN$.

Inversement, si l'on se donne à priori la longueur AM, et si l'on veut qu'elle représente le nombre 4, il faudra choisir AB de façon que cette longueur unité soit contenue 4 fois dans AM.

En général pour qu'une longueur L donnée représente le nombre N, il faut choisir pour module la longueur M donnée par l'équation $M = L/N$.

N° 4. – Importance du choix du module pour la précision à obtenir. – La longueur L qui représente le nombre N est proportionnelle à M. Un même nombre peut donc être représenté par une quantité d'autant plus grande que le module choisi est lui-même plus grand. Or, l'erreur que l'on commet en mesurant des longueurs relativement courtes (comme celles qu'on peut avoir à mesurer sur une épure, ou sur un abaque) est sensiblement constante et dans tous les cas elle croît beaucoup moins rapidement que la longueur mesurée. *La précision de la représentation d'un nombre croît donc avec le module de cette représentation.*

On conçoit d'ailleurs facilement qu'il soit possible de choisir le module de telle façon qu'on obtienne pour la représentation des nombres une précision déterminée et fixée à l'avance, d'après l'approximation que l'on veut obtenir dans les calculs.

N° 5. – Différents modes de représentation graphique des nombres. – Les éléments simples auxquels peut se ramener toute la géométrie plane sont

1° la distance de deux points

2° l'angle de deux directions.

Ces deux éléments peuvent être employés pour la représentation graphique des nombres.

On a vu au n° 3 comment on peut représenter les nombres par des longueurs comptées sur une droite, ou bien par les distances des différents points de la droite à un point choisi comme origine.

On peut remplacer la droite par une courbe quelconque. Un nombre est alors représenté par la longueur de l'arc de courbe correspondant, cette longueur étant proportionnelle au nombre. Il ne faut pas oublier que cette longueur représentative doit être comptée en suivant la courbe et non pas en ligne droite d'un point de la courbe à l'autre.

Les angles conduisent à une représentation graphique assez simple, si l'on admet que *chaque nombre est représenté par un angle proportionnel compté à partir d'une direction fixe.* Il ne saurait y avoir aucune limitation des nombres représentés par le fait que le tour d'horizon est limité à 4 angles droits ; on peut toujours admettre en effet que, pour passer d'une direction à une autre en tournant autour d'un point O, on effectue tout d'abord un nombre entier quelconque de révolutions complètes.

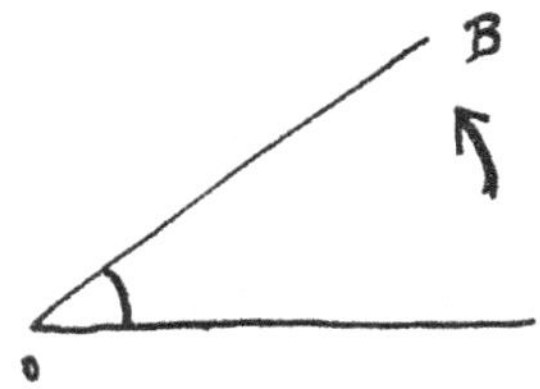

figure 2

Par exemple l'angle des deux directions OA, OB, compté de OA vers OB dans le sens de la flèche peut correspondre à 40 grades, ou à 440 grades, si on décrit un tour complet, ou à 840 grades si on décrit 2 tours complets, 1240 grades si on décrit 3 tours complets, etc. Il suffit donc, pour pouvoir représenter tous les nombres par des angles, d'avoir un moyen de distinguer l'intervalle auquel appartient le nombre représenté : par exemple il suffit de savoir si l'angle représentatif est compris entre 0 et 400, 400 et 800, 800 et 1200, etc., par période correspondant un tour complet.

N° 6. – Anamorphose de Mr Lalanne. – Jusqu'ici on a supposé que chaque nombre était représenté par une quantité proportionnelle. On obtient une représentation beaucoup plus générale en admettant qu'on fait correspondre au nombre N une longueur L donnée par l'équation $L = Mf(N)$, f étant une fonction de forme absolument quelconque.

C'est le principe de l'anamorphose de Mr Lalanne.

Un simple exemple suffira à établir dès maintenant l'importance de l'anamorphose.

Supposons qu'on prenne $f(N) = N^2$, c'est-à-dire que chaque nombre soit représenté par une quantité proportionnelle à son carré. Si on ajoute les longueurs représentatives de deux nombres N et N', on obtient une longueur proportionnelle à $N^2 + N'^2$ et qui correspond dans la représentation adoptée à $\sqrt{N^2 + N'^2}$.

Le calcul d'une quantité de la forme

$$\sqrt{N_1^2 + N_2^2 + N_3^2 + \cdots + N_k^2}$$

se trouve donc réduit à une simple addition.

N° 7. – Graduations linéaires. – Si l'on se bornait à admettre que tout nombre N doit être représenté par une longueur L donnée par la formule $L = Mf(N)$ comptée sur une courbe C, il n'en résulterait aucune simplification, puisque pour chaque nombre N on devrait faire le calcul numérique de la fonction f, de manière à avoir L. Il faut rendre la représentation effective, de façon qu'on ait en quelque sorte une image de la fonction f.

Supposons qu'à partir du point O choisi comme origine on porte les longueurs

$$\begin{aligned} L_1 &= Mf(1) \\ L_2 &= Mf(2) \\ &\cdots\cdots \\ L_n &= Mf(n) \\ &\cdots\cdots \end{aligned}$$

correspondant respectivement à tous les nombres entiers.

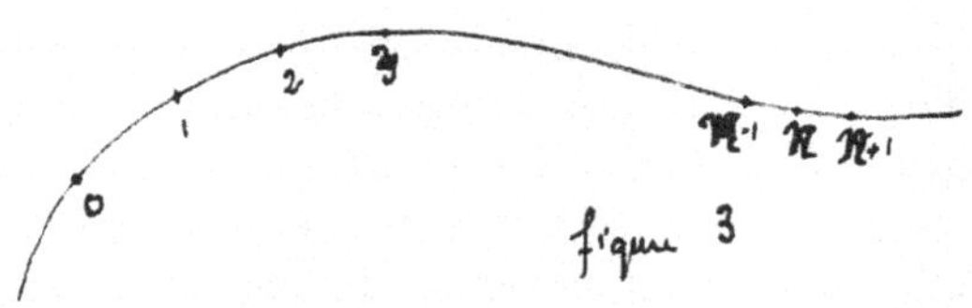

En marquant d'un trait l'extrémité de chacune des longueurs L_1, L_2, L_3, etc., et en inscrivant en regard de ce trait le nombre (1, 2, 3, etc.) auquel il correspond, on obtient une graduation linéaire correspondant à la représentation adoptée et qui peut remplacer complètement la fonction f.

Lorsque la fonction f se réduit au nombre à représenter, c'est-à-dire si l'on a $f(N) = N$, la graduation présente le maximum de simplicité. La distance comptée entre deux traits consécutifs est la même dans toute l'étendue de la graduation.

En effet

$$\begin{aligned} L_{n+1} &= M(n+1) \\ L_n &= Mn \\ \text{d'où} \qquad L_{n+1} - L_n &= M\left[(n+1) - n\right] = M. \end{aligned}$$

La distance de deux traits correspondant à des nombres entiers consécutifs est toujours égale au module.

Au contraire dans le cas de l'anamorphose, la distance de deux traits consécutifs varie avec n, c'est-à-dire avec la région de la graduation que l'on considère :

$$L_{n+1} - L_n = M\left[f(n+1) - f(n)\right].$$

N° 8. – Interpolation graphique. – Pour qu'une graduation remplisse complètement son but, il faut que l'on puisse y représenter un nombre quelconque différent de ceux qui correspondent aux traits de la graduation.

Il faut également qu'on puisse déterminer à quel nombre correspond un point quelconque de la graduation, en dehors des traits qui y sont normalement tracés.

C'est le but de l'opération appelée *interpolation graphique.*

Considérons tout d'abord la graduation correspondant à L = MN et proposons-nous de déterminer le point correspondant au nombre N.

N est compris entre 2 nombres entiers. Soient n et $n+1$ ces deux entiers. On a $n < \mathrm{N} < n+1$.

Par suite, sur la graduation, le point représentatif de N sera compris entre les traits correspondant à n et $n+1$.

On a, à partir de l'origine :

pour le nombre n	$\mathrm{OA} = \mathrm{L}_n = \mathrm{M}n$
.... N	$\mathrm{OK} = \mathrm{L_N} = \mathrm{MN}$
.... $n+1$	$\mathrm{OB} = \mathrm{L}_{n+1} = \mathrm{M}(n+1)$.

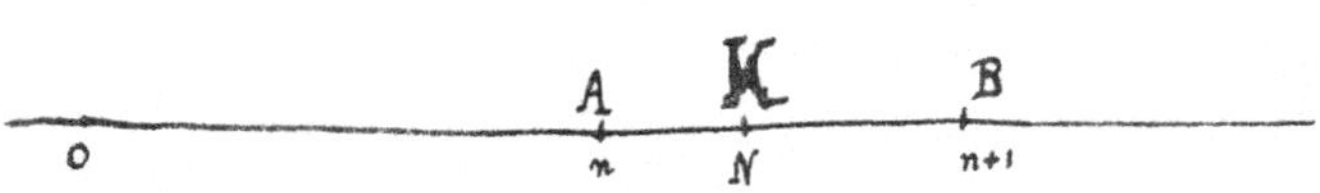

Figure 4

(On a représenté une droite comme support de la graduation. Le raisonnement serait le même avec une courbe quelconque à condition de compter la longueur sur la courbe elle-même.)

On a donc

$$\mathrm{AK} = \mathrm{M}(\mathrm{N} - n)$$
$$\mathrm{KB} = \mathrm{M}(n + 1 - \mathrm{N})$$
$$\mathrm{AB} = \mathrm{M}$$
$$\text{ou} \qquad \frac{\mathrm{AK}}{\mathrm{N} - n} = \frac{\mathrm{KB}}{n + 1 - \mathrm{N}} = \frac{\mathrm{AB}}{1} = \mathrm{M}.$$

Ces équations permettent de résoudre toutes les questions auxquelles peut donner lieu l'emploi de la graduation.

Si l'on veut déterminer le point K qui correspond à un nombre N donné on écrira

$$\mathrm{AK} = (\mathrm{N} - n)\,\mathrm{AB},$$

ce qui exprime qu'on obtient K en partageant AB dans le rapport de $(\mathrm{N} - n)$ à l'unité.

Si l'on cherche le nombre N qui correspond à un point M donné sur la graduation, on écrira

$$\mathrm{N} - n = \frac{\mathrm{AK}}{\mathrm{AB}}.$$

Le rapport AK/AB exprime donc la fraction d'unité qui forme l'excédent de N sur l'entier immédiatement inférieur n.

Dans le cas de l'anamorphose, il semble tout d'abord impossible d'employer le même procédé. En effet, les équations précédemment employées doivent être remplacées par les suivantes :

$$\frac{\mathrm{AK}}{f(\mathrm{N}) - f(n)} = \frac{\mathrm{KB}}{f(n+1) - f(\mathrm{N})} = \frac{\mathrm{AB}}{f(n+1) - f(n)} = \mathrm{M}.$$

Il serait donc nécessaire, en toute rigueur, de calculer $f(\mathrm{N})$ pour connaître AK et KB.

En réalité on s'arrange toujours de manière que, dans l'intervalle de deux traits consécutifs de la graduation, la variation de la fonction f puisse être considérée comme proportionnelle à la variation de N. On se trouve ainsi ramené au cas précédent et les différents problèmes qui peuvent se poser sont résolus de la même façon.

S'il ne suffit pas pour arriver à ce résultat de faire correspondre l'intervalle de 2 traits consécutifs de la graduation à l'unité, on augmentera le nombre de ces traits en faisant correspondre l'intervalle à une fraction convenablement choisie, c'est-à-dire suffisamment petite pour que dans chaque intervalle on puisse considérer la fonction $f(\mathrm{N})$ comme simplement proportionnelle à N.

Dans le cas où l'espacement AB des traits de la graduation est assez considérable (pratiquement dès que AB dépasse 5 millimètres, si l'on veut se mettre à l'abri d'erreurs graphiques atteignant 1/4 de millimètre), il est nécessaire de calculer exactement la longueur AK qui donne la position du point correspondant au nombre N.

Dans ces conditions la graduation ne remplace qu'imparfaitement les calculs numériques et il vaut mieux avoir recours à une construction géométrique.

Soit une graduation tracée sur la droite xx' (figure 5) et sur laquelle il s'agit de construire le point correspondant au nombre 3,453.

Supposons que par les traits 3 et 4 de la graduation on ait au préalable tracé 2 droites parallèles yy'.

Plaçons un double décimètre de manière que le trait 0 soit en A′ et le trait 10 en B′. Si par le point 4 centimètres 53 (en appréciant le 1/10e de millimètre), on fait passer une parallèle à yy', on obtient en K le point correspondant à 3,453. En effet on a partagé l'intervalle des traits 3 et 4 dans le rapport 4,53/10.

Pratiquement, pour éviter toute construction, on place par tâtonnement le décimètre dans la position AB telle que le point 4,53 soit exactement sur xx', axe de la graduation.

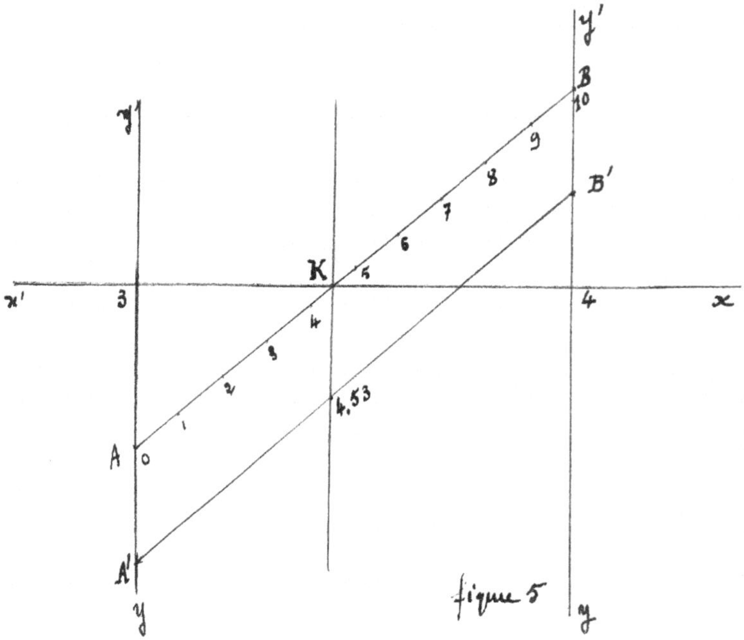

figure 5

Inversement, si l'on veut savoir à quel nombre correspond le point K de la graduation, on placera le décimètre de façon qu'il passe par K, et que, en même temps, les traits 0 et 10 soient sur les parallèles yy' passant par les traits 3 et 4.

Cette opération pourra être effectuée avec toute autre graduation qu'un décimètre. Il y a souvent avantage à utiliser une petite bande de papier de longueur appropriée à l'intervalle des traits de la graduation et partagée en un nombre convenable de parties.

N° 9. – Interpolation à vue. – La meilleure solution consiste toujours à réduire l'intervalle des traits de la graduation à quelques millimètres. Alors, on évalue à l'œil la fraction de l'intervalle correspondant à un nombre donné. On apprécie facilement 1/10e d'un intervalle variant de 1 à 5 millimètres. C'est ce qu'on appelle l'interpolation à vue. On voit ainsi qu'à la seule condition de réduire suffisamment l'intervalle de deux traits consécutifs, la graduation donne directement la correspondance entre le graphique et les nombres avec toute la précision dont le dessin est susceptible.

N° 10. – Équidistance. Intervalle. – On voit d'après ce qui précède qu'il y a toujours intérêt à ce que deux traits consécutifs d'une graduation soient à une distance de quelques millimètres.

L'intervalle graphique qui sépare deux traits consécutifs dépend de deux quantités :

1° le *module*, dont nous avons déjà vu la définition et qui caractérise en quelque sorte l'échelle du dessin de la graduation ;

2° l'*équidistance.*

Nous désignerons sous le nom d'*équidistance*, le nombre qui correspond à l'intervalle de deux traits consécutifs de la graduation.

Par exemple : si les traits de la graduation correspondent aux nombres entiers, l'équidistance est 1.

Si les traits correspondent aux centaines, 0, 100, 200, 300, etc., l'équidistance est 100.

Si au contraire les traits représentent les nombres de millième en millième, 0,001 - 0,002 - 0,003 - 0,004 - etc., l'équidistance est 0,001, ou 1/1000.

L'*équidistance graphique* n'est autre que l'équidistance réduite à l'échelle de la graduation.

C'est donc le produit de l'équidistance par le module :

$$\mathrm{E}_g = \mathrm{E} \times \mathrm{M}.$$

Exemple : une graduation en centièmes E = 0,01 doit être établie avec le module M = 300 millimètres.

L'équidistance graphique est

$$\mathrm{E}_g = 0{,}01 \times 300 = 3 \text{ millimètres}.$$

Les traits de la graduation seront donc écartés de 3 millimètres de l'un à l'autre.

Dans les graduations proportionnelles, l'écartement constant des traits n'est autre que l'équidistance graphique.

Dans le cas de l'anamorphose, l'écartement variable des traits dépend non seulement de l'équidistance graphique, mais encore de la forme de la fonction $f(\mathrm{N})$ *qui sert à la définition de la graduation.*

D'une façon générale, nous appellerons *intervalle*, I, la différence des deux valeurs de la fonction $f(\mathrm{N})$ pour des valeurs de N différant entre elles de l'équidistance.

L'*intervalle graphique* n'est autre que l'intervalle réduit à l'échelle du dessin, c'est-à-dire le produit de l'intervalle par le module :

$$\mathrm{I}_g = \mathrm{I} \times \mathrm{M}.$$

Par exemple : considérons la graduation définie par l'équation

$$\mathrm{L} = \mathrm{M.N}^2.$$

Comme plus haut, M = 300 millimètres, E = 0,01.

Pour N = 2, l'intervalle sera

$$\mathrm{I}_{(2)} = \left[(2,01)^2 - (2)^2\right]$$

$$\text{ou} \qquad \mathrm{I}_2 = 0,0401$$

et l'intervalle graphique pour la même valeur de N,

$$\mathrm{I}_{g(2)} = \mathrm{I}_2 \times \mathrm{M} = 300 \times 0,0401$$

$$\text{d'où} \qquad \mathrm{I}_{g(2)} = 12 \ {}^{\text{millimètres}}_{,}\ 03.$$

L'intervalle est donc en général variable dans l'étendue de la graduation et il ne se confond avec l'équidistance que dans le cas des graduations proportionnelles, c'est-à-dire lorsqu'on a $f(\mathrm{N}) \equiv \mathrm{N}$.

N° 11. – Graduations angulaires. – Tout ce qui vient d'être dit (depuis le N° 7) au sujet des graduations linéaires s'applique à la représentation des nombres par des angles ; il suffit de remplacer les longueurs par des angles.

La graduation angulaire est constituée par le tracé d'un certain nombre de directions correspondant à des nombres variant d'une quantité égale à l'équidistance. Chaque direction porte l'indication du nombre qui lui correspond.

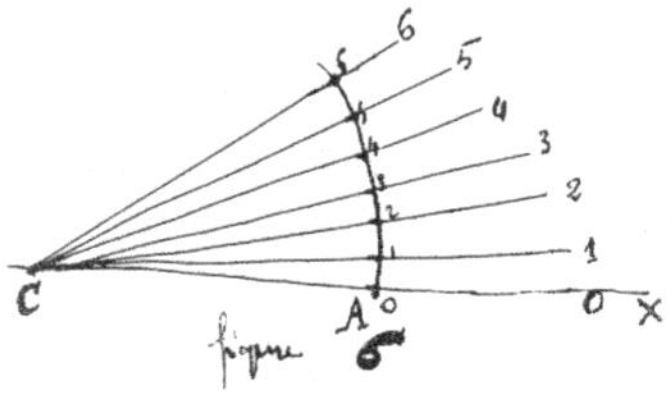

Il convient de remarquer que toute graduation angulaire peut être remplacée par une graduation linéaire qui est équivalente. En effet aux angles comptés à partir d'une direction CX autour de C, correspondent des arcs comptés sur un cercle de centre C à partir du point A de CX. On se trouve même pratiquement obligé d'en revenir à la graduation circulaire dès qu'on veut faire de l'interpolation graphique.

Courbes – Diagrammes – Équations à 2 variables

N° 12. – Représentation des équations par des courbes. – On sait que toute relation entre deux variables, qui peut être désignée dans sa forme la plus générale par

$$\mathrm{F}(x, y) = 0$$

peut être représentée par une courbe plane.

Il suffit pour cela de choisir deux axes de coordonnées Ox et Oy, parallèlement auxquels on compte les valeurs respectives de x et de y.

À chaque valeur de x, x_0 représentée par OA par exemple, on fait correspondre la valeur de y donnée par la résolution de l'équation

$$\mathrm{F}(x_0, y) = 0.$$

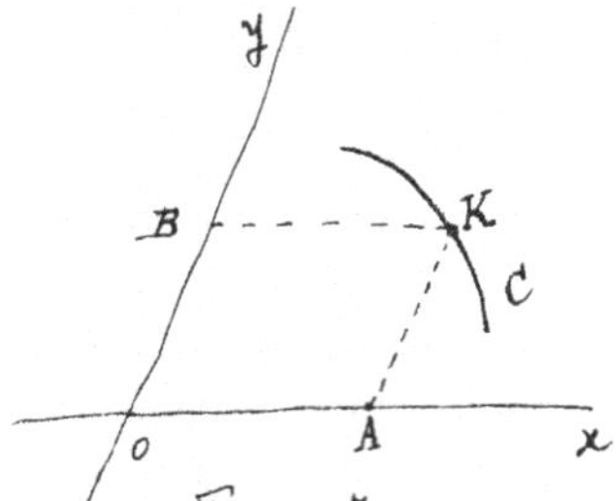

Figure 7

Cette valeur y_0 étant portée suivant AK parallèlement à Oy, le lieu du point K est une courbe C qui est la représentation de l'équation $F(x,y)=0$.

On a

$$\begin{aligned} \mathrm{BK} &= \mathrm{OA}, \quad \text{abscisse du point K} \\ \mathrm{OB} &= \mathrm{AK}, \quad \text{ordonnée} \ldots\ldots\ldots\ldots \end{aligned}$$

On peut dire, par suite, que les distances aux axes (comptées parallèlement à ces axes) de chaque point de la courbe C satisfont toujours à l'équation proposée $F(x,y)=0$.

On désigne souvent la courbe représentative de l'équation $F(x,y)=0$, sous le nom de *diagramme de l'équation* $F(x,y)=0$ ou encore de *diagramme de* y, y étant considéré comme une fonction de x donnée par l'équation représentée.

On s'est borné à dire précédemment que l'abscisse OA et l'ordonnée OB *représentent* les quantités x_0 et y_0 correspondant au point K et soumises à la relation $F(x_0,y_0)=0$.

En réalité, *comme on l'a vu pour les graduations*, il est nécessaire d'introduire les longueurs que l'on veut faire correspondre aux unités des quantités x et y, longueurs que nous avons déjà désignées sous le nom de *modules*.

Soit M_1 le module de x
M_2 le module de y.

L'abscisse OA (Figure 7) a une longueur égale à $X_0 = M_1 x_0$. De même l'ordonnée OB est $Y_0 = M_2\, y_0$.

Pour avoir l'équation qui lie l'ordonnée et l'abscisse de la courbe représentative, il suffira de remplacer x et y respectivement par leurs valeurs $\dfrac{X}{M_1}$ et $\dfrac{Y}{M_2}$ dans l'équation $F(x,y)=0$. On obtient ainsi

$$F\left(\frac{X}{M_1}, \frac{Y}{M_2}\right) = 0,$$

qui est *l'équation exacte du diagramme.*

En résumé x et y sont des *nombres* entre lesquels existe une certaine relation $F(x, y) = 0$.

Pour représenter cette relation nous adoptons respectivement pour x et y les modules M_1 et M_2.

Chaque valeur de x est alors représentée sur le dessin par X unités de longueur (millimètre par exemple).

Chaque valeur de y est représentée par Y unités de longueur et ces deux dernières quantités sont liées par la relation

$$F\left(\frac{X}{M_1}, \frac{Y}{M_2}\right) = 0,$$

qui résulte de la relation entre x et y.

N° 13. – Importance du choix des Modules. – Un simple exemple montrera l'importance du choix des modules.

Supposons qu'on veuille représenter dans l'intervalle de 0 à 100, l'équation

$$y^2 - \frac{x}{100} = 0.$$

On sait que cette équation est celle d'une parabole ayant pour axe Ox, pour sommet O et tangente à Oy.

Pour $x = 100$, on a $y = 1$.

y varie donc de 0 à 1 pendant que x varie de 0 à 100.

Si l'on prend pour module l'unité à la fois pour x et y on aura la courbe suivante :

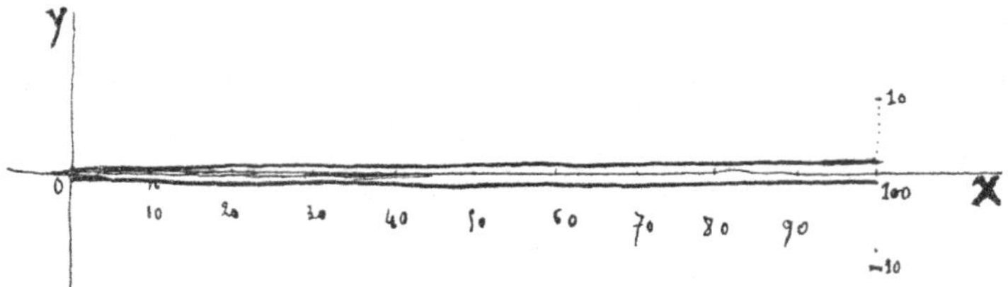

Figure 8

Il est bien évident qu'un tel diagramme ne pourrait être d'aucune utilité, à cause de la faible précision avec laquelle il permet d'obtenir la valeur de y correspondant à une valeur de x.

Supposons au contraire qu'on prenne

$$\begin{array}{ll} \text{pour } x & M_1 = 1 \quad \text{(millimètre)} \\ \text{pour } y & M_2 = 100. \end{array}$$

Les coordonnées du diagramme satisfont alors, comme on l'a vu précédemment, à l'équation

$$\frac{Y^2}{10.000} - \frac{X}{100} = 0$$

$$\text{ou} \qquad Y^2 - 100\,X = 0.$$

Cette équation représente encore une parabole possédant les mêmes propriétés générales que la précédente, mais cette fois, pour X = 100, on a Y = 100.

Y varie de 0 à 100 lorsque X varie de 0 à 100 et l'on obtient le diagramme suivant (limité aux valeurs positives de Y) :

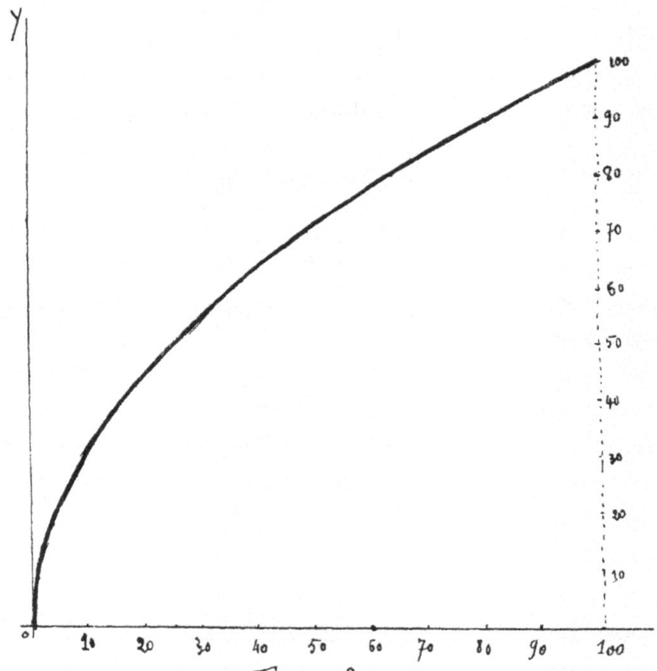

Figure 9.

Il est bien évident que ce diagramme donnera une précision plus grande que le précédent, ainsi pour $x = X = 60$, le diagramme (9) donne Y = 77,4, d'où

$$y = \frac{Y}{100} = 0,774,$$

tandis que le diagramme (8) aurait pu à peine donner l'approximation de $y = 0{,}8$.

Ainsi la précision à attendre de l'emploi d'un diagramme dépend essentiellement du choix des modules.

Il y a lieu de remarquer que les 2 courbes obtenues figures 8 et 9 ont les mêmes caractères, les mêmes particularités. La figure 9 est une simple déformation de la figure 8, les ordonnées étant amplifiées dans un certain rapport, alors que les abscisses restent les mêmes.

N° 14. – Diagrammes anamorphosés. – La déformation ainsi mise en évidence et due au choix des modules n'est qu'un cas particulier d'une déformation beaucoup plus générale résultant de la représentation par l'anamorphose des coordonnées x et y.

Supposons qu'au lieu de porter sur les axes OX et OY les quantités

$$\begin{aligned} X &= M_1\, x \\ Y &= M_2\, y, \end{aligned}$$

on emploie les quantités définies par les équations

$$\begin{aligned} X &= M_1\, \varphi_1(x) \\ Y &= M_2\, \varphi_2(y), \end{aligned}$$

où φ_1 et φ_2 sont des fonctions de x et y respectivement qui peuvent être choisies arbitrairement.

On obtiendra ainsi une courbe qui sera encore une représentation de l'équation $F(x, y) = 0$.

L'équation de la courbe représentative, qui constitue alors *un diagramme doublement anamorphosé*, s'obtient en éliminant x et y entre les 3 équations

$$\begin{cases} X = M_1\, \varphi_1(x) \\ Y = M_2\, \varphi_2(y) \\ F(x, y) = 0. \end{cases}$$

Soit $\Psi(X, Y) = 0$ la relation entre X et Y ainsi obtenue, on voit qu'elle dépend à la fois des modules M_1 et M_2 et de la forme des fonctions φ_1 et φ_2 employées pour l'anamorphose des coordonnées. Il est souvent possible de choisir ces fonctions φ_1 et φ_2 de manière à obtenir un diagramme beaucoup plus simple.

Nous allons montrer immédiatement *la corrélation qu'on peut établir entre les diagrammes et les graduations anamorphosées*, et d'autre part l'emploi de l'anamorphose pour la simplification des diagrammes.

N° 15. – Correspondance entre les graduations et les diagrammes. – *À chaque graduation anamorphosée on peut faire correspondre un diagramme.*

Considérons en effet la graduation définie à l'aide de l'équation $L = Mf(N)$ (voir N° 7).

Prenons deux axes de coordonnées OX et OY. Portons sur OX les valeurs de N et sur OY les valeurs de L. Le lieu des points tels que leurs coordonnées soient liées par l'équation précédente sera alors une courbe qui représente cette équation et est par suite équivalente à la graduation.

Inversement à toute courbe représentant une équation $F(X, Y) = 0$, on peut toujours faire correspondre une graduation qui est équivalente.

En effet cette équation peut être supposée résolue par rapport à Y et mise par suite sous la forme $Y = f(X)$.

Par suite en posant $Y = L$ et $X = N$, l'équation $L = f(N)$ définit une graduation anamorphosée de module 1 équivalente à la courbe proposée.

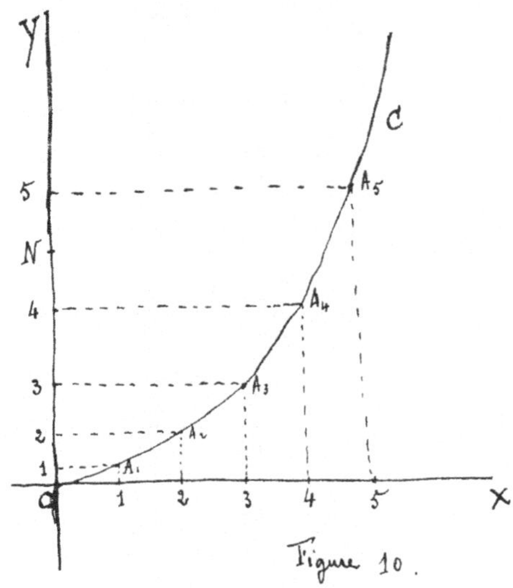

Figure 10.

Étant donné une courbe C rapportée à deux axes de coordonnées OX et OY, constituons sur OX une graduation proportionnelle d'équidistance E et de module 1. (Dans le cas de la figure 10 l'équidistance est égale à l'unité.) Par chaque trait de cette graduation menons l'ordonnée $1A_1$, $2A_2$, $3A_3$, etc. Enfin par les points A_1, A_2, A_3, A_4, etc., de la courbe ainsi obtenue menons des parallèles à OX, nous obtiendrons ainsi sur OY une graduation définie par l'équation $L = f(N)$ de la courbe C.

Cette corrélation des graduations et des diagrammes peut simplifier d'une façon notable la construction d'une graduation, lorsque le diagramme correspondant est une courbe de construction simple et rapide.

Supposons par exemple que l'on veuille construire une graduation correspondant à la relation

$$L = 5 \pm \sqrt{25 - N^2}.$$

En posant $Y = L$, $X = N$, et en rendant l'équation rationnelle, on obtient

$$X^2 + Y^2 - 10\,Y = 0.$$

C'est l'équation d'un cercle de rayon 5, tangent en O à OX. Le tracé de ce cercle permet de construire rapidement la graduation par projection sur OY.

Il résulte de ce qui précède qu'une graduation, qui n'exige l'emploi que d'une ligne, droite par exemple, peut toujours remplacer un diagramme, et par suite suffit toujours pour exprimer une relation entre 2 quantités.

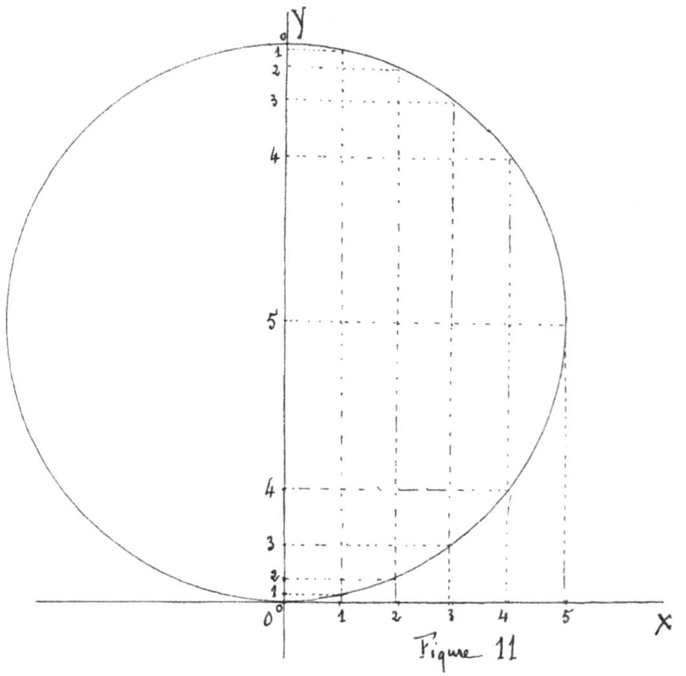

La courbe C (figure 10) représente l'équation F(X, Y) = 0. Si l'on marque sur la graduation de l'axe OY un point correspondant à une certaine valeur X = N par interpolation graphique sur cette graduation, il suffira de mesurer la longueur ON pour avoir la valeur de Y qui correspond à la valeur choisie de X.

En résumé, *toute graduation anamorphosée fournit non seulement une représentation particulière des nombres, mais encore, par comparaison avec une graduation donnant les longueurs, les différentes valeurs de la fonction qui sert à la définition de l'anamorphose.*

Exemple : une graduation anamorphosée est tracée sur l'équation

$$L = M \text{ Log } N.$$

La figure 12 est faite avec des logarithmes ordinaires pour M = 100 mm.

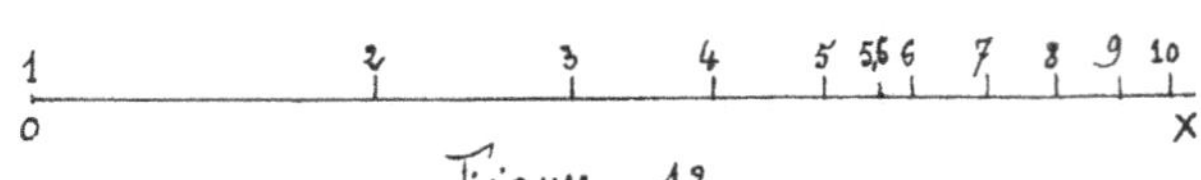

Le logarithme de 1 étant zéro, pour avoir le logarithme d'un nombre N, il suffit de placer ce nombre sur l'échelle et de mesurer la distance du point représentatif au point (0.1).

Le logarithme de 5,6 est approximativement 0,749.

N° 16. – Exemple de simplification d'un diagramme par l'anamorphose. – Considérons l'équation

$$x.y = 10.$$

On sait qu'elle correspond à une hyperbole ayant pour asymptote les axes de coordonnées. La représentation de l'équation proposée sans anamorphose conduira donc au diagramme suivant limité aux nombres positifs :

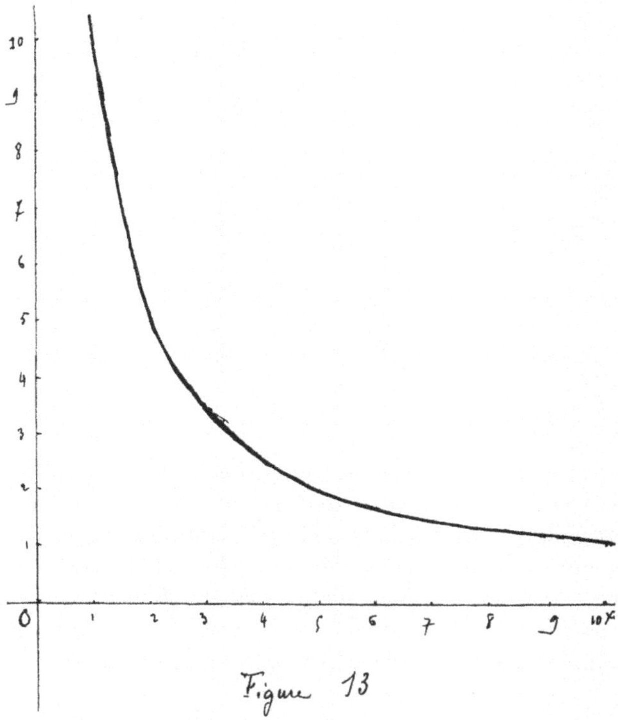

Figure 13

Le tracé d'un tel diagramme n'est pas sans difficulté. On peut fort heureusement le simplifier beaucoup en opérant de la façon suivante :

Représentons x et y par les fonctions logarithmiques

$$\mathrm{X} = \mathrm{Log}\, x$$
$$\mathrm{Y} = \mathrm{Log}\, y$$
$$\mathrm{Log}\, x.y = \mathrm{X} + \mathrm{Y} = \mathrm{Log}\, 10 = 1.$$

L'équation du nouveau diagramme est donc

$$\mathrm{X} + \mathrm{Y} = 1.$$

C'est-à-dire que le nouveau diagramme se réduit à une droite également inclinée sur les axes, à la condition de graduer ces deux axes suivant l'anamorphose logarithmique, c'est-à-dire d'y porter la graduation déjà employée à la figure 12, pour module = 100 millimètres.

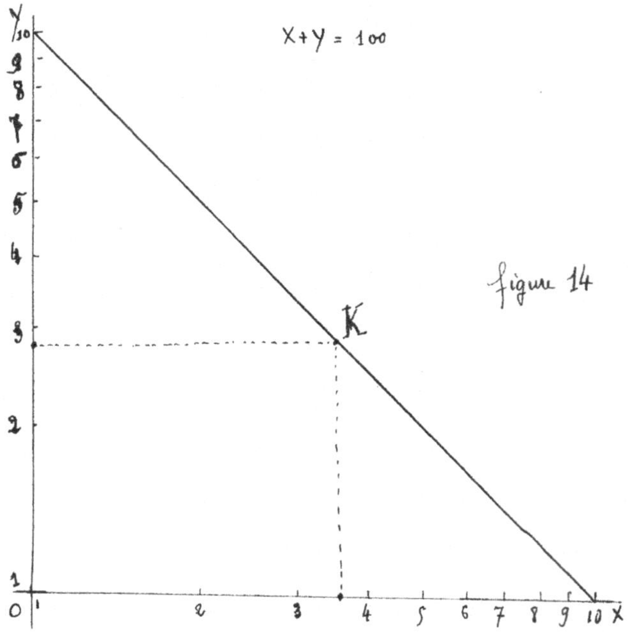

La construction de ce diagramme est donc particulièrement facile. Quant à son emploi, il n'est pas sensiblement plus compliqué que celui du précédent.

Une valeur de x étant donnée, on la place par interpolation graphique (ou à vue suivant les cas) sur la graduation OX, on trace l'ordonnée correspondante qui donne le point K. On mène par K une parallèle à OX et l'on obtient la valeur de y par interpolation sur la graduation de OY.

Les cas de simplification des diagrammes par l'anamorphose sont très fréquents. De là l'importance de ce procédé au point de vue du calcul graphique.

N° 17. – Utilisation des courbes pour la construction des diagrammes anamorphosés. – On peut comme on l'a vu au N° 15, utiliser les courbes pour construire les graduations des axes OX et OY dans le cas d'un diagramme doublement anamorphosé.

Considérons deux axes rectangulaires OX et OY, à l'aide desquels nous nous proposons de représenter la relation $f(x, y) = 0$, x étant représenté par l'anamorphose $X = \varphi_1(x)$, y par $Y = \varphi_2(y)$.

Soit Ox le prolongement de l'axe OY, Oy le prolongement de l'axe OX.

On peut construire le diagramme représentant $X = \varphi_1(x)$ en portant les quantités x et X respectivement sur les directions Ox et OX.

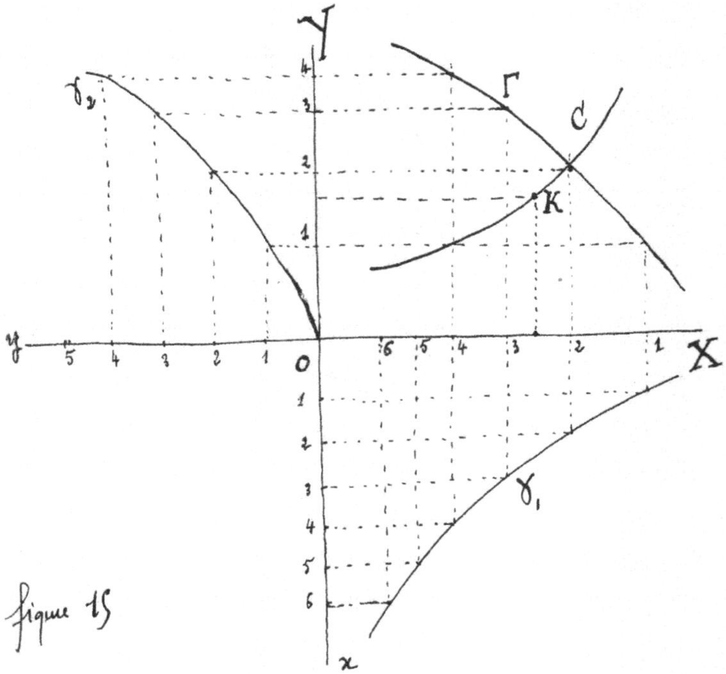

Nous pouvons ainsi faire correspondre, à l'aide de cette courbe γ_1, à la graduation de Ox représentant x sans anamorphose, la graduation OX représentant x à l'aide de l'anamorphose $X = \varphi_1(x)$.

On peut opérer de même pour l'anamorphose de y en rapportant les valeurs de y et Y aux deux axes Oy et OY respectivement.

Lorsqu'on a obtenu les deux graduations OX et OY, tout diagramme doublement anamorphosé tel que C doit <être rapporté> par interpolation à ces deux graduations.

Par exemple à $x = 2,6$ correspond le point K du diagramme C qui donne la valeur $y = 1,7$.

Remarque. – Si on mène par les points de même valeur des deux graduations des parallèles aux axes de coordonnées, on obtient des points pour lesquels on a $x = y$.

Le lieu de ces points n'est autre que le diagramme de cette équation. Ce diagramme, lorsqu'il peut être obtenu simplement, peut remplacer une des deux courbes γ_1 ou γ_2.

C'est la Courbe Γ de la figure 15. Sans anamorphose on sait que l'équation $x = y$ est représentée par la bissectrice de l'agle XOY. La courbe Γ n'est donc autre que l'équivalent de cette droite lorsqu'on emploie la double anamorphose.

N° 18. – Anamorphoses successives. – Il peut arriver qu'on soit conduit à représenter la variable x non plus par l'anamorphose $\mathrm{X} = \varphi_1(x)$, mais par une fonction de X, que nous pouvons désigner en général par $\Psi_1(\mathrm{X})$.

On fait ainsi deux anamorphoses successives.

On peut dire en réalité que la quantité qui représente x peut être mises sous la forme $\Psi[\varphi_1(x)]$ et qu'en calculant directement cette fonction de fonction on est ramené à une anamorphose simple.

Au point de vue pratique, il vaut mieux laisser subsister les 2 transformations. La construction des graphiques est généralement plus simple et on laisse bien en évidence la méthode qui a permis de les obtenir.

On est conduit à faire des anamorphoses successives, lorsque la première anamorphose qui a pour but de simplifier la construction du graphique, ne permet pas d'obtenir une précision suffisante pour certaines valeurs des quantités à calculer.

De même que l'on obtient une graduation anamorphosée à l'aide d'une courbe, on obtiendra facilement la graduation correspondant à 2 anamorphoses successives si l'on peut construire facilement les courbes correspondant aux fonctions $\varphi_1(x)$ et $\Psi_1(\mathrm{X})$.

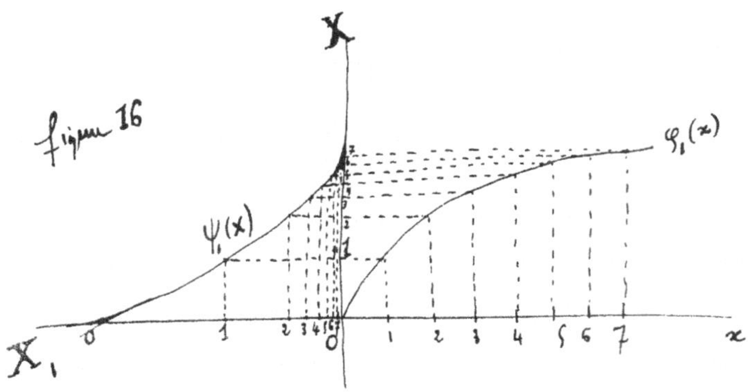

La figure 16 montre comment, en portant sur Ox les nombres x, on obtient sur OX la graduation correspondant à la première anamorphose et de même sur OX_1 la graduation correspondant aux deux anamorphoses successives.

N° 19. – Diagrammes gradués. – Considérons un diagramme C représentant la relation $f(x, y) = 0$ entre les quantités x et y. Lorsqu'on se donne une valeur de x, soit x_0, il y correspond un point K du diagramme, et par suite une valeur y_0 de y.

Considérons d'autre part une 2ème relation $f_1(x, z) = 0$ entre les quantités x et z. Elle fait correspondre à chaque valeur de x une valeur de z.

Inscrivons à côté du point K, la valeur z_0 qui correspond à x_0 et faisons de même pour toutes les valeurs de x.

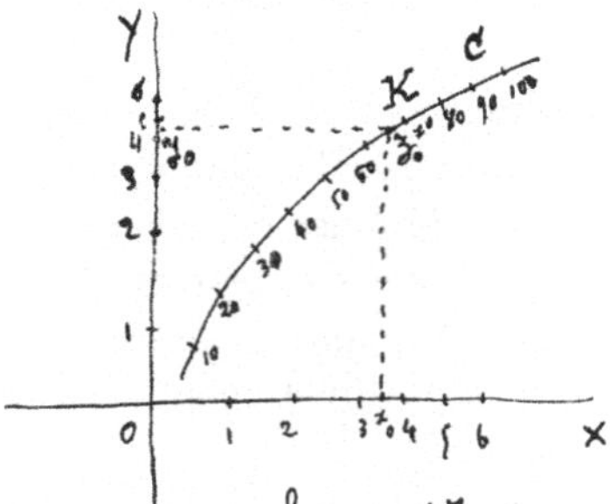

figure 17.

Nous obtiendrons ainsi sur le diagramme C une graduation donnant les valeurs de z, c'est-à-dire un *diagramme gradué.*

À l'aide de ce diagramme gradué, on pourra obtenir les valeurs de y et z qui correspondent à la même valeur de x.

Remarque. – Puisqu'on connaît les valeurs de y et z qui correspondent à une même valeur de x, on a par suite les valeurs de ces deux quantités qui se correspondent réciproquement, c'est-à-dire l'expression de la relation qui existe entre ces deux quantités.

Cette relation s'obtiendrait algébriquement en éliminant x entre les 2 équations

$$\begin{cases} f(x,y) = 0 \\ f_1(x,z) = 0. \end{cases}$$

On voit donc que le diagramme gradué tel qu'il vient d'être défini résout complètement le problème de l'élimination d'une inconnue entre deux équations. *C'est l'élimination graphique.*

N° 20. – Diagrammes conjugués. – On peut arriver au même résultat à l'aide de 2 diagrammes dans lesquels l'inconnue à éliminer est portée sur un axe commun.

Soit à éliminer la quantité x entre les 2 équations

$$\begin{cases} f(x,y) = 0 \\ f_1(x,z) = 0. \end{cases}$$

Représentons $f(x,y) = 0$ par un diagramme rapporté aux deux axes Ox et Oy.

Construisons de même le diagramme de l'équation $f_1(x,z) = 0$ en prenant le même axe Ox et pour axe Oz le prolongement de Oy (on pourrait d'ailleurs prendre le même axe Oy).

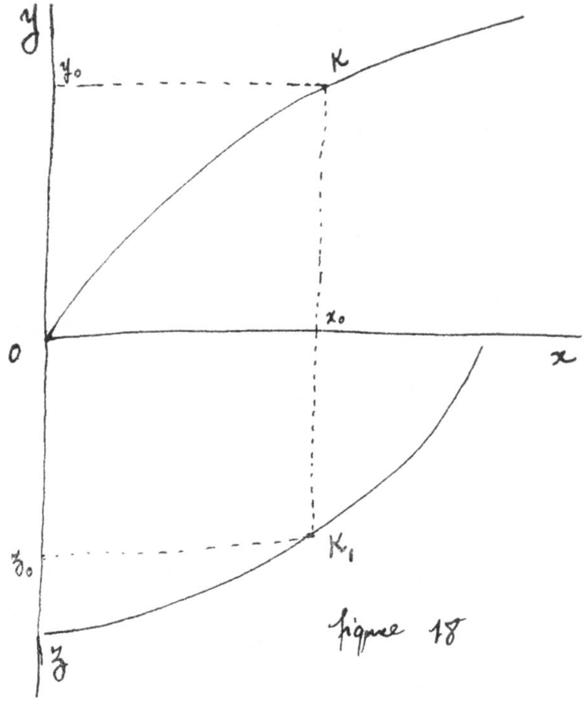

figure 18

Si l'on mène par le point x_0 de Ox une parallèle à zOy, on obtient en K et K_1 les points correspondants des deux diagrammes. Les valeurs y_0 et z_0 qui en résultent sont conjuguées, c'est-à-dire qu'à y_0 correspond forcément z_0 et inversement.

La construction montre d'ailleurs qu'en se donnant y_0 et en traçant les 3 droites y_0K, KK_1 et $\mathrm{K}_1 z_0$, on obtient la valeur correspondant à y_0.

Remarque. – On conçoit qu'on peut ainsi rapporter au même axe Ox plusieurs diagrammes permettant de déterminer autant de quantités en fonction de la même quantité x.

Représentation des surfaces. Équations à 3 variables

N° 21. – Sections planes des surfaces. Plan coté. – De même qu'on a vu au N° 12 qu'à toute équation à 2 variables on peut faire correspondre une courbe rapportée à deux axes de coordonnées, de même à chaque équation à 3 variables on peut faire correspondre dans l'espace une surface rapportée à 3 axes de coordonnées.

Cette surface est la représentation géométrique de la relation qui existe entre les 3 variables.

Soit $f(x, y, z) = 0$ cette relation.

Chaque point pris sur la surface S correspondante a pour coordonnées des quantités x_0, y_0, z_0 qui satisfont à l'équation, c'est-à-dire telles que $f(x_0, y_0, z_0) = 0$.

Inversement tout point tel que ses 3 coordonnées satisfont à l'équation se trouve forcément sur la surface S.

Mais alors qu'il est facile et commode d'utiliser un diagramme tracé sur une feuille de papier et qui peut remplacer une équation algébrique à deux variables, il serait très incommode d'avoir recours à la représentation complète d'une surface dans l'espace. La construction de cette surface présenterait déjà de grosses difficultés ; mais sontemploi serait encore plus compliqué, sinon impraticable.

Il est donc nécessaire de chercher un mode de représentation des surfaces, se réduisant à la forme d'un plan ou dessin, et susceptible d'être utilisé dans les mêmes conditions.

Le procédé le plus simple est le *plan coté*, analogue aux plans topographiques dans lesquels le figuré du terrain est obtenu à l'aide de courbes de niveau.

Considérons la surface S, correspondant à l'équation $f(x, y, z) = 0$ rapportée à 3 axes de coordonnées Ox, Oy et Oz.

Coupons cette surface par un plan parallèle au plan xOy, soit $z = z_0$.

Nous obtenons ainsi une certaine courbe C_0.

Projetons cette courbe sur le plan xOy, en menant par chacun de ses points une parallèle à Oz. Nous obtenons ainsi une courbe égale à C_0 et que nous pourrons appeler "*courbe* z_0".

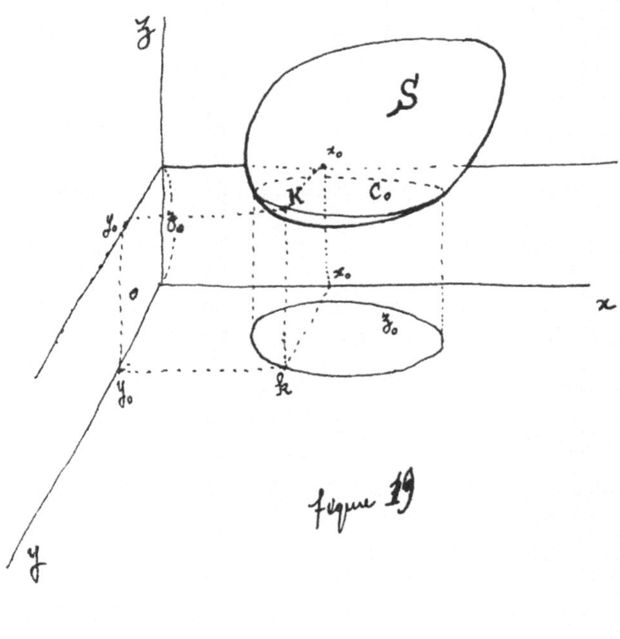

figure 19

En effet chaque point k de cette courbe z_0 a les mêmes coordonnées x_0 et y_0 que le point correspondant K de la courbe C_0.

Nous pouvons donc dire que la courbe z_0 du plan xOy représente exactement l'ensemble des points de la surface S *dont la coordonnée z a la même valeur z_0.*

Si nous faisions la même opération pour toutes les valeurs de z, nous pourrions tracer sur Ox un ensemble de courbes correspondant chacune à une valeur particulière de z.

Cet ensemble est absolument équivalent à la surface.

Supposons en effet que l'on donne les coordonnées x_0 et y_0 d'un point de la surface. En construisant le point du plan xOy qui a ces mêmes coordonnées, on trouve qu'il appartient à une courbe cotée z_0, c'est-à-dire telle que le point de la surface a pour coordonnées

$$\begin{cases} x = x_0 \\ y = y_0 \\ z = z_0. \end{cases}$$

On conçoit qu'on ne peut songer à construire une infinité de sections correspondant à toutes les valeurs de z, on se limite donc aux sections correspondant à des valeurs convenablement choisies de z, par exemple à des valeurs variant suivant une progression arithmétique. On inscrit sur le plan à côté de chaque courbe obtenue la valeur de z qui lui correspond.

On obtient ainsi un *plan coté* de la surface S.

N° 22. – Modules. Dilatation des ordonnées. – Comme on l'a déjà vu à propos des graduations linéaires (N° 3), puisqu'il s'agit d'une représentation, d'un dessin, il importe de préciser la longueur (c'est-à-dire le nombre d'unités de longueur du dessin) par laquelle on représentera l'unité de chacune des coordonnées d'un point de la surface, ou de chacune des quantités x, y, z qui figurent dans la relation $f(x, y, z) = 0$.

En d'autres termes, on conviendra que l'on représente

x	par la longueur	$X = M_1\, x$
y	....	$Y = M_2\, y$
z	....	$Z = M_3\, z$

M_1, M_2, M_3 étant les modules correspondant respectivement aux quantités x, y et z.

Dans le cas le plus simple et le plus général comme on le verra plus loin, où on adopte le même module M pour les 3 quantités, M n'est autre que le rapport des dimensions du dessin aux dimensions correspondantes de la surface à représenter (à condition toutefois d'employer la même unité de longueur pour la surface réelle et pour sa représentation). M n'est donc autre chose que ce qu'on appelle en topographie l'*échelle d'un plan.*

Dans le cas où l'on adopte des modules différents, on voit que la relation qui existe entre les 3 quantités X, Y et Z qui servent à représenter x, y, z, n'est autre que

$$f\left(\frac{X}{M_1}, \frac{Y}{M_2}, \frac{Z}{M_3}\right) = 0.$$

Dans la représentation, la surface subit donc une certaine déformation que l'on doit mettre à profit pour donner au plan coté une plus grande précision.

Cette déformation est souvent désignée sous le nom de *dilatation des ordonnées.*

N° 23. – Anamorphose. – La déformation précédente n'est qu'un cas particulier de la déformation qu'on peut obtenir à l'aide de l'anamorphose.

On peut appliquer l'anamorphose à une seule, à deux ou aux trois quantités qui interviennent dans la définition de la surface. *On a ainsi l'anamorphose simple, double, ou triple.*

Si l'on représente une surface

$$f(x, y, z) = 0 \qquad (1)$$

en faisant subir aux trois coordonnées les anamorphoses définies respectivement par les 3 équations

$$\left.\begin{aligned} X &= M_1\,\varphi_1(x) \\ Y &= M_2\,\varphi_2(y) \\ Z &= M_3\,\varphi_3(z) \end{aligned}\right\} \qquad (2)$$

l'opération équivaut à remplacer la surface à représenter par une autre surface dont l'équation serait

$$F(X, Y, Z) = 0,$$

cette équation étant obtenue par l'élimination de x, y, z entre les 4 équations précédentes (1) et (2).

Autrement dit, s'il était possible de résoudre les équations (2) par rapport à x, y,z et par conséquent de les mettre sous la forme

$$x = \Psi_1\left(\frac{X}{M_1}\right)$$
$$y = \Psi_2\left(\frac{Y}{M_2}\right)$$
$$z = \Psi_3\left(\frac{Z}{M_3}\right),$$

l'équation de la surface représentatrice F(X, Y, Z) = 0 ne serait autre que

$$f\left[\Psi_1\left(\frac{X}{M_1}\right), \Psi_2\left(\frac{Y}{M_2}\right), \Psi_3\left(\frac{Z}{M_3}\right)\right] = 0,$$

obtenue en remplaçant x, y et z par leurs expressions respectives, dans l'équation $f(x, y, z) = 0$ de la surface à représenter.

Il est important de remarquer que dans les équations (2) il y a deux éléments distincts et également importants dans la représentation :

1° la fonction φ qui définit l'anamorphose, ou en quelque sorte la loi de transformation de la coordonnée considérée;

2° le module auquel la longueur représentatrice (X par exemple) est proportionnelle et qui, par suite, caractérise la grandeur de la surface représentatrice, ou si l'on veut l'échelle du dessin du plan coté.

Exemple I. – Un exemple simple montrera mieux les avantages que l'on peut retirer de l'emploi de l'anamorphose.

Considérons la relation

$$x^2 + y^2 + z^2 - \mathrm{R}^2 = 0.$$

Elle correspond à une sphère de rayon R.

On peut représenter cette surface à l'aide d'un plan. Posons en effet $\mathrm{X} = x^2$, $\mathrm{Y} = y^2$, $\mathrm{Z} = z^2$ (les 3 modules sont égaux à l'unité de longueur).

En remplaçant x^2, y^2, z^2 par leurs valeurs dans l'équation de la sphère, on obtient

$$\mathrm{X} + \mathrm{Y} + \mathrm{Z} - \mathrm{R}^2 = 0,$$

équation d'un plan également incliné sur les axes de coordonnées.

Ceci veut dire que, si au lieu de porter sur les axes de coordonnées des graduations proportionnelles aux coordonnées x, y, z, on y porte des graduations anamorphosées suivant la loi des carrés des coordonnées, chaque plan également incliné sur les axes devient la représentation d'une sphère.

Dans ces conditions on obtient, en projetant sur le plan XOY, l'un des plans cotés ci-dessous, suivant qu'on représente la sphère sans anamorphose, ou qu'on emploie l'anamorphose triple étudiée plus haut.

Les 2 plans cotés représentent l'équation

$$x^2 + y^2 + z^2 - 100 = 0.$$

<Figure 20 :> Module commun pour x, y et z : M = 1 centimètre.

Chaque courbe représente l'intersection de la sphère avec les plans parallèles à xOy correspondant aux valeurs entières de z, de 0 à 10. On a laissé subsister les constructions qui indiquent comment ces cercles sont obtenus.

Les axes Ox et Oy sont gradués en centimètres, par suite de 0 à 10.

La graduation de z se trouve sur les cercles en suivant la bissectrice de l'angle des axes.

<Figure 21 :> $\mathrm{X} = \mathrm{M}x^2$, $\mathrm{Y} = \mathrm{M}y^2$, $\mathrm{Z} = \mathrm{M}z^2$; M = 1 millimètre, ce qui détermine les graduations de Ox et Oy.

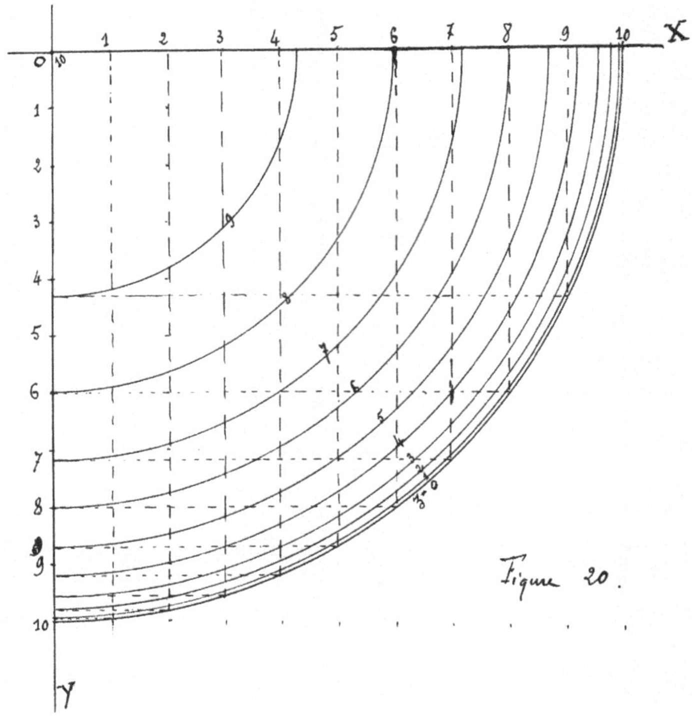

Figure 20.

Les cercles du plan coté précédent sont transformés en droites également inclinées sur les axes.

Ici la construction de ces droites est très simple. En effet la surface représentatrice est le plan

$$X + Y + Z - 100 = 0.$$

Pour avoir les points d'intersection des droites avec OX, il suffit de prendre l'intersection du plan avec le plan de coordonnées XOZ, ce qui revient à faire $Y = 0$, d'où $X + Z - 100 = 0$, ou encore $Z = 100 - X$.

Il en résulte que si l'on fait une graduation de z sur l'axe OX, cette graduation sera exactement égale à celle de X, mais dirigée en sens inverse et avec son origine au point 100.

Par exemple pour avoir l'intersection de OX avec la droite $z = 4$, il faut résoudre l'équation $X + Y + Z - 100 = 0$ en faisant $Y = 0$, $Z = z^2 = 16$ millimètres. On a donc $X + 0 + 16 - 100 = 0$, $X = 100 - 16$. Or 16 représente précisément la distance du point O au point 4 de la graduation OX.

Comme il en est ainsi pour toutes les valeurs de z, quand on a construit la graduation OX, il suffit de la retourner bout pour bout et l'on a sur ce même axe OX une graduation de z.

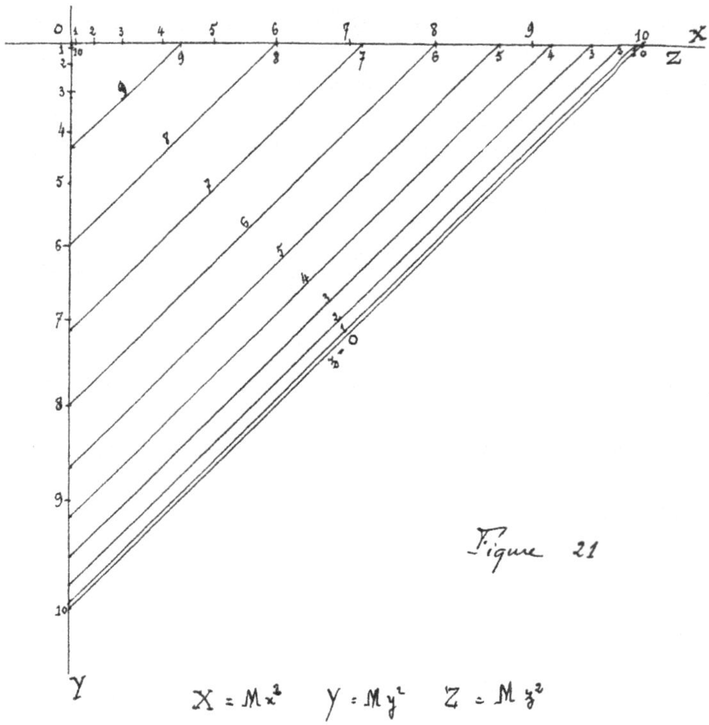

Remarquons, pour compléter cet exemple, que rien n'oblige à prendre le même module pour les 3 coordonnées ; si l'on veut obtenir une précision différente pour chacune d'elles, on choisira des modules convenables, l'échelle du dessin variant proportionnellement au module.

En prenant dans l'exemple précédent

$$\mathrm{X} = \mathrm{M}_1\, x^2$$
$$\mathrm{Y} = \mathrm{M}_2\, y^2$$
$$\mathrm{Z} = \mathrm{M}_3\, z^2,$$

la sphère serait représentée par le plan

$$\frac{\mathrm{X}}{\mathrm{M}_1} + \frac{\mathrm{Y}}{\mathrm{M}_2} + \frac{\mathrm{Z}}{\mathrm{M}_3} - \mathrm{R}^2 = 0$$

dont l'inclinaison n'est plus la même par rapport aux 3 axes.

Exemple II. – Dans l'exemple précédent, la construction du plan coté sans anamorphose (Figure 20) était facile, puisque les sections planes de la sphère se projetaient suivant des cercles faciles à construire.

Il n'en est plus ainsi lorsque les sections de la surface à représenter ne se projettent pas suivant des droites ou des arcs de cercle. S'il faut construire des courbes point par point, l'opération est extrêmement longue et l'anamorphose devient d'autant plus avantageuse, si elle permet de supprimer la construction de ces courbes, en les remplaçant par des droites ou en général par d'autres courbes plus faciles à tracer.

Considérons par exemple l'équation

$$\frac{x^2}{a^2} + \frac{y^2}{b^2} + \frac{z^2}{c^2} - 1 = 0.$$

On sait qu'elle représente un ellipsode rapporté à ses axes dont les $\frac{1}{2}$ longueurs sont respectivement a, b et c.

Les sections parallèles au plan $x\mathrm{O}y$ sont des ellipses.

L'établissement d'un plan coté sans anamorphose comporterait donc la construction d'un certain nombre d'ellipses, construction déjà beaucoup plus délicate que celle des cercles.

Employons l'anamorphose triple définie par les équations

$$\mathrm{X} = \mathrm{M}_1 \frac{x^2}{a^2}$$
$$\mathrm{Y} = \mathrm{M}_2 \frac{y^2}{b^2}$$
$$\mathrm{Z} = \mathrm{M}_3 \frac{z^2}{c^2}.$$

La surface de l'ellipsode se trouve alors représentée par le plan

$$\frac{\mathrm{X}}{\mathrm{M}_1} + \frac{\mathrm{Y}}{\mathrm{M}_2} + \frac{\mathrm{Z}}{\mathrm{M}_3} - 1 = 0.$$

Les ellipses sont remplacées par des droites et l'on obtient un plan coté analogue à celui de la figure 21.

N° 24. – Interpolation. – On a vu dans ce qui précède que tout plan coté représentant une surface ou bien une relation entre trois variables telle que $f(x, y, z) = 0$, se présente sous la forme d'une série de courbes correspondant chacune à une valeur de z qui est inscrite à côté de la courbe correspondante.

L'ensemble de ces courbes est rapporté à 2 axes de coordonnées portant les graduations relatives aux nombres x et y.

Pour déterminer la valeur de z qui correspond à des valeurs données des deux autres variables, x_0 et y_0 par exemple, il faut tout d'abord construire le point du plan dont les coordonnées correspondent à x_0 et y_0, c'est-à-dire qu'il faut déterminer par interpolation sur les graduations de Ox et Oy, les points x_0 et y_0. En menant par ces deux points des parallèles aux axes de coordonnées, on obtient à leur point de rencontre le point du plan correspondant au couple de valeurs (x_0, y_0).

Si ce point se trouve exactement sur une des courbes du plan coté, la solution du problème est simple, il suffit de lire la valeur de z inscrite sur cette courbe. Soit z_0 ce nombre, c'est celui qui d'après la relation $f(x, y, z) = 0$ correspond aux valeurs x_0 et y_0 des deux autres variables.

Mais en général, le point du plan correspondant au couple de valeurs (x_0, y_0) ne se trouve pas sur une des courbes z. Il se trouve entre deux de ces courbes correspondant respectivement aux nombres z_1 et z_2.

Il est alors nécessaire d'avoir recours à l'interpolation pour déterminer le nombre z_0 qui correspond à (x_0, y_0).

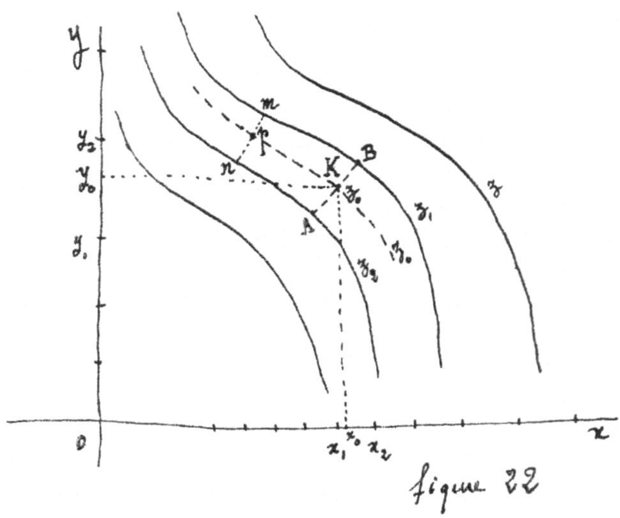

L'interpolation est basée sur ce fait que les courbes z représentent des sections d'une même surface par des plans généralement très rapprochés les uns des autres (au moins à l'échelle du dessin) ; leur forme se modifie progressivement de l'un à l'autre, de telle sorte que la courbe z_2 est sensiblement parallèle à la courbe voisine z_1 en toutes ses parties.

Il en résulte que si en un point quelconque m de z_1 on mène une normale à cette courbe, elle est aussi sensiblement normale en n à la courbe voisine z_2.

Supposons qu'en tous les point de z_1 on mène les normales telles que mn et que, sur chacune de ces normales, dans l'intervalle mn, on place un point p tel que

$$\frac{mp}{mn} = r.$$

En joignant tous ces points p on aura une courbe partageant l'intervalle entre z_1 et z_2 dans le rapport $r/1$ et que l'on peut par suite faire correspondre au nombre z_0 qui partage l'intervalle de z_1 à z_2 dans le même rapport, c'est-à-dire tel que

$$\frac{z_0 - z_1}{z_2 - z_1} = r$$

$$\text{ou} \qquad z_0 = z_1 + r(z_2 - z_1).$$

Tout point K situé sur cette courbe correspondra à la valeur z_0.

Inversement, pour avoir la valeur z_0 qui correspond à un point K, il suffit de mener par K la normale AB aux deux courbes voisines et chercher dans quel rapport le point K partage le segment AB. Soit r ce rapport, on doit avoir

$$\frac{\text{KB}}{\text{AB}} = \frac{z_0 - z_1}{z_2 - z_1} = r$$

$$\text{d'où} \qquad z_0 = z_1 + r(z_2 - z_1).$$

Si l'intervalle de deux courbes consécutives, c'est-à-dire $(z_2 - z_1)$ est pris comme unité, on voit que l'on a par interpolation $z_0 = z_1 + r$.

Remarque. – On voit immédiatement que ce qui précède suppose que le déplacement d'une courbe z est exactement proportionnel à la variation de z ou KB proportionnel à $(z_0 - z_1)$.

Il n'en est pas ainsi en général. Aussi n'a-t-on le droit d'interpoler que si l'écartement des courbes z est suffisamment petit pour que dans l'intervalle de deux courbes consécutives on puisse considérer le déplacement de la courbe comme proportionnel à la variation de z.

Pour préciser cette question, l'intervalle de deux courbes voisines n'étant pas absolument proportionnel à la différence des valeurs de z qui leur correspondent, on commet toujours en interpolant une petite erreur variable dans chaque cas, mais que l'on peut toujours évaluer. Cette erreur marque la limite de l'approximation que l'on peut demander à l'interpolation.

En effet, même si les dimensions du dessin permettaient d'obtenir une précision plus grande dans la détermination du nombre z_0 par exemple, on obtiendrait la valeur de z correspondant aux valeurs données (x_0, y_0) non pas dans la relation $f(x, y, z) = 0$ représentée par le plan cotée, mais dans une autre relation qui lui est substituée par le fait de l'interpolation. Cette valeur z_0' de z ne se confond avec z_0 que jusqu'à l'ordre d'approximation où l'interpolation est permise.

Cette remarque peut être appliquée à l'interpolation dans le cas des graduations anamorphosées quelconques.

Remarque II. – Si l'on se reporte à la figure 21 on voit qu'on sait également construire par interpolation la courbe qui correspond à une valeur z_0 ou calculer la valeur z_0 correspondant à une courbe ou un point donné ; le plan coté peut donc indifféremment servir à déterminer z si l'on se donne x et y, ou bien y si l'on se donne x et z.

Dans le premier cas, l'intersection des deux parallèles aux axes passant par les points x_0 et y_0 donne le point K et par suite z_0.

Dans le $2^{\text{ème}}$ cas, le point K est obtenu par l'intersection de l'ordonnée x_0 et de la courbe z_0. En menant par K une parallèle à Ox on obtient le point y_0 sur Oy.

On opérerait d'une façon analogue si on se donnait y et z, pour déterminer x.

Remarque III. – Il y a tout intérêt, comme on l'a vu pour les graduations, à déterminer le plan coté de façon que l'écartement des courbes soit toujours assez faible pour qu'on puisse se contenter de faire l'interpolation sans aucune construction, c'est-à-dire *que l'interpolation à vue puisse toujours être suffisante.*

N° 25. – Équidistance. Intervalle. Interligne. – Dans le cas où z n'est soumis à aucune anamorphose, on construit en général les sections de la surface par des plans parallèles à xOy et correspondant à des valeurs de z variant en progression arithmétique.

Ces plans sont équidistants les uns des autres. La différence $z_2 - z_1$ des deux valeurs de z correspondant respectivement à deux sections consécutives est l'*équidistance* relative à la variable z.

On voit immédiatement l'analogie qui existe entre la quantité ainsi définie et celle qu'on appelle équidistance dans les plans topographiques.

Si l'on a tracé sur Ox et Oy des graduations correspondant aux deux variables x et y, on a de même pour chacune de ces 2 graduations une équidistance, qui peut être différente de celle de z.

Comme on l'a vu à propos des graduations, *l'équidistance graphique* est *l'équidistance réduite à l'échelle du dessin, c'est-à-dire le produit de l'équidistance par le module* :

$$\mathrm{E}_g = \mathrm{E} \times \mathrm{M}.$$

Ceci semble en défaut dans le cas prévu au N° 22, où l'on veut employer des modules différents pour les différentes variables, de manière à produire une dilatation des ordonnées. Mais, comme on l'a déjà remarqué, la dilatation des ordonnées n'est qu'un cas particulier d'anamorphose.

En conséquence si on veut employer la représentation définie par

$$\begin{aligned} \mathrm{X} &= \mathrm{M}x \\ \mathrm{Y} &= \mathrm{M}'y \\ \mathrm{Z} &= \mathrm{M}''z, \end{aligned}$$

on pourra la mettre sous la forme

$$X = M\,x$$
$$Y = M.\left(\frac{M'}{M}\right) y$$
$$Z = M.\left(\frac{M''}{M}\right) z.$$

On voit ainsi que les 3 variables peuvent être considérées comme ayant le même module M, $\left(\frac{M'}{M}y\right)$ et $\left(\frac{M''}{M}\right)z$ représentant les anamorphoses de y et z et pouvant d'ailleurs rentrer dans des anamorphoses plus complexes.

Nous considérerons donc toujours dans ce qui suit que toutes les variables ont le même module.

Dans le cas de l'anamorphose, comme on l'a vu au sujet des graduations, l'espacement des traits sur Ox et Oy, de même que l'écartement des courbes z, n'est plus constant ; l'équidistance et l'équidistance graphique sont remplacées respectivement par l'*intervalle* et l'*intervalle graphique* avec la relation

$$I_g = I \times M.$$

Mais dans l'étude d'un plan coté intervient un autre élément. En effet, si l'on reproduisait exactement dans l'espace la surface représentée par l'équation $f(x, y, z) = 0$, la graduation correspondant à la variable z serait tracée sur l'axe Oz, sur lequel on pourrait déterminer l'intervalle ou l'équidistance graphique relative à z.

Lorsqu'on remplace la représentation exacte de la surface par le plan coté correspondant sur le plan Oxy, la graduation de l'axe Oz se trouve remplacée par l'ensemble des sections parallèles au plan xOy et qui correspondent aux valeurs successives de z.

L'intervalle de deux traits de la graduation de Oz, z_1 et z_2 par exemple, est donc remplacé par l'écartement des deux courbes correspondantes, z_1 et z_2. Si l'on se reporte à la figure 22, on voit que c'est cet écartement AB qui sert à l'interpolation entre les valeurs z_1 et z_2, et par suite caractérise la détermination de z à l'aide du plan coté.

Nous désignerons sous le nom d'*interligne* l'écartement de 2 courbes consécutives sur le plan coté.

L'interligne correspond exactement à l'intervalle d'une graduation ; plus il est grand, plus la longueur qui représente la différence entre deux valeurs consécutives de z est considérable, et, par suite, plus on peut obtenir de précision par l'interpolation, au moins dans les limites où l'interpolation est permise.

Considérons deux courbes z_1 et z_2 d'un plan coté qui sont les projections sur le plan xOy (figure 23) des sections de la surface S parallèles au plan xOy et correspondant aux valeurs z_1 et z_2 de z.

La droite ab que l'on peut considérer comme normale à la fois aux deux courbes z_1 et z_2 mesure l'interligne dans la région considérée du plan coté.

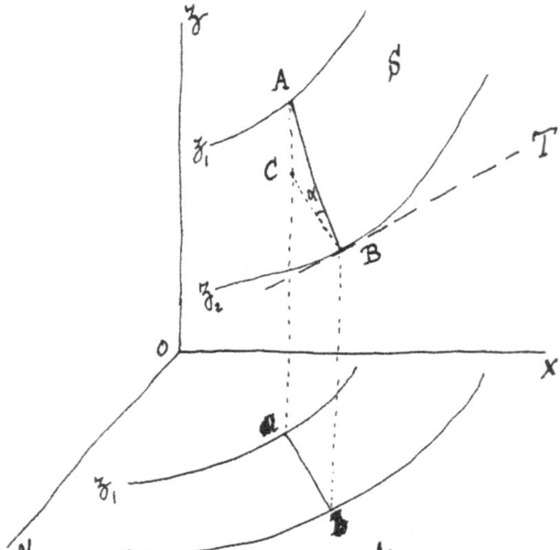

Ce segment *ab* est la projection sur le plan $x\text{O}y$ du segment AB que l'on peut également considérer comme normal en même temps aux deux sections de la surface, les sections étant généralement très rapprochées les unes des autres et par suite sensiblement parallèles, comme on l'a déjà fait remarquer.

Si par le point B on mène une parallèle à *ab*, c'est-à-dire une parallèle au plan $x\text{O}y$, cette droite est contenue dans le plan de la section z_2, également parallèle à $x\text{O}y$.

On a donc en AC l'*intervalle graphique* de la graduation de z, puisque AC représente la différence des 2 valeurs consécutives z_1 et z_2.

Dans le cas des coordonnées rectangulaires, le triangle ABC est rectangle en C.

$$\text{AC} = \text{Intervalle graphique} = \text{I}_g$$

$$\text{BC} = ab = \text{Interligne} \qquad = \text{I}_l.$$

L'angle en B, α peut être considéré comme l'angle de la surface avec le plan $x\text{O}y$, ou inclinaison de la surface en B sur le plan de projection. En effet, les points A et B étant voisins sur la surface S, AB peut être assimilé à une tangente en B à cette surface. D'autre part AB est perpendiculaire en B à la courbe z_2 et par suite à sa tangente en B, qui est dans un plan parallèle à $x\text{O}y$. ABC est donc l'angle plan du dièdre formé par le plan tangent à la surface en B et le plan parallèle à $x\text{O}y$ et passant par B. α est donc bien l'inclinaison du plan tangent en B sur $x\text{O}y$ ou, autrement dit, l'inclinaison de la surface sur le plan de projection qui contient le plan coté.

Le triangle ABC donne immédiatement la relation

$$\text{BC} = \text{AC cotg } \alpha$$

$$\text{d'où} \qquad \text{I}_l = \text{I}_g \text{ cotg } \alpha.$$

Lorsqu'on s'est fixé une valeur de l'intervalle graphique pour la graduation de z, le plan coté est d'autant plus avantageux que l'interligne correspondant est plus grand, ou que cotg α a une plus grande valeur. Il y a donc tout intérêt à ce que la surface soit aussi peu inclinée que possible sur le plan de projection. Les parties de la surface qui donnent le maximum de précision pour la détermination de z sont les moins inclinées sur le plan de projection.

Il résulte de ce qui précède que si l'on veut augmenter la précision d'un plan coté, ce n'est pas sur la coordonnée z qu'il faut agir. En effet, la dilatation des ordonnées appliquée à z sans toucher à x et y, déplacerait simplement tous les points de la surface respectivement sur des parallèles à Oz telles que Aa, Bb ; par suite les sections parallèles à xOy se déplaceraient sur des cylindres à génératrices parallèles à Oz, et elles conserveraient toujours la même projection sur le plan xOy.

C'est donc sur les coordonnées x et y qu'il faut agir pour obtenir une valeur convenable de l'interligne, l'équidistance de z étant alors choisie de façon à permettre l'interpolation dans toute l'étendue du plan coté.

N° 26. – Différence entre les diagrammes gradués ou conjugués et les plans cotés. – On a vu aux N°s 19 et 20 qu'il est possible de représenter 3 variables à l'aide d'un diagramme gradué ou de diagrammes conjugués. il y a cependant une différence essentielle entre ces diagrammes et les plans cotés qui représentent également 3 variables.

Dans le cas des diagrammes, les 3 variables doivent nécessairement être assujetties à 2 relations. Il suffit de connaitre x pour que les valeurs correspondantes de y soient immédiatement déterminées.

Dans le cas du plan coté, au contraire, les 3 variables ne sont liées que par une seule relation ; il faut connaître 2 de ces variables, x et y par exemple, pour déterminer la troisième z.

On peut dire que le diagramme gradué est l'équivalent d'une courbe de l'espace, tandis que le plan coté est l'équivalent d'une surface.

N° 27. – Plans cotés conjugués. Élimination graphique. – Supposons que l'on ait deux relations telles que

$$f_1(x, y, z) = 0$$
$$f_2(x, y, t) = 0.$$

Rapportons aux mêmes axes de coordonnées Ox et Oy les deux plans cotés correspondants, en employant la même représentation pour x et y dans les 2 cas.

En chaque point du plan correspondant aux valeurs (x_0, y_0) de x et y, passeront deux courbes donnant respectivement les valeurs z_0 et t_0 de z et t, qui avec x_0 et y_0 satisfont en même temps aux deux relations

$$f_1(x_0, y_0, z_0) = 0$$
$$f_2(x_0, y_0, t_0) = 0.$$

Inversement, si l'on se donne les valeurs y_0 et z_0, on pourra trouver les valeurs x_0 et t_0 correspondantes. On peut faire abstraction de x et considérer seulement la valeur de t qui correspond au couple de valeurs y_0, z_0. Ce qui revient à dire que l'on détermine graphiquement la relation qui existe nécessairement entre les 3 quantités y, z et t.

Cette relation s'obtiendrait algébriquement en éliminant x entre les 2 équations

$$f_1(x, y, z) = 0$$
$$f_2(x, y, t) = 0.$$

Les plans cotés conjugués permettent donc d'obtenir l'élimination graphique d'une inconnue commune à deux relations contenant chacune 3 variables. C'est la généralisation de ce principe d'élimination qui conduit à la notion des *abaques*.

Chapitre II. Abaques

Abaques à deux variables

N° 28. – Généralisation de la représentation des nombres. Graduations planaires. – Considérons une relation de la forme

$$f(x, y, \mathrm{N}) = 0 \qquad (1)$$

qui contient une quantité ou paramètre N dont la valeur n'est pas déterminée et à laquelle on peut, par suite, attribuer toutes les valeurs comprises entre deux limites déterminées ou même toutes les valeurs possibles de $-\infty$ à $+\infty$.

Si l'on donne à N une valeur particulière N_0, la relation (1) devient $f(x, y, \mathrm{N}_0) = 0$; elle peut être représentée, sur un plan rapporté à 2 axes de coordonnées quelconques, par une courbe dont la position sur le plan et même souvent la forme dépendent de la valeur N_0 attribuée au paramètre N.

À chaque valeur de N correspondra de même une courbe du plan.

Inversement si toutes les courbes étaient tracées sur le même plan, en les rapportant aux mêmes axes de coordonnées, à chaque courbe correspondrait une valeur du paramètre N.

Il y a correspondance complète entre les courbes et les valeurs de N.

D'autre part, considérons le plan rapporté à deux axes de coordonnées Ox et Oy, et sur lequel nous supposons tracées toutes les courbes correspondant aux différentes valeurs du paramètre N.

Un point M du plan correspond aux coordonnées x_0, y_0. Si on porte ces valeurs dans la relation (1), on obtient l'équation $f(x_0, y_0, \mathrm{N}) = 0$ dans laquelle tout est connu sauf N. Elle permet donc de déterminer la valeur de N.

Par conséquent, à tout point du plan correspond une valeur de N.

De plus, si on désigne par N_0 la valeur de N correspondant au point M, (x_0, y_0), on a nécessairement $f(x_0, y_0, \mathrm{N}_0) = 0$ et il est évident que (x_0, y_0) est un système particulier de solutions de l'équation $f(x, y, \mathrm{N}_0) = 0$.

D'où l'on peut conclure que la valeur de N *qui correspond au point* M *du plan, n'est autre que la valeur* N_0 *du paramètre qui caractérise la courbe passant par le point* M.

En résumé, l'équation (1) permet d'établir sur le plan un ensemble de courbes, tel que l'on puisse faire correspondre

1° à chaque point du plan, une valeur de N ;

2° à chaque valeur de N, une courbe particulière du plan.

Il ne peut être question de tracer sur un plan toutes les courbes possibles, leur nombre étant infini et les courbes correspondant à des valeurs très voisines de N étant pratiquement confondues.

On doit donc se contenter de tracer sur le plan, en les rapportant à un même système de coordonnées, l'ensemble des différentes courbes correspondant à des valeurs convenablement choisies du paramètre N, par exemple à des valeurs de N variant en progression arithmétique.

L'ensemble des courbes ainsi tracées sur le plan peut être désigné sous le nom de *graduation planaire*, par suite de l'analogie avec les graduations linéaires dont il a été question plus haut.

On voir en effet que si, précédemment la graduation était limitée à des points situés sur une ligne, elle se trouve dans ce dernier procédé étendue à l'ensemble du plan.

N° 29. – Interpolation. – On a donc tracé sur le plan, par rapport à deux axes de coordonnées Ox, Oy, les différentes courbes correspondant aux valeurs N_0, N_1, N_2, …, N_k du paramètre N de la relation $f(x, y, \mathrm{N}) = 0$. À côté de chacune de ces courbes, on inscrit la valeur de N qui lui correspond.

Pour établir la correspondance entre un point quelconque du plan, généralement en dehors des courbes tracées, et les valeurs de N, on a recours à l'interpolation telle qu'elle a été décrite eu N° 24, au sujet des plans cotés.

On peut en effet considérer la quantité variable N comme une troisième coordonnée. Alors la relation $f(x, y, \mathrm{N}) = 0$ représente une surface et la graduation planaire, telle qu'elle vient d'être définie, n'est autre qu'un plan coté de cette surface.

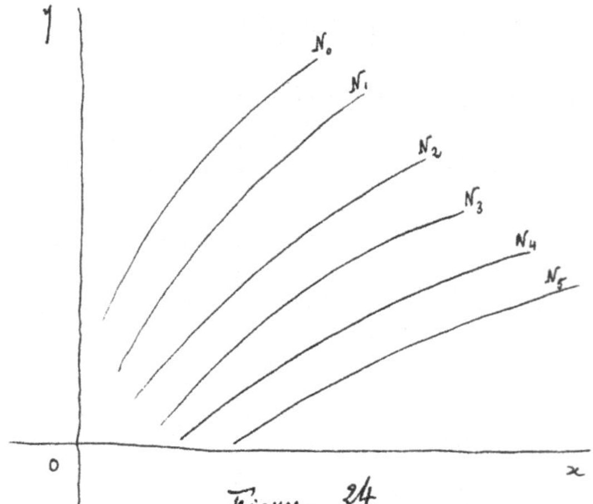

Figure 24

N° 30. – Équidistance. Intervalle. Interligne. – De même la comparaison des plans cotés et des graduations planaires permet de conserver les éléments définis au N° 25 à propos des plans cotés, avec les définitions qui en ont été données.

L'*équidistance* n'est autre que la différence qui sépare les valeurs du paramètre N correspondant à deux courbes consécutives de la graduation, lorsque cette différence est constante, c'est-à-dire lorsque les valeurs de N qui sont inscrites sur la graduation N_0, N_1, N_2, N_3, ...sont en progression arithmétique.

L'*équidistance graphique* est cette même quantité multipliée par le module, c'et-à-dire réduite à l'échelle du dessin.

L'*intervalle* remplace l'équidistance, lorsque la différence des valeurs consécutives de N n'est pas constante, c'est-à-dire lorsque les valeurs successives N_0, N_1, N_2, ...suivent toute autre loi que celle de la progression arithmétique. l'intervalle est variable avec les valeurs de N à laquelle il se rapporte.

L'*intervalle graphique* est l'intervalle réduit à l'échelle du dessin.

L'*interligne*, en un point quelconque de la graduation planaire, est la distance de deux courbes consécutives de la graduation, distance comptée normalement aux deux courbes, généralement assez voisines l'une de l'autre, comme forme et comme position, pour qu'on puisse les regarder comme parallèles dans le voisinage du point considéré.

L'interligne est relié à l'intervalle graphique comme on l'a vu au N° 25 [1].

N° 31. – Comparaison des graduations planaires et des graduations linéaires. – Il existe un lien très étroit entre les graduations planaires et les graduations linéaires.

[1] On a vu au N° 25 que l'on a $I_l = I_g$ cotg α. α est l'angle du plan tangent à la surface représentée avec le plan des xy, ou bien l'angle de la normale à la surface avec le troisième axe qui serait ici l'axe ON. Or, dans le cas des coordonnées rectangulaires, les cosinus des angles que fait avec les axes la normale au point (x, y, N) à la surface représentée par $f(x, y, \mathrm{N}) = 0$ sont

Considérons tout d'abord une graduation linéaire rectiligne tracée sur un axe Ox.

Si l'on rapporte le plan à deux axes de coordonnées Ox et Oy, on sait que tous les points d'une droite parallèle à Oy correspondent à la même valeur x_0 de la coordonnée x ; x_0 étant l'abscisse du point de rencontre de l'axe Ox avec la parallèle considérée à l'axe Oy.

En menant ainsi des parallèles à Oy par tous les points de la graduation de l'axe Ox, on détermine une graduation planaire que l'on peut substituer à la graduation linéaire considérée.

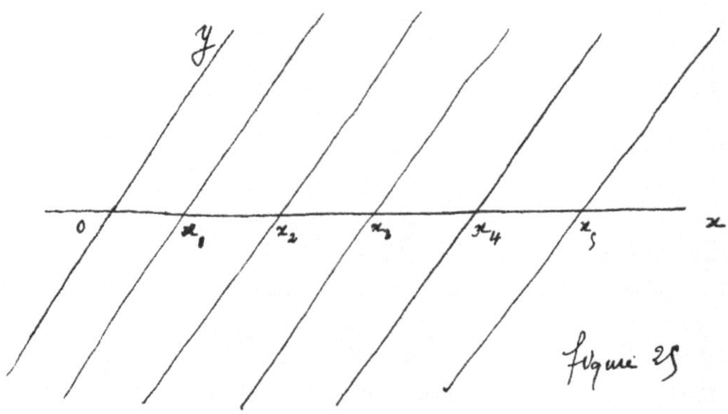

figure 25

Inversement si l'on a une graduation planaire constituée par des droites parallèles, on peut, en la coupant par une transversale Ox, la remplacer par une graduation linéaire dont les différents points sont obtenus par l'intersection de la droite Ox et des différentes droites de la graduation planaire.

Il en est ainsi toutes les fois que la graduation planaire est constituée par des courbes égales, autrement dit par une courbe unique se déplaçant dans le plan.

En effet, lorsqu'une courbe C de forme invariable se déplace dans un plan d'un mouvement continu, chacun de ses points décrit une courbe Γ coupant sous un angle α la courbe C dans chacune de ses différentes positions. Si donc on connaît la courbe Γ et le point M de la courbe C qui la décrit, on pourra facilement replacer la courbe C dans toutes les positions qu'elle occupe successivement, à la seule condition de connaître l'angle α pour chaque position du point M. On peut en effet ainsi, d'une part mettre en place un point de la courbe C, et d'autre part orienter cette courbe sur le plan en la faisant tourner autour de la position du point M jusqu'à ce qu'elle fasse l'angle α avec la courbe Γ lieu de M.

proportionnels à $\frac{\partial f}{\partial x}, \frac{\partial f}{\partial y}, \frac{\partial f}{\partial \mathrm{N}}$, dérivées partielles de f. On en déduit facilement

$$\cot g\, \alpha = \frac{\frac{\partial f}{\partial \mathrm{N}}}{\sqrt{(\frac{\partial f}{\partial x})^2 + (\frac{\partial f}{\partial y})^2}}, \quad \text{d'où} \quad \mathrm{I}_l = \mathrm{I}_g \frac{\frac{\partial f}{\partial \mathrm{N}}}{\sqrt{(\frac{\partial f}{\partial x})^2 + (\frac{\partial f}{\partial y})^2}}.$$

Dans la pratique, la substitution de la graduation linéaire à la graduation planaire n'est réellement pratique, que *si l'angle α est constant* et en particulier si cet angle est droit. Dans ce dernier cas, il existe un point M_0 de la courbe mobile C, qui décrit une courbe Γ_0 normale en chacun de ses points à la courbe C qui y passe [2].

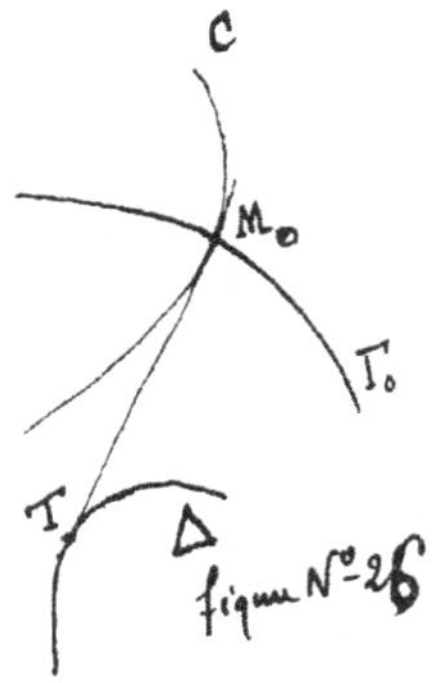

Considérons alors sur la courbe Γ_0 les points qui correspondent aux positions de la courbe C données par les valeurs N_1, N_2, N_3, ..., N_k, ...du paramètre, on obtient ainsi une graduation linéaire.

Traçons d'autre part sur une feuille de papier à calquer la courbe C mobile. Marquons sur cette courbe le point M_0 et la normale n au point M_0. *Cette feuille de papier à calquer ainsi préparée constitue ce qu'on appelle un* transparent.

Pour avoir la position de la courbe C correspondant dans la graduation planaire à la valeur N du paramètre, il suffit de prendre le point N de la graduation linéaire tracée sur Γ_0, puis de placer la feuille de papier à calquer de façon que M_0 soit sur N de Γ_0, et que la normale n soit tangente à Γ_0.

En effet M_0 occupe bien la valeur qui correspond au paramètre N et la courbe C est bien normale en M_0 à Γ_0.

Au point de vue analytique, la graduation linéaire n'est qu'un cas particulier de la graduation planaire.

En effet toute graduation planaire correspond à une relation $f(x, y, \mathrm{N}) = 0$ de forme absolument quelconque.

Dans le cas où cette fonction peut être mise sous la forme

$$[x - \varphi_1(\mathrm{N})]^2 + [y - \varphi_2(\mathrm{N})]^2 = 0,$$

[2] Considérons la tangente à la courbe mobile C, au point M_0 qui décrit la courbe Γ_0 constamment normale à C. Si l'on suppose cette tangente M_0T entraînée dans le mouvement de la courbe C, elle enveloppe une courbe Δ. Or M_0T est constamment normale à Γ_0 ; par suite Δ est la développée de Γ_0. Pour que le point M_0 de cette tangente décrive la courbe Γ_0, il faut que la droite M_0T roule sans glissement sur la développée Δ. Il suffit donc, pour que la courbe Γ_0 existe, que le mouvement de la courbe C soit tel qu'une de ses tangentes roule sur son enveloppe. En d'autres termes, le centre instantané de rotation du mouvement de C est toujours sur la tangente au point M_0 à cette courbe.

où φ_1 et φ_2 sont des fonctions réelles pour les valeurs de N considérées, les seules valeurs réelles de x et y qui satisfont à cette équation sont données par le système

$$\begin{cases} x = \varphi_1(\mathrm{N}) \\ y = \varphi_2(\mathrm{N}). \end{cases}$$

Lorsqu'on donne à N les valeurs successives N_1, N_2, N_3, ...on obtient une série de points qui sont sur une courbe Γ.

La relation proposée permet donc l'établissement d'une graduation linéaire.

C'est un cas particulier de courbe mobile de forme constante : ici la courbe mobile est un cercle de rayon nul ou réduit à un point. La courbe Γ_0 n'est autre que le lieu des positions de ce point suivant les différentes valeurs de N.

Il est facile de montrer que la *graduation angulaire* (voir N° 11) est aussi un cas particulier de graduation planaire. En effet toutes les droites passant par un point de coordonnées a et b dans un plan rapporté à deux axes de coordonnées rectangulaires, peuvent être représentées par la relation

$$y - b = (x - a)\,\mathrm{tg}\,\mathrm{N},$$

N étant le nombre de grades (ou degrés) de l'angle que fait la droite avec l'axe Ox.

Toute graduation angulaire peut donc être représentée par une équation de cette forme et inversement.

La graduation planaire a l'avantage

1° *de montrer immédiatement la façon dont la courbe mobile se déplace* ;

2° *de donner immédiatement la valeur du paramètre qui correspond à un point du plan.*

Avec la graduation planaire, les deux opérations :

chercher un point du plan d'après la valeur du paramètre,

trouver le paramètre correspondant à un point du plan,

sont sensiblement équivalentes.

La graduation planaire parle aux yeux.

Elle a l'inconvénient de nécessiter un dessin souvent long et compliqué.

En effet pour permettre l'interpolation, il faut que les courbes successives soient suffisamment rapprochées (voir N° 24). Il faut donc tracer un grand nombre de courbes, ce qui est toujours une opération délicate et très longue, toutes les fois qu'il s'agir de lignes autres que des droites.

Les graduations linéaires substituées aux graduations planaires ont l'avantage de simplifier beaucoup le dessin, en remplaçant le tracé d'un grand nombre de courbes par celui de 3 lignes seulement, dont une droite.

De plus, l'interpolation entre deux points d'une graduation linéaire est un peu plus facile que l'interpolation dans l'interligne d'une graduation planaire, de telle sorte que les graduations linéaires sont généralement considérées comme plus commodes et plus exactes pour le calcul des différentes valeurs du paramètre.

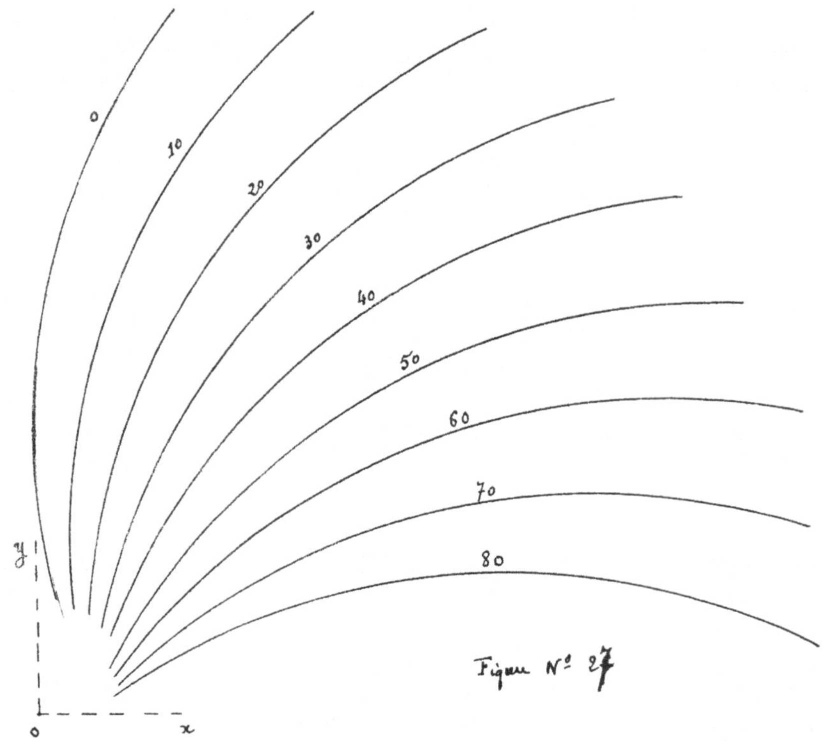

Exemple. – Considérons la graduation planaire correspondant à la relation

$$(x - r \sin \mathrm{N} - \mathrm{R} \cos \mathrm{N})^2 + (y - r \cos \mathrm{N} + \mathrm{R} \sin \mathrm{N})^2 - \mathrm{R}^2 = 0.$$

La forme même de cette relation indique que les différentes courbes sont des cercles égaux de rayon R.

En traçant ces cercles par rapport aux axes Ox, Oy, on obtient la graduation planaire représentée dans la figure N° 27.

N est le nombre de grades de l'angle qui figure dans la relation ; r = 5 centimètres ; R = 12 centimètres.

Les axes Ox, Oy ne sont utiles que pour la construction ; quand toutes les courbes sont tracées et numérotées, on peut les faire disparaître.

Il est facile de montrer que tous les cercles de la graduation planaire coupent normalement le cercle de rayon r décrit de O comme centre [3].

[3] En effet la puissance du centre du cercle mobile par rapport à ce cercle de rayon r est :

$$(r \sin \mathrm{N} + \mathrm{R} \cos \mathrm{N})^2 + (r \cos \mathrm{N} - \mathrm{R} \sin \mathrm{N})^2 - r^2 = \mathrm{R}^2.$$

Donc la tangente menée au cercle de centre O et de rayon R du centre du cercle mobile a toujours pour longueur le rayon de ce dernier cercle, ce qui exprime bien que les 2 cercles se coupent à angle droit.

Par suite, la graduation planaire de la figure 27 peut être remplacée par la graduation linéaire de la figure 28, qui n'est autre qu'une graduation en grades sur un cercle de 5 centimètres de rayon.

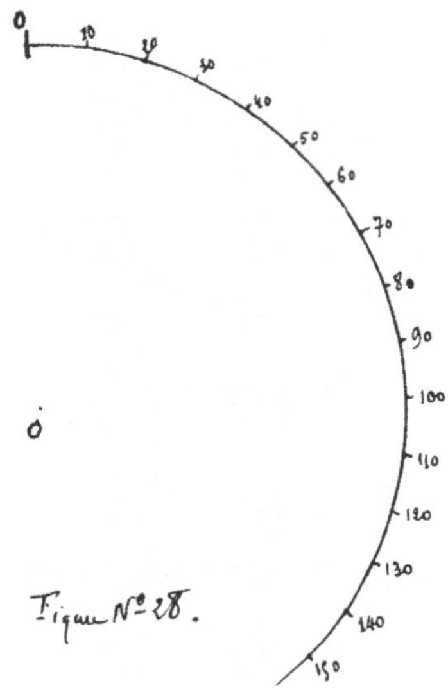

Figure N° 28.

Le transparent contiendra un des cercles de la figure 27 avec une tangente : par exemple l'ensemble du cercle coté 0 et de la tangente Oy. (Le point de contact M_0 de cette tangente étant marqué d'une façon précise, à l'aide du rayon perpendiculaire à cette tangente.)

Pour avoir la position du cercle mobile correspondant à N = 65 :

1° Déterminer sur la graduation 28 le point 65.

2° Amener en ce point le point M_0 du transparent.

3° Faire tourner le transparent sans bouger le point M_0, de façon que la tangente passe par O.

Ici, il est plus commode d'employer une tangente à la courbe mobile, parce qu'elle est normale au cercle fixe et par suite doit toujours passer par le centre O de ce cercle.

32. – Double graduation. Représentation des relations à deux variables. Généralisation de l'anamorphose. – Considérons les deux graduations définies par les équations

$$\left.\begin{array}{l} f_1(x, y, \mathrm{N}_1) = 0 \\ f_2(x, y, \mathrm{N}_2) = 0 \end{array}\right\} \qquad (1)$$

les éléments des deux graduations étant supposés déterminés par rapport aux mêmes axes Ox et Oy.

D'après ce que nous avons déjà vu, N° 27, si l'on fait $N_1 = N_2$, on peut considérer les deux graduations comme des plans cotés conjugués et on peut les utiliser pour l'élimination de N entre les deux relations $f_1(x, y, N) = 0$, $f_2(x, y, N) = 0$.

Considérons dans les deux graduations les courbes correspondant aux valeurs égales des paramètres ; on aura ainsi des paires de courbes conjuguées. À chaque valeur donnée simultanément aux deux paramètres, correspondra un point du plan, intersection de deux courbes conjuguées. L'ensemble de ces points donnera une courbe du plan.

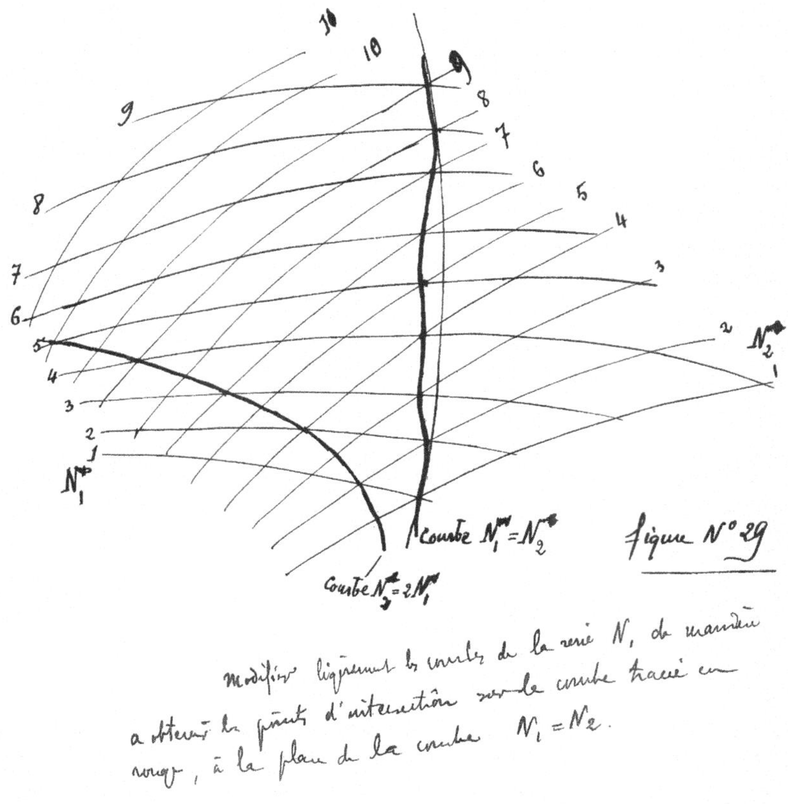

Cette courbe n'est autre que la représentation de l'équation $N_1 = N_2$ (Figure N° 29) *dans le système de double graduation défini par la relation* (1).

Plus généralement ce système de double graduation peut servir à représenter une relation quelconque entre N_1 et N_2 que nous désignerons par

$$F(N_1, N_2) = 0. \qquad (2)$$

En effet, cette équation fait correspondre une valeur de N_2 à chaque valeur de N_1 ; par suite on peut, comme on l'a fait précédemment, grouper par paire les courbes des deux graduations et les points d'intersection des courbes de chaque paire décrivent une courbe qui est la représentation graphique de l'équation (2). (Figure N° 29. Courbe représentant l'équation $N_2 = 2\,N_1$.)

D'ailleurs, si l'on joint l'équation (2) et les 2 équations (1), on obtient un système de 3 équations entre lesquelles on peut éliminer deux des quantités variables, par exemple N_1 et N_2.

Le résultat de l'élimination est une équation contenant x et y, $\Phi(x, y) = 0$, qui n'est autre que l'équation de la courbe représentative de la relation (2), dans le système de double graduation (1).

Il n'est pas sans intérêt de remarquer à ce sujet que *chaque graduation planaire n'est que le résultat d'une généralisation de l'anamorphose.*

En effet, on peut toujours imaginer que l'équation $f_1(x, y, N_1) = 0$ qui définit la graduation planaire soit résolue par rapport à N_1, et mise par suite sous la forme $N_1 = \Psi_1(x, y)$.

Or lorsqu'il s'agit d'une graduation linéaire sur l'axe Ox par exemple, l'anamorphose consiste à représenter N_1, de telle façon que l'on puisse écrire par exemple $N_1 = \Psi(x)$.

Ici nous avons les deux quantités x et y à notre disposition et nous représentons le nombre N_1, à l'aide d'une fonction de ces deux quantités.

La double graduation peut donc être définie, au lieu du système (1), par les deux relations

$$\left.\begin{aligned} N_1 &= \Psi_1(x, y) \\ N_2 &= \Psi_2(x, y). \end{aligned}\right\} \qquad (3)$$

Ce nouveau système est plus commode pour la recherche des graduations les plus commodes pour la représentation d'une relation de la forme (2), $F(N_1, N_2) = 0$.

Remplaçons en effet N_1 et N_2 par leurs valeurs tirées du système (3), nous obtenons pour l'équation de la courbe représentative de la relation (2) :

$$F[\Psi_1(x, y), \Psi_2(x, y)] = 0. \qquad (4)$$

On sait déjà qu'on a toute facilité pour représenter la relation (2) à l'aide d'un simple diagramme, en la rapportant à deux axes de coordonnées ON_1, ON_2 sur lesquels on compte respectivement les valeurs de N_1 et N_2, même au besoin par l'intermédiaire d'une anamorphose. Par suite, le système de double graduation ne peut être avantageux, que si on peut l'employer successivement à la représentation de plusieurs relations ne différant que par les données numériques, et si les courbes représentatrices qui en résultent, se trouvent simplifiées.

Il y a donc intérêt à rechercher les graduations telles que la courbe représentatrice (4) soit simple.

N° 33. – Représentation d'une relation par une ligne droite. Exemple. – Cherchons par exemple la condition pour que la courbe représentatrice soit une droite.

Si nous revenons au système définissant la double graduation

$$\left.\begin{array}{l} f_1(x, y, \mathrm{N}_1) = 0 \\ f_2(x, y, \mathrm{N}_2) = 0 \end{array}\right\} \qquad (1)$$

nous voyons que ces deux équations peuvent fournir les valeurs de deux des inconnues en fonction des deux autres. On peut donc exprimer x et y en fonction de N_1 et N_2 et écrire :

$$\left.\begin{array}{l} x = \varphi_1(\mathrm{N}_1, \mathrm{N}_2) \\ y = \varphi_2(\mathrm{N}_1, \mathrm{N}_2). \end{array}\right\} \qquad (5)$$

Nous voulons que ce système soit tel que la courbe représentatrice soit une droite dont l'équation peut s'écrire

$$ax + by + c = 0.$$

Si nous y remplaçons x et y par leurs expressions tirées de (5), nous avons la relation

$$a\,\varphi_1(\mathrm{N}_1, \mathrm{N}_2) + b\,\varphi_2(\mathrm{N}_1, \mathrm{N}_2) + c = 0,$$

qui doit être identique à la relation qu'on se propose de représenter, c'est-à-dire

$$\mathrm{F}(\mathrm{N}_1, \mathrm{N}_2) = 0. \qquad (2)$$

Donc, cette relation (2) étant mise sous la forme

$$a\,\varphi_1(\mathrm{N}_1, \mathrm{N}_2) + b\,\varphi_2(\mathrm{N}_1, \mathrm{N}_2) + c = 0,$$

elle peut être représentée par une ligne droite, à la condition d'employer la double graduation définie par le système

$$\left.\begin{array}{l} x = \varphi_1(\mathrm{N}_1, \mathrm{N}_2) \\ y = \varphi_2(\mathrm{N}_1, \mathrm{N}_2). \end{array}\right\}$$

Exemple. – La relation

$$\mathrm{N}_1^4 + \mathrm{N}_2^4 - \mathrm{R}^2 = 0,$$

où R est une constante, peut se représenter par une droite, où l'on prend pour définir la double graduation les équations

$$x = \mathrm{N}_1^4$$
$$y = \mathrm{N}_2^4.$$

La droite représentatrice n'est autre que

$$x + y - \mathrm{R}^2 = 0,$$

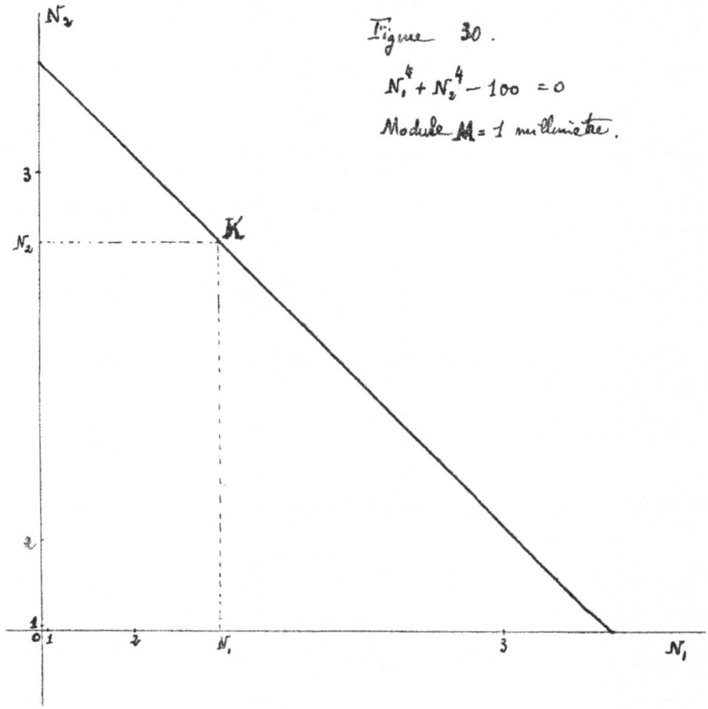

parallèle à une bissectrice de l'angle des axes.

De plus, on voit que chacune des deux graduations est constituée par des droites parallèles respectivement à chacun des axes de coordonnées. Le tracé de ces graduations est donc très simple et très rapide, et il en est de même de la droite représentatrice.

Au contraire, le diagramme de la relation proposée est une courbe peu différente du cercle, mais qu'il serait toujours difficile d'obtenir avec précision.

N° 34. – Représentation d'une relation à deux variables par un cercle. – Si nous cherchons à représenter la relation proposée par un cercle dont l'équation peut être mise sous la forme

$$(x-a)^2+(y-b)^2-\mathrm{R}^2=0$$

(a et b étant les coordonnées du centre et R le rayon), il suffit, d'après ce qui précède, que la relation proposée puisse se mettre sous la forme d'une somme algébrique de carrés

$$[\varphi_1(\mathrm{N}_1,\mathrm{N}_2)]^2+[\varphi_2(\mathrm{N}_1,\mathrm{N}_2)]^2-[\varphi_3(\mathrm{N}_1,\mathrm{N}_2)]^2=0.$$

On peut en effet la ramener immédiatement à la forme suivante :

$$\left[\mathrm{R}\,\frac{\varphi_1(\mathrm{N}_1,\mathrm{N}_2)}{\varphi_3(\mathrm{N}_1,\mathrm{N}_2)}\right]^2+\left[\mathrm{R}\,\frac{\varphi_2(\mathrm{N}_1,\mathrm{N}_2)}{\varphi_3(\mathrm{N}_1,\mathrm{N}_2)}\right]^2-\mathrm{R}^2=0$$

en multipliant par R^2 et divisant par $[\varphi_3(\mathrm{N}_1,\mathrm{N}_2)]^2$, et la double graduation est définie par le système

$$\begin{aligned} x-a&=\mathrm{R}\,\frac{\varphi_1(\mathrm{N}_1,\mathrm{N}_2)}{\varphi_3(\mathrm{N}_1,\mathrm{N}_2)}\\ y-b&=\mathrm{R}\,\frac{\varphi_2(\mathrm{N}_1,\mathrm{N}_2)}{\varphi_3(\mathrm{N}_1,\mathrm{N}_2)}. \end{aligned}$$

Exemple – Reprenons l'exemple précédent $\mathrm{N}_1^4+\mathrm{N}_2^4-\mathrm{R}^2=0$. La double graduation se trouve immédiatement définie par

$$\begin{aligned} x-a&=\mathrm{N}_1^2\\ y-b&=\mathrm{N}_2^2. \end{aligned}$$

Supposons que le centre du cercle soit à l'origine des coordonnées : $a=0$, $b=0$. Par suite la double graduation devient

$$\begin{aligned} x&=\mathrm{N}_1^2\\ y&=\mathrm{N}_2^2. \end{aligned}$$

On voit qu'on a encore deux graduations composées de droites. Ce dernier système présente sur le précédent l'avantage que l'établissement des graduations nécessite seulement le calcul des carrés de N_1 et N_2, tandis que dans le premier système il fallait calculer les $4^{\text{èmes}}$ puissances.

On pourrait de même chercher à représenter la relation proposée

$$\mathrm{F}(\mathrm{N}_1,\,\mathrm{N}_2)=0$$

par une courbe quelconque dont l'équation peut s'écrire

$$\varphi(x,y)=0.$$

Cette recherche présenterait en général peu d'intérêt au point de vue pratique.

N° 35. – Théorie des abaques à 2 variables. – Pour que la théorie puisse s'appliquer facilement aux différents cas, nous la généraliserons de la façon suivante :

Soit la double graduation définie par les équations

$$\left.\begin{aligned} f_1(x,y,\mathrm{N}_1)&=0\\ f_2(x,y,\mathrm{N}_2)&=0. \end{aligned}\right\}\qquad(1)$$

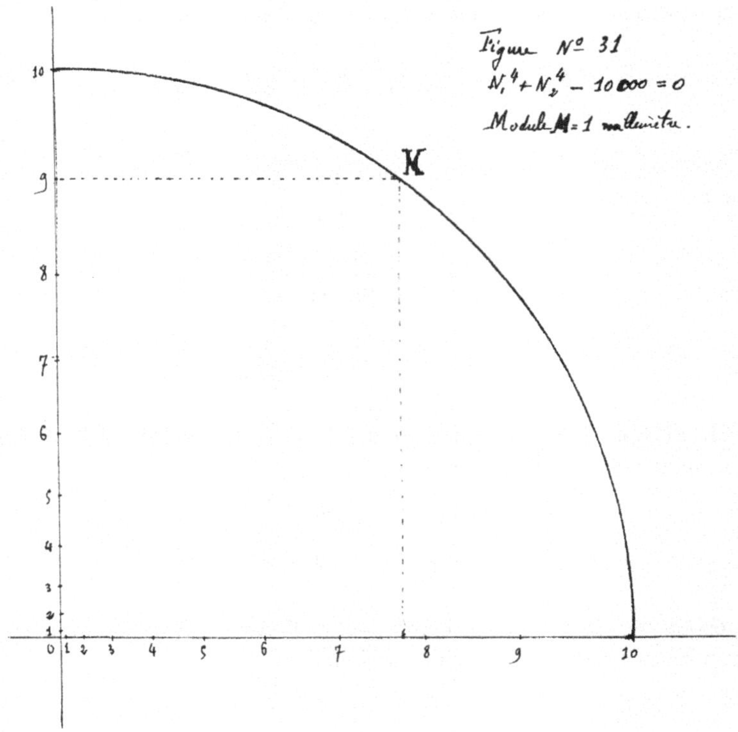

Soit d'autre part

$$\Phi(x, y, \mathrm{N}_1, \mathrm{N}_2) = 0 \qquad (2)$$

l'équation d'une courbe (qui peut dépendre des deux variables N_1 et N_2, ou en être indépendante), choisie à priori pour représenter la relation donnée

$$\mathrm{F}(\mathrm{N}_1, \mathrm{N}_2) = 0 \qquad (3)$$

Il est clair que, quelles que soient ces relations, l'équation (3) fournit toujours des couples de valeurs de N_1 et N_2 auxquels correspondent des couples d'éléments de la double graduation (1) que l'on peut dire conjugués, et l'équation $\mathrm{F}(\mathrm{N}_1, \mathrm{N}_2) = 0$ exprime qu'entre chacun de ces couples d'éléments [4] de la double graduation et la courbe (2), il y a une relation géométrique qu'il suffit de mettre en évidence pour pouvoir faire correspondre à chaque valeur de N_1 une valeur de N_2 satisfaisant à l'équation (3).

Par exemple, on a employé jusqu'à maintenant la courbe (2) passant par l'intersection des deux courbes déterminées dans les graduations planaires (1) par

[4] Nous employons à dessein le terme "élément" qui s'appliquera également à tous les genres de graduation ; il désignera indifféremment le point d'une graduation linéaire, que la ligne correspondant à une valeur du paramètre dans une graduation planaire.

les couples de valeurs de N_1 et N_2 satisfaisant à l'équation (3). Mais on pourrait aussi bien exprimer que les deux courbes correspondantes de la double graduation sont tangentes à une même droite qui enveloppe la courbe (2) (Figure 32).

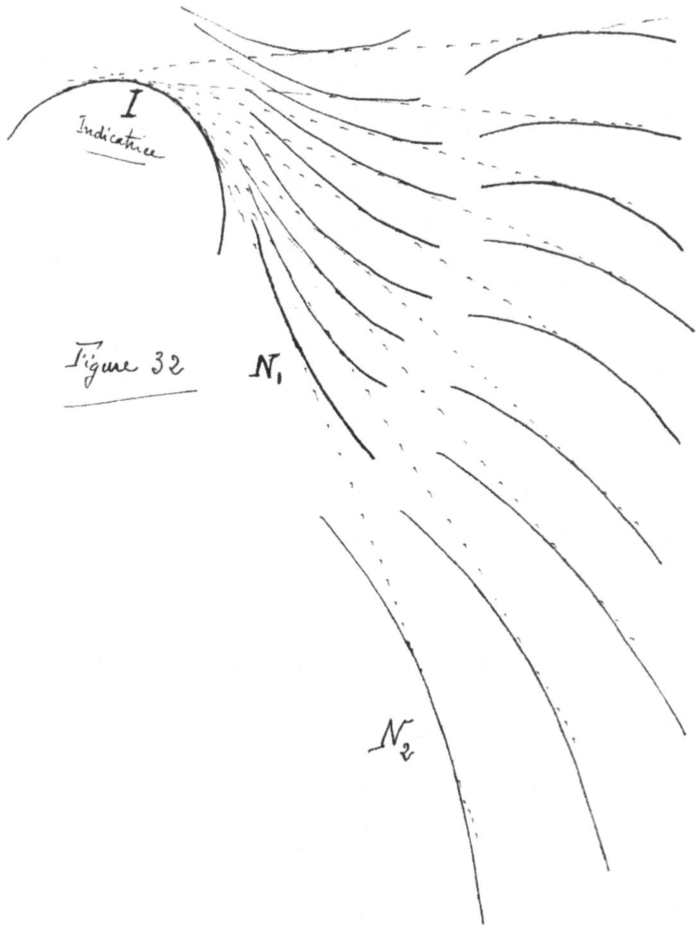

On pourrait encore supposer que de deux points fixes, on mène des tangentes aux courbes conjuguées de la double graduation et que la courbe (2) est le lieu des points d'intersection de ces tangentes.

Il est évident que le nombre des combinaisons que l'on peut faire est, au moins théoriquement, illimité.

On voit donc que, pour la représentation d'une relation entre les 2 nombres N_1 et N_2, on peut adopter une double graduation absolument quelconque et, de même, une courbe représentatrice quelconque, à la seule condition de pouvoir mettre en évidence les relations géométriques entre la courbe représentatrice et les courbes conjuguées de la double graduation.

Nous appellerons *indicatrice* la courbe $\Phi(x, y, \mathrm{N}_1, \mathrm{N}_2) = 0$ choisie dans chaque cas pour représenter la relation proposée.

On doit conclure de ce qui précède que l'indicatrice doit être simple, point, droite ou cercle, rarement une autre courbe. Mais il faut aussi que la relation géométrique entre l'indicatrice et chacune des graduations planaires soit simple, pour que la détermination des couples de courbes conjuguées et par suite des couples de valeurs correspondantes de N_1 et N_2 soit aussi facile que possible.

Lorsque l'indicatrice dépend des variables N_1 et N_2, sa position et même sa forme peuvent varier d'après les valeurs des variables. Théoriquement on peut toujours supposer que l'indicatrice dépend de N_1 et N_2. Pratiquement il est indispensable qu'elle ne dépende que d'une variable, N_1 par exemple. En effet la connaissance de N_1 doit permettre de déterminer la position correspondante de l'indicatrice et, par suite, à l'aide de la relation géométrique connue avec la graduation N_2, la courbe de cette graduation qui est conjuguée de celle qui a servi de point de départ dans la première graduation, et par suite la valeur de N_2 conjuguée de N_1.

Nous allons montrer sur quelques exemples simples l'application de cette théorie.

N° 36. – Exemples d'abaques à 2 variables. – 1°) Considérons tout d'abord deux graduations linéaires proportionnelles égales, parallèles et dirigées en sens inverses (Figure 33).

Joignons les origines O et O′ des deux graduations.

Si l'on mène une sécante KAK′, on voit que l'on a dans les triangles semblables OAK et O′AK′ :

$$\frac{\mathrm{OK}}{\mathrm{O'K'}} = \frac{\mathrm{OA}}{\mathrm{O'A'}}.$$

Par suite, si le point A a été choisi de manière que l'on ait

$$\frac{\mathrm{OA}}{\mathrm{O'A'}} = k,$$

à toute sécante passant par le point A correspondra la relation

$$\mathrm{OK} = k \times \mathrm{O'K'} \qquad \text{ou} \qquad \mathrm{N}_1 = k\mathrm{N}_2.$$

L'indicatrice se réduit ici au point A.

Quant à la relation entre l'indicatrice et les graduations, elle s'exprime par ce fait que les points conjugués des deux graduations sont en ligne droite avec le point A.

La figure 33 *est l'équivalent d'un compas de proportion.*

Dans le cas de la figure, $k = 0,484$. Pour $\mathrm{N}_1 = 4,17$, on trouve $\mathrm{N}_2 = 8,63$ à 0,02 près environ, d'après l'échelle de la figure.

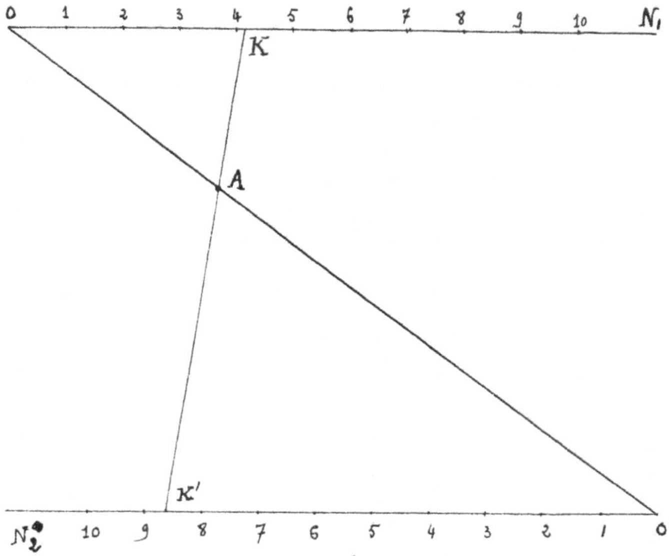

Figure 33
Module M = 1 centimètre.

2°) Soit à représenter une relation telle que

$$\operatorname{tg} N_1 = K \operatorname{tg} N_2 \qquad (1)$$

où N_1 et N_2 sont des nombres de grades par exemple.

On sait qu'une tangente peut toujours être considérée comme le coefficient angulaire d'une droite rapportée à deux axes de coordonnées rectangulaires. Il est donc naturel de chercher à représenter l'équation proposée à l'aide de deux faisceaux de droites concourantes :

$$\begin{cases} y = \operatorname{tg} N_1 . x \\ y = \operatorname{tg} N_2 \, (x - d). \end{cases}$$

On a ainsi deux graduations angulaires, la première autour de l'origine, la deuxième autour d'un point de l'axe Ox situé à une distance d de l'origine.

Les équations précédentes peuvent se mettre sous la forme

$$\begin{cases} \operatorname{tg} N_1 = \dfrac{y}{x} \\ \operatorname{tg} N_2 = \dfrac{y}{x - d} \end{cases}$$

qui permettent d'éliminer N_1 et N_2 dans la relation à représenter (1). On obtient ainsi

$$\frac{y}{x} = K \frac{y}{x - d},$$

$$\text{d'où} \qquad x - d = \mathrm{K}x \qquad \text{ou} \qquad x = \frac{d}{1-\mathrm{K}}.$$

Ce qui veut dire que si l'on adopte le système de double graduation angulaire, l'équation proposée peut être représentée à l'aide d'une *indicatrice rectiligne* parallèle à Oy.

Ici la relation géométrique entre la double graduation et l'indicatrice est la condition de se couper en un même point ; l'indicatrice est le lieu des points de rencontre des courbes conjuguées des deux graduations.

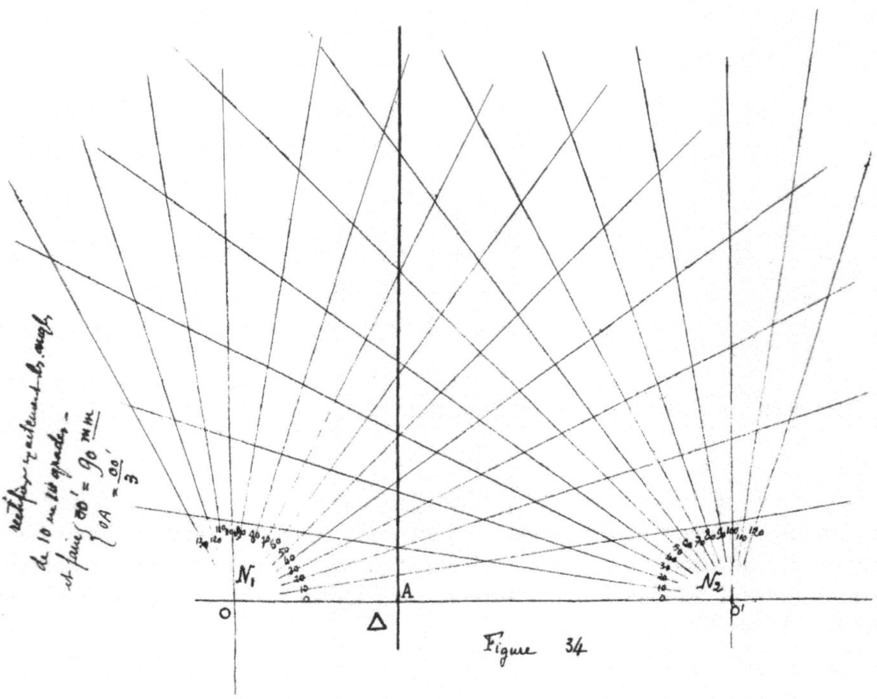

Les graduations correspondent aux angles de 10 en 10 grades.

La droite Δ représente sensiblement la relation

$$\operatorname{tg} \mathrm{N}_1 = 2 \operatorname{tg} \mathrm{N}_2.$$

Il est à remarquer que l'on a fait la graduation de N_2 en sens inverse de celle de N_1 ; cela revient à changer le signe de K, par suite, en se reportant à ce qui précède, on doit avoir

$$\mathrm{OA} = \frac{\mathrm{OO}'}{1+\mathrm{K}}$$

et par suite pour toutes les valeurs positives de K, le point A est entre O et O′.

Remarque. – Toutes les fois qu'il s'agit de relations entre des lignes trigonométriques, il est indiqué à priori d'avoir recours à des graduations angulaires ou circulaires.

3°) Nous emprunterons à la pratique de la navigation un troisième exemple.

Lorsqu'un navire se rend d'un point à un autre, il suit autant que possible une ligne faisant un angle constant avec les différents méridiens, qu'on appelle *loxodromie*.

Chaque jour, des déterminations astronomiques permettent de faire le point, c'est-à-dire de déterminer les coordonnées géographiques de la position du navire à midi.

Il importe donc, pour vérifier la route, de pouvoir calculer rapidement les coordonnées géographiques d'un point de la loxodromie ou, en d'autres termes, de déduire la longitude du point lorsque l'on a déterminé sa latitude.

À cet effet, les marins emploient des cartes dressées d'après la projection de Mercator, dans laquelle les méridiens sont représentés par des droites parallèles et équidistantes, et les parallèles par des droites parallèles, perpendiculaires aux méridiens, mais dont l'intervalle va en augmentant de l'équateur au pôle, qui d'ailleurs se trouve rejeté à l'infini.

La figure 35 montre le canevas du *planisphère* obtenu d'après ce système de projection.

L'intervalle des parallèles est calculé de telle façon que les angles de la surface de la terre soient conservés dans la projection.

Il est évident, par suite, que la loxodromie est représentée par une ligne qui fait des angles égaux avec les droites représentant les méridiens : cette ligne ne peut être qu'une droite.

Si donc on veut aller du point A au point B en décrivant une loxodromie, on devra décrire la courbe représentée par la droite AB et par conséquent parcourir la route qui fait en chacun de ses points l'angle α (figure 35) avec la direction du méridien.

Il est facile à l'aide de la droite AB de calculer les deux coordonnées d'un point quelconque de la route :

par exemple on voit immédiatement que pour la latitude 50° Nord, on devra se trouver à la longitude 5° Est ;

inversement, on traversera le méridien 0° à la latitude 48° Nord.

Nous n'insisterons pas sur le calcul de la graduation des latitudes, cet exemple ayant été choisi principalement pour montrer *l'intérêt qu'il peut y avoir très souvent à aborder le problème à l'aide de considérations géométriques, qui peuvent mettre sur la voie de la solution la plus simple pour la représentation et simplifier le plus souvent les calculs nécessaires à l'établissement des graduations.*

Quels que soient les points A et B, l'indicatrice est toujours la droite joignant A et B.

La relation géométrique entre l'indicatrice et les deux graduations des méridiens et parallèles est encore la condition de se couper en un même point.

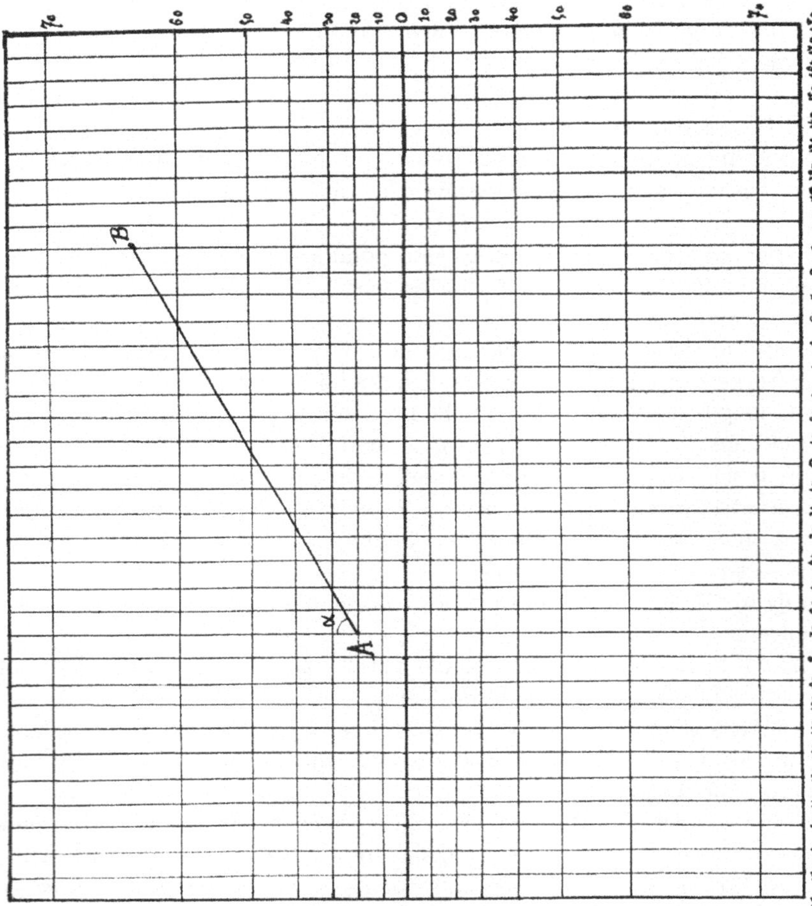

Figure 35
Planisphère de Mercator — Module $0^m,0004$ pour 1°

Nous ne nous étendrons pas davantage sur la représentation des relations à deux variables ; tout ce qui sera dit par la suite au sujet des relations à trois variables s'applique évidemment comme cas particulier aux relations à deux variables, puisqu'il suffit de rendre la $3^{\text{ème}}$ variable constante.

4°) *Indicatrice mobile.* Proposons-nous de calculer l'angle sous lequel on voit un cercle de rayon R d'un point O de son plan, en fonction de la distance du centre C du cercle au point O.

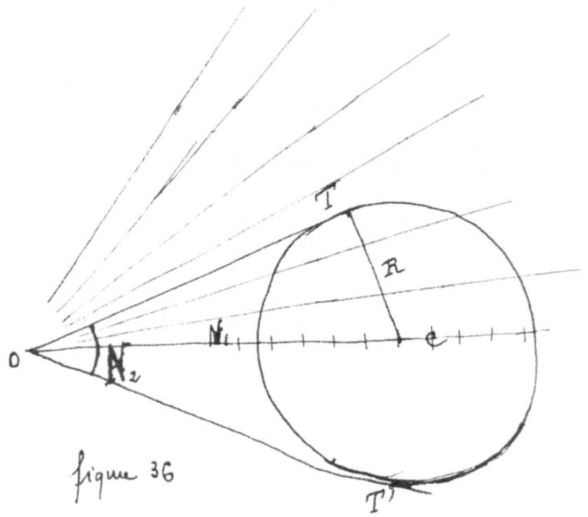

figure 36

Désignons par N_1 la distance OC et par N_2 l'angle $\widehat{TOT'}$. Le triangle rectangle COT donne immédiatement la relation qu'il s'agit d'exprimer, c'est-à-dire

$$R = N_1 \sin \frac{N_2}{2},$$

les longueurs R et N_1 étant exprimées à l'aide de la même unité.

La figure 36 montre immédiatement comment on peut traduire cette relation en abaque, en utilisant directement la relation géométrique qui sert de point de départ.

Il suffit de tracer sur la droite OC une graduation des longueurs, et autour du point O une graduation angulaire. Pour avoir la valeur de N_2 correspondant à une valeur donnée de N_1, on prend le point de OC correspondant à N_1, on trace autour de ce point un cercle de rayon R ; la valeur de N_2 est alors fournie par la droite OT de la graduation angulaire qui se trouve tangente au cercle ainsi tracé.

L'indicatrice est un cercle de rayon R, dont le centre se déplace sur OC. Si l'on prend comme axes de coordonnées OC et la perpendiculaire en O, l'équation de l'indicatrice est :

$$(x - N_1)^2 + y^2 - R^2 = 0.$$

La double graduation est définie par les deux équations

$$\begin{cases} x = N_1 \\ y = x \operatorname{tg} N_2. \end{cases}$$

La relation à représenter exprime que l'indicatrice ayant son centre en N_1 (relation avec la première graduation qui donne immédiatement la position de l'indicatrice), la droite correspondante de la deuxième graduation lui est tangente (relation avec la deuxième graduation).

Si l'on était obligé de tracer l'indicatrice pour chaque nouvelle valeur de N_1, un tel abaque serait peu pratique. En réalité on trace le cercle sur un transparent que l'on peut déplacer à volonté sur l'abaque comprenant la double graduation.

On doit en conclure que l'emploi d'une indicatrice courbe n'est réellement pratique que si cette courbe a une forme invariable, ce qui permet l'emploi du transparent.

On pourrait en effet admettre que l'on trace sur le transparent les différentes formes que prend l'indicatrice, ce qui reviendrait à y tracer une graduation planaire, mais on conçoit qu'on arrive ainsi à un système compliqué, d'emploi difficile et par conséquent à éviter, ce qui est toujours possible.

La relation proposée dans l'exemple précédent peut être représentée d'une façon plus simple ; la solution indiquée n'a pas un grand intérêt pratique, elle n'a été donnée que pour préparer à ce qui va suivre, en précisant la notion de l'indicatrice mobile.

Théorie générale des abaques

N° 37. – Représentation des relations à un nombre quelconque de variables. – On peut immédiatement conclure de ce qui précède qu'il est possible de représenter une relation quelconque

$$F(N_1, N_2, \ldots, N_p) = 0 \qquad (1)$$

contenant p quantités variables $N_1, N_2, \ldots, N_p$, à l'aide d'un abaque dans lequel chaque quantité variable est représentée en général par une graduation.

L'abaque à p variables se compose donc tout d'abord d'un système de p graduations que nous pouvons représenter par les équations

$$\left.\begin{array}{l} f_1(x,y,N_1) = 0 \\ f_2(x,y,N_2) = 0 \\ f_3(x,y,N_3) = 0 \\ \ldots\ldots\ldots\ldots\ldots \\ f_p(x,y,N_p) = 0. \end{array}\right\} \qquad (2)$$

Pour pouvoir résoudre l'équation (1), c'est-à-dire déterminer graphiquement N_p par exemple lorsqu'on se donne les $(p-1)$ autres variables, il faut disposer d'une relation entre les p éléments correspondants des p graduations.

On y arrive facilement à l'aide de *l'indicatrice* qui permet d'utiliser des relations géométriques aussi simples que possible.

N° 38. – Indicatrice. Conditions qu'elle doit remplir. – Soit

$$\Phi(x, y, N_1, N_2, \ldots, N_p) = 0 \qquad (3)$$

l'équation de l'indicatrice.

Sous cette forme, elle paraît dépendre des p paramètres variables. En réalité, en utilisant l'équation (1) on peut remplacer une des variables par son expression en fonction des autres. Par suite l'indicatrice ne dépend effectivement que de $(p-1)$ des quantités variables.

Si l'on se donne des valeurs de $(p-1)$ variables, la $p^{\text{ème}}$ se trouve déterminée par suite de la relation (1). L'indicatrice est déterminée à l'aide de ses $(p-1)$ relations géométriques avec les éléments des $(p-1)$ graduations qui correspondent aux variables connues. Par suite, la relation géométrique de l'indicatrice et de la dernière graduation permet de fixer la valeur de la dernière variable, qui se trouve ainsi déterminée en fonction des autres.

Pour que la détermination de la $p^{\text{ème}}$ variable puisse être faite comme il vient d'être dit, il est nécessaire, tout d'abord, que l'indicatrice soit déterminée dès qu'on connaît $(p-1)$ variables. L'équation

$$\Phi(x, y, \mathrm{N}_1, \mathrm{N}_2, \ldots, \mathrm{N}_p) = 0$$

ne peut donc contenir aucune autre quantité indéterminée que les p variables N, dont l'une est déterminée en fonction des $(p-1)$ autres par l'équation (1).

Par contre, l'équation de l'indicatrice doit dépendre effectivement des $(p-1)$ paramètres variables. Supposons en effet qu'il suffise de connaître les q premiers paramètres pour déterminer complètement l'équation de l'indicatrice $(q < p-1)$. La connaissance des éléments des q premières graduations permettrait alors de fixer la position de l'indicatrice, et par suite de ses relations avec les autres graduations, les $(p-q)$ autres variables se trouveraient déterminées en même temps. Ces $(p-q)$ variables devraient donc être considérées comme des fonctions des q premières ; il y aurait en réalité $(p-q)$ relations entre les p variables, ce qui est contraire à l'hypothèse, puisqu'il s'agit de représenter la relation unique

$$\mathrm{F}(\mathrm{N}_1, \mathrm{N}_2, \mathrm{N}_3, \ldots, \mathrm{N}_p) = 0.$$

Pour fixer les idées, si l'équation de l'indicatrice peut être mise sous la forme d'un polynôme en x et y, elle doit dépendre de $(p-1)$ coefficients différents et de $(p-1)$ seulement (p sous la forme homogène).

Soient A_1, A_2, A_3, …, A_{p-1} ces $(p-1)$ coefficients ; ce sont des fonctions des variables N_1, N_2, …, $\mathrm{N}_{(p-1)}$ (N_p ayant été éliminé comme on l'a déjà vu). Les conditions à remplir par ces fonctions sont telles, que si on se donne les valeurs des $(p-1)$ variables N, les coefficients A sont complètement déterminés ; inversement, la condition que toutes les quantités N interviennent effectivement équivaut à dire que si l'on se donne les valeurs des $(p-1)$ coefficients A, les valeurs des $(p-1)$ variables N seront également déterminées.

Dans sa forme géométrique la plus générale, le problème se présente de la façon suivante : la connaissance de $(p-1)$ éléments des $(p-1)$ premières graduations devant permettre de fixer l'indicatrice, cette courbe doit dépendre exactement de $(p-1)$ conditions.

Par exemple : si l'on a p graduations planaires, on emploie une indicatrice qui est déterminée par $(p-1)$ tangentes ; si l'on a p graduations linéaires, l'indicatrice doit être déterminée dans tous les cas par $(p-1)$ points appartenant aux $(p-1)$ premières graduations. L'intersection de l'indicatrice et de la $p^{\text{ème}}$ graduation détermine la valeur de la variable à calculer.

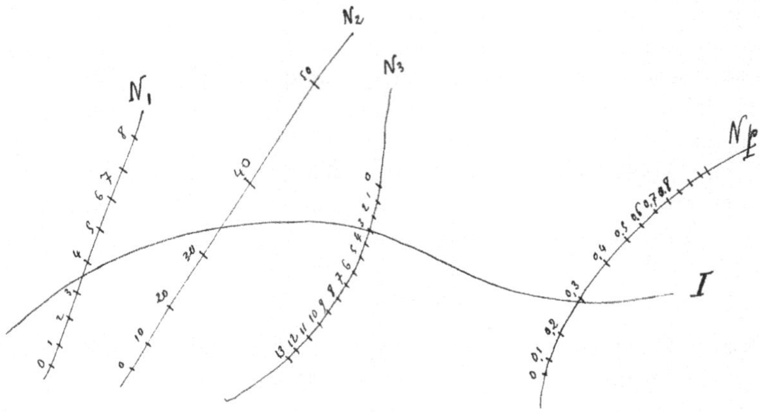

Figure 37

Remarque. – Il est clair qu'on peut toujours utiliser une indicatrice dépendant de plus de $(p-1)$ coefficients pour représenter une relation à p variables. Il suffit en effet de donner arbitrairement des valeurs constantes à certains des coefficients de manière qu'il en reste seulement $(p-1)$ à déterminer.

Appendix D

Chronology of the life of André-Louis Cholesky

- **1875**
  - *15 October:* Birth of André–Louis Cholesky in Montguyon.
- **1893**
  - *24 July:* Bachelor's degree.
- **1895**
  - *15 October:* Admission to *École Polytechnique.*
- **1897**
  - *1 October:* Admission to *École d'Application de l'Artillerie et du Génie.*
- **1899**
  - *1 October:* Lieutenant en second at the 22th Artillery Regiment.
- **1902**
  - *17 January - 27 June:* Mission in Tunisia.
  - *21 November 1902 - 1 May 1903:* Mission in Algeria.
- **1903–1904**
  - *31 December 1903 - 6 June 1904:* Mission in Algeria.
- **1905**
  - *24 June:* Assigned to the Geographical Service of the Army.
  - *July - September:* Mission in the Rhône valley, in Isère, and in Cevennes.

- *26 September:* Promoted premier lieutenant.

- **1907**
  - *10 May:* Marriage with Anne Henriette Brunet.
  - *June - September:* Surveying of the meridian of Lyon.

- **1907–1908**
  - *7 November 1907 - 25 June 1908:* Mission in Crete.
  - *16 June 1908:* Birth of his first son René.

- **1909**
  - *25 March:* Promoted capitaine en second at the 27th Artillery Regiment.
  - *28 August:* Removed from the Geographical Service.
  - *14 September:* Assigned to the 13th Artillery Regiment for a statutory period of two years as a battery commander.
  - *30 November:* Birth of his first daughter Françoise.
  - *December:* Professor at *École Spéciale des Travaux Publics, du Bâtiment et de l'Industrie.*

- **1910**
  - *2 December:* Manuscript of Cholesky on his method for solving systems of linear equations.

- **1911**
  - *4 June:* Birth of his second daughter Hélène.
  - *24 September:* Assigned to the particular general staff of artillery.
  - *13 October:* Assigned to the Geographical Service of the Army as head of leveling in Algeria and Tunisia.

- **1911–1912**
  - *27 October 1911 - 24 April 1912:* Leveling in Algeria.

- **1912–1913**
  - *8 October 1912 - 17 April 1913:* Leveling in Tunisia.

- **1913–1914**
  - *25 May 1913:* Appointed head of the Topographical Service of the Regency of Tunisia.

- *2 August 1914:* Mobilization. He joined the 7th artillery group in Bizerte.
- *18 September:* Assigned to the 16th Artillery Regiment in Issoire.
- *24 September:* Appointed as commander of the 9th battery of the 23rd Artillery Regiment.

- **1915**
  - *3 January:* Detached to the general commanding the artillery of the 17th Army Corps for organizing the firing.
  - *11 February:* Assigned to the Geographical Service.
- **1916–1918**
  - *25 September 1916 - February 1918:* Technical Director of the Geographical Service of the Romanian Army.
  - *6 July 1917:* Promoted chef d'escadron.
- **1918**
  - *5 June:* Assigned to the 202nd Field Artillery Regiment.
  - *11 June:* Commander of the 2nd group of artillery.
  - *15 August:* Fights against the Hindenburg line.
  - *23 August:* Fighting on the Ailette river.
  - *31 August:* André–Louis Cholesky died at 5 am near Bagneux from injuries received on the battlefield. Buried in the military cemetery of Chevillecourt.
- **1919**
  - *19 January:* Birth of his second son André–Louis.
- **1921**
  - *24 October:* Transfer of his body to the military cemetery of Cuts.
- **1924**
  - *April - June:* Benoît's paper on Cholesky's method.
- **1946**
  - Lectures on Cholesky's method by John Todd at King's College in London.

Acknowledgments

I am grateful to many people who helped me during the preparation of this work: Antonio Arricó, Assistant Professor at the University of Naples (Italy), Olivier Azzola, Archivist of *École Polytechnique* (France), Jean–Pierre Barriot, Engineer at CNES (France), Zsombor Bartos–Elekes, Senior Lecturer, Babeş–Bolyai University, Cluj–Napoca (Romania), Bruno Belhoste, Professor at the University of Paris I–Panthéon–Sorbonne (France), Claudine Billoux, former Archivist of *École Polytechnique* (France), Konstantinos Chatzis, Professor at *École Nationale des Ponts et Chaussées* (France), Richard W. Cottle, Professor at the University of Stanford (USA), Florence Darmon, Director of the *École Spéciale des Travaux Publics, du Bâtiment et de l'Industrie* (France), Frédéric De Ligt, President of the APMEP of the *Région Poitou–Charentes* (France), Jean–Paul Devilliers, Editor–in–Chief of *SABIX* (France), Iain Duff, Head of Numerical Analysis Group at Rutherford Appleton Laboratory (United Kingdom) and Project Leader of Parallel Algorithms Group at CERFACS (France), Yves Dumont, Researcher at CIRAD (France), Jérome Dyment, Research fellow at the CNRS, *Institut de Physique du Globe de Paris* (France), Robert Erra, Professor at the *École Supérieure d'Informatique, Électronique, Automatique* (France), Richard William Farebrother, Professor at the University of Manchester (United Kingdom), Laurence Fioux, *Responsable des Renseignements Bibliographiques à Distance*, BDIC (France), Jean Fromentin, Member of APMEP of the *Région Poitou–Charentes* (France), Seiji Fujimo, Professor at Kyushu University (Japan), Walter Gautschi, Professor Emeritus, Purdue University, West Lafayette (USA), Magdalena Głodek, Librarian of the *Bibliothèque Polonaise de de Paris* (France), Roland Glowinski, Professor at the University of Houston (USA), Christian Gout, Professor at INSA Rennes, Editor–in–Chief of MATAPLI (France), Joseph F. Grcar, Professor at University of Illinois at Urbana–Champaign (USA), Fred G. Gustavson, Emeritus IBM Watson Research (USA) and Umeå University (Sweden), Freddy Haeghens (Belgium), Nicholas J. Higham, Professor at the University of Manchester (United Kingdom), Mircea Ivan, Professor at the University of Cluj (Romania), Al Jiménez, Professor at the California Polytechnic State University (USA), Zbigniew Judycki, *Institut des Recherches Biographiques de Vaudricourt* (France), William Kahan, Professor at the University of California et Berkeley (USA), Michel Kasser, President of the *Association Française de Topographie* (France), Catherine Langlais, Secretary of SABIX (France), Véronique Magnol–Malhache, Collections manager, Library André Desguine, Departmental Archives of Val de Marne (France), Jean–Pierre Maillard, Editor–in–Chief of *XYZ* (France), John Maindonald, Professor at the Australian National University, Canberra (Australia), Therese A. McCarty, Vice President of Academic Affairs, Professor of Economics, Union College, Schenectady, NY (USA), Gérard Meurant, former Research Director at CEA/DIF/DSSI (France), Marilena Mitrouli, Professor at the University of Athens (Greece), Alexandre Moatti, President of SABIX (France), Hiroshi Murakami, Professor at the Tokyo Metropolitan University (Japan), Gerhard Opfer,

Professor at the University of Hamburg (Germany), Serge Parpay, Member of AP-MEP of the *Région Poitou–Charentes* (France), Stefan Paszkowski, Professor at the University of Wrocław (Poland), David H. Pattison, Social Security Administration (USA), Sophie Pecquerie, Librarian at CNES (France), Jeanne Peiffer, Research Director at CNRS, former Editor–in–Chief of the *Revue d'Histoire des Mathématiques* (France), François Robert, former Professor at IMAG in Grenoble (France), Giuseppe Rodriguez, Professor at the University of Cagliari (Italy), Dominique Samper, Director of the Municipal Library of Cachan (France), Scott Sanders, Antioch College, Yellow Springs, Ohio (USA), Ridgway Scott, Professor at the University of Chicago (USA), Roger Serre, former Chief Engineer at the *Institut Géographique National* and Professor at the *École Nationale des Sciences Géographiques* (France), George P. H. Styan, Professor at McGill University (Canada), Cécile Tardy, Responsible of the *Département des Services aux Publics*, BDIC (France), Marie–Christine Thooris, Head of the Library of *École Polytechnique* (France), Alain Tavernese (France), Alain Vienne, Director of the Observatory of Lille and Professor at the *Université des Sciences et Technologies de Lille* (France), Robert Vincent, Honorary President of the *Association Française de Topographie* (France).

I am indebted to Raymond Nuvet, vice–major of Montguyon (France), who found most of the documents concerning the family of Cholesky, and to the municipal council of Montguyon for deciding to give the name of André–Louis Cholesky to the cultural center of the city. This center was inaugurated on 1st September 2012, and I would like to thank its staff and Frédéric De Light for their active involvement in this project. Three grandchildren of Cholesky attended this event.

I am grateful to the family of André–Louis Cholesky, in particular his grandson Michel Gross and his granddaughter Jacqueline, for providing me all the informations and the documents in their possession and allowing me to use them.

A special thank is due to Michela Redivo–Zaglia, Professor at the University of Padua (Italy), for her help in searches on internet, in formatting the illustrations, and for her active participation to the final layout of the book.

I also thank the staff of the *Bibliothèque de l'Université Charles–de–Gaulle - Lille III, Bibliothèque de l'Institut Henri Poincaré, Archives Départementales du Val de Marne, Bibliothèque de la Ville de Cachan, Archives et Bibliothèque de l'École Polytechnique, Bibliothèque MIR de l'Université de Paris VI, Bibliothèque de Documentation Internationale Contemporaine, Service Historique de la Défense, Biblioteca del Seminario Matematico dell'Università di Padova.*

Dominique Tournès not only had the difficult task to reproduce the unpublished handwritten book (and the figures) of Cholesky given in the Appendix C, but he wrote the very documented Chapter 7. Moreover, he accompanied me since the beginning of this project, he always encouraged me, and he also provided interesting references and quotations. I greatly appreciated his help. An important part of the value of this book is due to his work and his scientific qualifications.

We are both indebted to Richard Kennedy for the excellent quality of his translation from French of Chapter 7, and to Edwin Beschler, our copy editor, who had the daunting task of correcting the English.

We don't forget the archivists of the *École Polytechnique*, Mrs. Claudine Billoux and Mr. Olivier Azzola for helping us and giving us the right to use the material the archives possess, and Mrs. Marie–Christine Thooris, head of the Library.

Finally, we would like to thank Anna Mätzener and Karin Neidhart, our editors, and the staff of Birkhäuser for assistance in the final preparation of this book.

André–Louis Cholesky around 1910
© Cholesky's family

Bibliography

[1] Dossier Benoît, No. 7YE 2.925, Service Historique de l'Armée de Terre, Fort Vieux de Vincennes.

[2] Dossier Cholesky, No. 5YE 126.454, Service Historique de l'Armée de Terre, Fort Vieux de Vincennes.

[3] *Annuaire des anciens élèves du navire école Jacques Cartier 1919–1932*, Compagnie Générale Transatlantique, 1961.

[4] *Historique du 202e Régiment d'Artillerie de Campagne pendant la Guerre 1914–1918*, Imprimerie Allier Frères, Grenoble, 1920.

[5] *L'Enseignement par Correspondance*, École Spéciale des Travaux Publics, du Bâtiment et de l'Industrie, Paris, 1918.

[6] *La nouvelle carte de France*, Service Géographique de l'Armée, 1923.

[7] *Le Régime foncier aux Colonies. 3e série, tome III : Tunisie–Érytrée–Philippines*, Institut colonial international, Bruxelles, 1899.

[8] *Le Service Géographique de l'Armée. Son Histoire - Son Organisation - Ses Travaux*, Ministère de la Défense Nationale et de la Guerre, Imprimerie du Service Géographique de l'Armée, Paris, 1938.

[9] *Matériaux d'Étude Topologique pour l'Algérie et la Tunisie*, Service Géographique de l'Armée, No. 23, 1906.

[10] *Topographie d'Exploration*, Cahiers du Service Géographique de l'Armée, No. 27, 1907.

[11] *Rapport sur les Travaux Effectués en 1906*, Cahiers du Service Géographique de l'Armée, No. 28, 1907.

[12] *Matériaux d'Étude Topologique pour l'Algérie et la Tunisie*, Service Géographique de l'Armée, No. 29, 1908.

[13] *Rapport sur les Travaux Effectués en 1907*, Cahiers du Service Géographique de l'Armée, No. 30, 1908.

[14] *Rapport sur les Travaux Effectués en 1908*, Cahiers du Service Géographique de l'Armée, No. 31, 1909.

[15] *Rapport sur les Travaux Effectués en 1911*, Cahiers du Service Géographique de l'Armée, No. 34, 1912.

[16] *Rapport sur les Opérations du Nivellement de Précision d'Algérie et de Tunisie pendant les campagnes 1910–1911, 1911–1912, 1912–1913 par le Capitaine Cholesky*, Cahiers du Service Géographique de l'Armée, No. 35, 1913, 34 pp.

[17] *Rapport sur les Travaux Effectués en 1912*, Cahiers du Service Géographique de l'Armée, No. 36, 1913.

[18] SHAPE Officer's Association Newsletter, 124, April 2004, http://www.nato.int/shape/community/soa/members/nl124_files/NL124.-pdf.

[19] C. Abbe, A historical note on the method of least squares, Amer. J. Science and Arts, 1 (1871) 411–415.

[20] O.S. Adams, *Application of Theory of Least Squares to the Adjustment of Triangulation*, Special publication No. 28, Government Printing Office, Washington, 1915.

[21] R. Adrain, Research concerning the probabilities of the errors which happen in making observations, The Analyst or Mathematical Museum, 1 (1808) 93–108.

[22] R. Adrain, Investigation of the figure of the earth and of the gravity in different latitudes, Trans. Amer. Phil. Soc., 1 (1818) 119–135.

[23] R. Adrain, Research concerning the mean diameter of the earth, Trans. Amer. Phil. Soc., 1 (1818) 353–366.

[24] H.W. Ahlmann, L. Rosenbaum, Scientific results of the Swedish–Norwegian Arctic expedition in the Summer of 1931, Part I–III, Geografiska Annaler, 15 (1933) 1–68.

[25] A.C. Aitken, On the evaluation of determinants, the formation of their adjugates and the practical solution of simultaneous linear equations, Proc. Edinb. Math. Soc. (2), 3 (1932) 207–219.

[26] J. Aldrich, Doing least squares: perspectives from Gauss and Yule, Internat. Stat. Rev., 66 (1998) 61–81.

[27] I.H. Alexander, On "The first method to adjust a traverse based on statistical considerations" by Herbert W. Stoughton, Surveying and Mapping, 35 (1975) 174–176.

[28] R. Allmendinger, N. Cardozo, D.M. Fisher, *Structural Geology Algorithms. Vectors and Tensors*, Cambridge University Press, Cambridge, 2012.

[29] S.C. Althoen, R. McLaughlin, Gauss–Jordan reduction: a brief history, Amer. Math. Monthly, 94 (1987) 130–142.

[30] S. Arend, Voies nouvelles dans le calcul scientifique, Ciel et Terre, 57 (1941) 497–513.

[31] Arthur–Lévy, *Le Service Géographique de l'Armée 1914–1918*, Berger–Levrault, Nancy–Paris–Strasbourg, 1926.

[32] V.L. Assur, A.M. Filatov, *Practical Guise to Surveying*, Mir Publishers, Moscow, 1988.

[33] R. Aublet, Historique des écoles d'artillerie, Bulletin d'Information et de Liaison des Officiers d'Artillerie d'Active et de Réserve, 6 (1953) – 13 (1955).

[34] M. Baboulin, L. Giraud, S. Gratton, A parallel distributed solver for large dense symmetric systems: applications to geodesy and electromagnetism problems, Int. J. of High Performance Computing Applications, 19 (2005) 353–363.

[35] T. Banachiewicz, Principes d'une nouvelle technique de la méthode des moindres carrés. Méthode de résolution numérique des équations linéaires, du calcul des déterminants et des inverses, et de réduction des formes quadratiques, Bulletin international de l'Académie polonaise des sciences et des lettres, Classe des sciences mathématiques, Série A, 1938, pp. 134–135, 393–404.

[36] T. Banachiewicz, *Études d'Analyse Pratique*, Cracow Observatory Reprint 22, University of Cracow, 1938.

[37] T. Banachiewicz, An outline of the Cracovian algorithm of the method of least squares, Astronom. J., 50 (1942) 38–41.

[38] R. Barrett, M. Berry, T.F. Chan, J. Demmel, J.M. Donato, J. Dongarra, V. Eijkhout, R. Pozo, C. Romine, H. Van der Vorst, *Templates for the Solution of Linear Systems. Building Blocks for Iterative Methods*, 2nd ed., SIAM, Philadelphia, 1994.

[39] Zs. Bartos–Elekes, I. Rus, G. Timár, The geodetic basis of the Romanian historical topographic maps in Lambert–Cholesky projection, Geophysical Research Abstracts, 10 (2008) EGU2008–A–00428.

[40] J. Bauschinger, H. Andoyer, Théorie des erreurs, in *Encyclopdédie des Sciences Mathématiques Pures et Appliquées*, Édition française, J. Molk et Ch. Lallemand éds., Tome I, Volume 4, Gauthier–Villars, Paris et B.G. Teubner, Leipzig, 1908; réédition par Jacques Gabay, Paris, 1993.

[41] Cdt. Benoît, Le Commandant Cholesky, Union Géodésique et Géophysique Internationale, Première Assemblée Générale, Rome, Mai 1922, Section de Géodésie, Bulletin Géodésique, 1 (1922) 159–161.

[42] Cdt. Benoît, Note sur une méthode de résolution des équations normales provenant de l'application de la méthode des moindres carrés à un système d'équations linéaires en nombre inférieur à celui des inconnues, (Procédé du Commandant Cholesky), Bulletin Géodésique, 2 (1924) 67–77.

[43] Lieutenant-colonel E. Benoît, Formules pratiques pour le calcul des coordonnées géodésiques; application dans le cas de l'ellipsoïde de référence international, avec une Introduction par le colonel G. Perrier, des exemples et des tables, calculés par M. Périn dans le système de la division sexagésimale de la circonférence, Bulletin Géodésique, 12 (octobre–novembre–décembre 1926), p. 187.

[44] Lieutenant-colonel E. Benoît, Formules pratiques pour le calcul des coordonnées géodésiques; application dans le cas de l'ellipsoïde de référence international (suite). Exemples et tables calculés par M. Périn, dans le système de la division centésimale de la circonférence, Bulletin Géodésique, 16 (octobre–novembre–décembre 1927), p. 135.

[45] F.W. Bessel, J.J. Baeyer, *Gradmessung in Ostpreussen und ihre Verbindung mit preussischen und russischen Dreieckskettten*, Dümmler, Berlin, 1838.

[46] C. Billoux, *"In Extremis"*. L'arrivée des archives d'André Louis Cholesky à l'École Polytechnique, Bull. Soc. Amis Bib. Éc. Polytech., 39 (2005) 5.

[47] Å. Björck, *Numerical Methods for Least Squares Problems*, SIAM, Philadelphia, 1996.

[48] E. Bodewig, Bericht über die vershiedenen Methoden zur Lösung eines Systems linear Gleichungen mit reellen Koffizienten. I–V, Indagationes Mathematicae, Part I, 9(4):441–452 (1947); Part II, 9(5): 518–530 (1947); Part III, 9(5):611–621 (1947); Part IV, 10(1):24–35 (1948); Part V, 10(1):82–90 (1948). Also published in Koninklijke Nederlandsche Akademie van Wetenschappen, Proceedings, Part I, 50(7–8):930–941 (1947); Part II, 50(9–10):1104–1116 (1947); Part III, 50(9–10):1285–1295 (1947); Part IV, 51(1–2):53–64 (1948); Part V, 51(1–2):211-219 (1948).

[49] E. Bodewig, *Matrix Calculus*, Noth–Holland, Amsterdam, 1956.

[50] A.D. Booth, *Numerical Methods*, Butterworths Scientific Publications, London, 1955.

[51] C.B. Boyer, *A History of Mathematics*, 2nd edition, Wiley, New York, 1968.

[52] C. Brezinski, André–Louis Cholesky, in *Numerical Analysis, A Numerical Analysis Conference in Honour of Jean Meinguet*, Bull. Soc. Math. Belg., 1996, pp. 45–50.

[53] C. Brezinski, Géodésie, topographie et cartographie, Bull. Soc. Amis. Bib. Éc. Polytech., 39 (2005) 33–68.

[54] C. Brezinski, La méthode de Cholesky, Rev. Hist. Math., 11 (2005) 205–238.

[55] C. Brezinski, The life and work of André Cholesky, Numer. Algorithms, 43 (2006) 279–288.

[56] C. Brezinski, André-Louis Cholesky: his life and works, in *Transgressive Computing 2006*, Jean–Guillaume Dumas ed., Granada, Spain, pp. 31–35.

[57] C. Brezinski, *Les images de la Terre. Cosmographie, géodésie, topographie et cartographie à travers les siècles*, Éditions L'Harmattan, Paris, 2010.

[58] C. Brezinski, André–Louis Cholesky, topographe et mathématicien (1875–1918), XYZ, 129 (2011) 51–62.

[59] C. Brezinski, M. Gross–Cholesky, La vie et les travaux d'André Cholesky, Bull. Soc. Amis Bib. Éc. Polytech., 39 (2005) 7–32.

[60] C. Brezinski, M. Gross–Cholesky, R. Nuvet, La famille d'André–Louis Cholesky, Matapli, 97 (2012) 69-86.

[61] C. Brezinski, L. Wuytack, Numerical analysis in the twentieth century, in *Numerical Analysis: Historical Developments in the 20th Century*, C. Brezinski and L. Wuytack eds., North–Holland, Amsterdam, 2001, pp. 1–40.

[62] R.A. Buckingham, *Numerical Methods*, Pitman, London, 1957.

[63] J.L. Burgess, *The Origins of the* Banlieue Rouge *: Politics, Local Government and Communal Identity in Arcueil and Cachan, 1919–1958*, Ph.D. Thesis, University of New South Wales, Australia, 2005.

[64] J.–P. Callot, *Histoire de l'École Polytechnique*, Les Presses Modernes, Paris, 1959.

[65] G. Carey, Cholesky problems, Behav. Genet., 35 (2005) 653–665.

[66] A. Celmiņš, The method of Gauss in 1799, Statistical Science, 13 (1998) 123–135.

[67] J.–L. Chabert et al., *Histoire d'Algorithmes, du caillou à la puce*, Belin, Paris, 1994; English translation by C. Weeks, *A History of Algorithms. From the Pebble to the Microship*, Springer, Berlin-New York, 1999.

[68] A.L. Cholesky, *Levés d'Études à la Planchette*, 3e éd., révisée par M. Lafosse, École Spéciale des Travaux Publics, Paris, 1923.

[69] A.L. Cholesky, *Cours de Topographie. 2e Partie, Topographie Générale*, 7e éd., École Spéciale des Travaux Publics, Paris, 1937.

[70] B.–I. Clasen, Sur une nouvelle méthode de résolution des équations linéaires et sur l'application de cette méthode au calcul des déterminants, Ann. Soc. Sci. Bruxelles (2), 12 (1888) 251–281 ; reprint Mathesis, 9 (1889), suppl. 2, 1–31.

[71] B.-É. Cousinéry, *Le calcul par le trait*, Carilian–Gœury & Dalmont, Paris, 1839.

[72] V. Crăciunescu, S. Constantinescu, I. Ovejanu, I. Rus, Project eHarta: a collaborative initiative to digitally preserve and freely share old cartographic documents in Romania, e-Perimetron, 6 (2011) 261–269.

[73] G. Cramer, *Introduction à l'analyse des lignes courbes algébriques*, Geneva, 1750.

[74] S.H. Crandall, *Engineering Analysis. A Survey of Numerical Procedures*, McGraw–Hill, New York, 1956.

[75] P.D. Crout, A short method for evaluating determinants and solving systems of linear equations with real or complex coefficients, Transactions of the American Institute of Electrical Engineers, 60 (1941) 1235–1240.

[76] P.D. Crout, *A short method for evaluating determinants and solving systems of linear equations with real or complex coefficients*, Marchant Methods MM–182, Sept. 1941, Marchant Calculating Machine Co., Oakland, California.

[77] C. Culmann, *Die Graphische Statik*, Meyer & Zeller, Zürich, 1866 : 2e éd. refondue, vol. 1, 1875 ; Trad. fr. de la 2e éd. allemande par G. Glasser, J. Jacquier & A. Valat, *Traité de statique graphique*, Dunod, Paris, 1880.

[78] J.A. Da Cunha, *Principes Mathématiques de Feu Joseph–Anastase Da Cunha Traduits Littéralement du Portugais par J.M. D'Abreu*, Imprimerie d'André Racle, Bordeaux, 1911; Reprint by Departamento de Matemática, Faculdade de Ciências e Tecnologia, Universidade de Coimbra, 1987.

[79] J. Daguillon, *Le Sol est fait de nos Morts*, Nouvelles Éditions Latines, Paris, 1987.

[80] R. Danger, *Les Carrières dans la Topographie*, Eyrolles, Paris, 1924 (2nd ed. with a foreword by Maréchal Lyautey, 1932).

[81] P.J. Davis, My early experiences with scientific computation, in *The Birth of Numerical Analysis*, A. Bultheel and R. Cools eds., World Scientific, Singapore, 2010, pp. 187–192.

[82] W. De Sitter, Determination of the mass of Jupiter and elements of the orbits of its satellites from observations made with the Cape heliometer by Sir David Gill, K.C.B., and W.H. Finlay, M.A., part I, Annals of the Cape Observatory, 12 (1915) 160–173.

[83] F. Debergh, A. Gaillard, *1918. Les Chemins de l'Armistice*, Éditions France–Empire, Paris, 1998.

[84] R. Descartes, *La géométrie*, appendice au *Discours de la méthode*, Leyde, 1637 ; réed., Dover, New York, 1954.

[85] A. Desguine, *En souvenir de M. Léon Eyrolles, 1861–1945*, Hors commerce, Ch. Saintard, Cachan, 1979.

[86] J.–P. Devilliers, Éditorial, Bull. Soc. Amis Bib. Éc. Polytech., 39 (2005) 3.

[87] J. Dommanget, In Memoriam Sylvain Arend, Ciel et Terre, 108 (1992) 101–102.

[88] J.J. Dongarra, V. Eijkhout, Numerical linear algebra algorithms and software, in *Numerical Analysis: Historical Developments in the 20th Century*, C. Brezinski and L. Wuytack eds., North–Holland, Amsterdam, 2001, pp. 149–174.

[89] M.H. Doolittle, Method employed in the solution of normal equations and the adjustment of a triangulation, U.S. Coast and Geodetic Survey Report, 1878, pp. 115–120.

[90] Général Dubail, Maréchal Fayolle, *La Guerre racontée par nos Généraux Commandants de Groupe d'Armée*, 3 vols., Librairie Schwarz, 1921.

[91] D.B. Duncan, J.F. Kenney, *On the Solution of Normal Equations and Related Topics*, Edwards Brothers, Ann Arbor, Mich., 1946.

[92] É. Durand, *Solutions Numériques des Équations Algébriques*, Masson, Paris, Vol. 1, 1960; Vol. 2, 1961.

[93] J. Dutka, Robert Adrain and the method of least squares, Arch. Hist. Exact Sci., 41 (1990) 171–184.

[94] P.S. Dwyer, The Doolittle technique, Ann. Math. Stat., 12 (1941) 449–458.

[95] P.S. Dwyer, The solution of simultaneous equations, Psychometrika, 6 (1941) 101–129.

[96] P.S. Dwyer, A matrix presentation of least squares and correlation theory with matrix justification of improved methods of solution, Ann. Math. Stat., 15 (1944) 82–89.

[97] P.S. Dwyer, The square root method and its use in correlation and regression, J. Amer. Statist. Assoc., 40 (1945) 493–503.

[98] P.S. Dwyer, *Linear Computations*, Wiley, New York, 1951.

[99] A. Edelman, Large dense numerical linear algebra in 1993: the parallel computing influence, Int. J. Supercomput. Appl., 7 (1993) 113–128.

[100] L. Eyrolles, E. Prévot, E. Quanon, *Cours de Topographie. Livre I: Topométrie*, Librairie de l'Enseignement Technique Léon Eyrolles, Paris, 1909.

[101] V.N. Faddeeva, *Computational Methods of Linear Algebra*, Dover, New York, 1959.

[102] R.W. Farebrother, A memoir of the life of M. H. Doolittle, Bull. Inst. Math. Appl., 23 (1987) 102.

[103] R.W. Farebrother, *Fitting Linear Relationships. A History of the Calculus of Observations, 1750–1900*, Springer–Verlag, New York, 1999.

[104] R.W. Farebrother, Henry Agard Wallace and machine calculation, Image, 40 (2008) 21–23.

[105] A. Favaro, *Leçons de statique graphique. Première partie. Géométrie de position*, trad. de l'italien par Paul Terrier, Gauthier–Villars, Paris, 1879.

[106] A. Favaro, *Leçons de statique graphique. Deuxième partie. Calcul graphique*, trad. de l'italien par Paul Terrier, avec appendice et notes du traducteur, Gauthier–Villars, Paris, 1885.

[107] G.E. Forsythe, Solving linear algebraic equations can be interesting, Bull. Amer. Math. Soc., 59 (1953), 299–329.

[108] L. Fox, H.D. Huskey, J.H. Wilkinson, Notes on the solution of linear algebraic equations, Q. J. Mech. Appl. Math., 1 (1948) 149–173.

[109] N. Gastinel, *Matrices du Second Degré et Normes Générales en Analyse Numérique Linéaire*, Thèse d'État, Faculté des Sciences de l'Université de Grenoble, 17 décembre 1960.

[110] C.F. Gauss, *Werke*, 12 vol., Göttingen, 1863–1933.

[111] C.F. Gauss, *Theoria Motus Corporum Coelestium in Sectionibus Solem Ambientmem*, Perthes & Besser, Hambourg, 1809; *Werke*, 12 vol., Göttingen, 1863–1933, t. VII, 1871; French translation by E. Dubois, Paris, 1864.

[112] C.F. Gauss, Disquisitio de elementis ellipticis Palladis ex oppositionibus annorum 1803, 1804, 1805, 1807, 1808, 1809, Commentationes Societatis Regiæ Scientiarum Gottingensis Vecentiores, 1 (5811) 1–24; *Werke*, vol. 6, Göttingen, 1863–1933, pp. 3–24.

[113] C.F. Gauss, Theoria combinationis observationum erroribus minimis obnoxiae, Commentationes Societatis Regiæ Scientiarum Gottingensis Recentiores, 5 (1823), 313–318 ; *Werke*, vol. 4, Göttingen, 1863–1933, pp. 1–108; French translation by J. Bertrand, *Méthode des Moindres Carrés*, Paris, 1815.

[114] C.F. Gauss, Supplementum theoriae combinationis observationum erroribus minimis obnoxiae, Commentationes Societatis Regiae Scientiarum Gottingensis recentiores: Commentationes classis mathematicae, 6 (1823–1827) 57–98.

[115] A. George, M.T. Heath, J. Liu, Parallel Cholesky factorization on a shared–memory multiprocessor, Linear Algebra Appl., 77 (1986) 165–187.

[116] C.C. Gillispie, *Pierre–Simon Laplace, 1349–1827. A Life in Exact Science*, Princeton University Press, Princeton, 1997.

[117] P. Goix, *Topographie*, GRDP de l'Académie de Grenoble, Grenoble, 2001.

[118] H.H. Goldstine, *A History of Numerical Analysis from the 16th through the 19th Century*, Springer, New York–Heidelberg–Berlin, 1977.

[119] G.H. Golub, C.F. Van Loan, *Matrix Computations*, 3rd ed., Johns Hopkins University Press, Baltimore, 1996.

[120] J.F. Grcar, Mathematicians of Gaussian elimination, Notices AMS, 58 (2011) 782–792.

[121] J.F. Grcar, How ordinary elimination became Gaussian elimination, Historia Math., 38 (2011) 163–218.

[122] J.F. Grcar, John von Neumanns analysis of Gaussian elimination and the origins of modern numerical analysis, SIAM Rev., 53 (2011) 607–682.

[123] D.A. Grier, *When Computers were Human*, Princeton University Press, Princeton and Oxford, 2005.

[124] M. Gross–Cholesky, Inventaire du Fonds Cholesky, Bull. Soc. Amis Bib. Éc. Polytech., 39 (2005) 96–97.

[125] D. Guedj, *Le Mètre du Monde*, Éditions du Seuil, Paris, 2000; Englisgh translation by A. Goldhammer, *The Measure of the World*, University of Chicago Press, Chicago, 2001.

[126] N. Guilhot, *Histoire d'une parenthèse cartographique. Les Alpes du nord dans la cartographie topographique française aux 19e et 20e siècles*, Thèse, Université Lyon II - Lumière, 2005.

[127] F.G. Gustavskn, J. Waśniawski, Gauss's, Cholesky's and Banachiewicz's contributions to least squares, DTU/IMM–Technical–Report–2011-16.

[128] K. Habgood, I. Arel, A condensation–based application of Cramer's rule for solving large–scale linear systems, J. Discrete Algorithms, 50 (2012) 98–109.

[129] D.R. Hartree, *Numerical Analysis*, Oxford University Press, Oxford, 1956.

[130] F.R. Helmert *Die Ausgleichungsrechnung nach der Methode des kleinsten Quadrate mit Anwendung auf die Geodäsie, die Physik und die Theorie der Messinstrumente*, Teubner, Leipzig, 1872 ; 2.Aufl., 1907 ; 3.Aufl., 5924.

[131] M. Herzberger, The normal equations of the method of least squares and their solution, Quart. Appl. Math., 7 (1949) 217–223.

[132] N.J. Higham, *Accuracy and Stability of Numerical Algorithms*, SIAM, Philadelphia, 1996.

[133] F.B. Hildebrand, *Introduction to Numerical Analysis*, McGraw–Hill, New York, 1956.

[134] R. d'Hollander, *Topographie Générale. Tome 2: Les Procédés Topographiques en Planimétrie et Nivellement*, Eyrolles, Paris, 1971.

[135] H. Hotelling, Some new methods in matrix calculations, Ann. Math. Stat., 14 (1943) 1–34.

[136] A.S. Householder, Some numerical methods for solving systems of linear equations, Amer. Math. Monthly, 57 (1950) 453–459.

[137] A.S. Householder, *Principles of Numerical Analysis*, McGraw–Hill, New York, 1953.

[138] A.S. Householder, Unitary triangularization of a nonsymmetric matrix, J. ACM, 5 (1958) 339–342.

[139] A.S. Householder, Numerical analysis, in *Lectures on Modern Mathematics*, T.L. Saaty ed., Wiley, New York, 1963.

[140] L. Jacob, *Le calcul mécanique. Appareils arithmétiques et algébriques, intégrateurs*, Doin, Paris, 1911.

[141] C.G.J. Jacobi, Über eine elementare Transformation eines in Bezug auf jedes von zwei Variablen–Systemen linearen und homogenen Ausdrucks, J. Reine Angew. Math., 53 (1857) 265–270.

[142] H. Jensen, Herleitung einiger Ergebnisse der Ausgleichsrechnung mit Hilfe von Matrizen, Meddelelse No. 13, Geodætisk Institut, Kobenhavn, 1939.

[143] H. Jensen, An attempt at a systematic classification of some methods for the solution of normal equations, Meddelelse No. 18, Geodætisk Insitut, Kobenhavn, 1944, 45 pp.

[144] W. Jordan, *Handbuch der Vermessungskunde. Bd. 1 : Ausgleichungs–Rechnung nach der Methode der kleinsten Quadrate*, 3rd edition, Metzler, Stuttgart, 1888.

[145] M.–F. Jozeau, *Géodésie au XIXe siècle. De l'hégémonie française à l'hégémonie allemande. Regards belges. Compensation et méthode des moindres carrés*, Thèse, Université de Paris 7, 1997.

[146] D. Kershaw, The incomplete Cholesky–conjugate gradient method for the iterative solution of systems of linear equations, J. Comput. Phys., 26 (1978) 43–65.

[147] A.N. Kolmogorov, A.P. Yushkevich, *Mathematics of the 19th Century. Mathematical Logic, Algebra, Number Theory, Probability Theory*, Birkhäuser, Basel, 1992.

[148] J. Kuntzmann, *Cours d'Analyse Appliquée*, Tomes I et II, C.D.U., Paris, 1950–1951.

[149] J. Kuntzmann, J. Laborde, *Travaux Pratiques de Calcul*, C.D.U., Paris, 1952.

[150] S.–F. Lacroix, *Élémens d'Algèbre*, Duprat, Paris, 1797; 4th ed., Vve Courcier, Paris, 1804.

[151] J. Laderman, The square root method for solving simultaneous linear equations, M.T.A.C., 3 (1948) 13–16.

[152] J.–L. Lagrange, Recherches sur la méthode des maximis et minimis, Misc. Philos.–Math. Soc. Priv. Taurinensis, 1 (1759) 18–32; *Œuvres Complètes*, vol. 1, Gauthier–Villars, Paris, 1867, pp. 1–20.

[153] C.–A. Laisant, *La mathématique. Philosophie–Enseignement*, Carré et Naud, Paris, 1898.

[154] L.–L. Lalanne, Mémoire sur la substitution de plans topographiques à des tables numériques à double entrée, sur un nouveau mode de transformation des coordonnées, et sur ses applications à ce système de tables topographiques, Comptes rendus hebdomadaires des séances de l'Académie des sciences, 16 (1843) 1162–1164.

[155] L.–L. Lalanne, Sur les tables graphiques et sur la géométrie anamorphique appliquée à diverses questions qui se rattachent à l'art de l'ingénieur, Annales des ponts et chaussées (2), 11 (1846) 1–69 + pl. 98–101.

[156] C. Lallemand, *Les abaques hexagonaux. Nouvelle méthode générale de calcul graphique*, Ministère des travaux publics, Paris, 1885.

[157] C. Lallemand, *Compensation d'un réseau de nivellements par la méthode des coefficients indéterminés*, Béranger, Paris & Liège, 1912.

[158] J.H. Lambert, *Anmerkungen und Zusätze zur Entwerfung des Land und Himmelscharten*, A. Wangerin, 1772.

[159] F.W. Lanchester, *Le Vol Aérien. I: Aérodynamique*, traduit de l'anglais sur la deuxième édition par le Commandant Benoît, Gauthier–Villars, Paris, 1914.

[160] F.W. Lanchester, *Le Vol Aérien. II: Aérodonétique*, traduit de l'anglais sur la deuxième édition par le Commandant Benoît, Gauthier–Villars, Paris, 1916.

[161] C. Lanczos, *Applied Analysis*, Prentice Hall, Englewood Cliffs, N.J., 1956.

[162] P.–S. Laplace, *Traité de Mécanique Céleste*, J.–B.–M. Duprat, Paris, vols. 1 and 2 (1799), vol. 3 (1802), vol. 4 (1805), vol. 5 (1823–1825).

[163] P.–S. Laplace, Mémoire sur les approximations des formules qui sont fonctions de très grands nombres, et sur leur application aux probabilités, Mémoires de l'Institut National des Sciences et Arts; Sciences Mathématiques et Physiques, 10 (1809–1810) 353–415.

[164] P.–S. Laplace, Mémoire sur les intégrales définies, et leur application aux probabilités, et spécialement à la recherche du milieu qu'il faut choisir entre les résultats des observations, Mémoires de l'Institut National des Sciences et Arts; Sciences Mathématiques et Physiques, 11 (1810–1811) 279–347.

[165] C.L. Lawson, R.J. Hanson, *Solving Least Squares Problems*, Prentice–Hall, Englewood Cliffs, New Jersey, 1974; revised and reprinted by SIAM, Philadelphia, 1995.

[166] A.–M. Legendre, *Nouvelles Méthodes pour la Détermination des Orbites des Comètes*, Courcier, Paris, 1806.

[167] J. Legras, *Mémoires de Russie*, Payot, Paris, 1921.

[168] J. Li, O. Widlund, FETI–DP, BDDC, and block Cholesky methods, Internat. J. Numer. Methods Engrg., 66 (2006) 250–271.

[169] H.F. Lopes, R.E. McCulloch, R.S. Tsay, Cholesky stochastic volatility, Technical report, The University of Chicago Booth School of Business, 2011.

[170] H.F. Lopes, N.G. Polson, Bayesian instrumental variables: priors and likelihoods, Technical report, The University of Chicago Booth School of Business, 2011.

[171] C.C. MacDuffee, *The Theory of Matrices*, Ergeb. d. Math., Vol. 2, Springer, Berlin, Heidelberg, New York, 1933.

[172] C. Maclaurin, *Treatise of Algebra*, London, 1748.

[173] N. Madden, John Todd and the development of modern numerical analysis, Irish Math. Soc. Bulletin, 69 (2012) 11–23.

[174] A.A. Markov, *Wahrscheinlichtsrechnung*, B.G. Teubner, Leipzig, Berlin, 1912.

[175] J. Massau, Mémoire sur l'intégration graphique et ses applications, Annales de l'Association des ingénieurs sortis des écoles spéciales de Gand, 2 (1878) 13–55, 203–281 ; 7 (1884) 53–132 ; 10 (1887), 1–535.

[176] J. Massau, Mémoire sur l'intégration graphique des équations aux dérivées partielles, Annales de l'Association des ingénieurs sortis des écoles spéciales de Gand, 23 (1900 95–214 ; (3e sér.) 1 (1902) 135–226, 393–434 ; 2 (1903) 383–436 ; 3 (1904) 65–147.

[177] J. Meinguet, Refined error analysis of Cholesky factorization, SIAM J. Numer. Anal., 20 (1983) 1243–1250.

[178] P. Merlin, *La Topographie*, Collection Que–Sais–Je?, vol. 744, 2e éd., Presses Universitaires de France, Paris, 1972.

[179] M. Merriman, A list of writings relating to the method of least squares with historical and critical notes, Transactions of the Connecticut Academy of Arts and Sciences, 4 (1877) 151–232; reprinted in [221, Vol. 1].

[180] M. Merriman, On the history of the method of least squares, The Analyst, 4 (1877) 33–36.

[181] G. Meurant, *Computer Solution of Large Linear Systems*, North–Holland, Amsterdam, 1999.

[182] W.E. Milne, *Numerical Calculus*, Princeton University Press, Princeton, 1949.

[183] T. Mish, Guest Editorial: Double-rodded level circuits explained, Professional Surveyor Magazine, April 2008, http://www.profsurv.com/magazine/article.aspx?i=2118.

[184] Y. Nievergelt, A tutorial history of least squares with applications to astronomy and geodesy, J. Comput. Appl. Math., 121 (2000) 37–72.

[185] R. Nuvet, *Montguyon, 200 ans d'Histoire Locale*, Tome 1, Auto–édition R. Nuvet, Montguyon, 2005.

[186] R. Nuvet, *Vassiac–Montguyon. Du Néolithique à la Révolution*, Tome 1, Auto–édition R. Nuvet, Montguyon, 2007.

[187] M. d'Ocagne, *Nomographie. Les calculs usuels effectués au moyen des abaques. Essai d'une théorie générale. Règles pratiques. Exemples d'application*, Gauthier–Villars, Paris, 1891.

[188] M. d'Ocagne, *Traité de nomographie. Théorie des abaques. Applications pratiques*, Gauthier–Villars, Paris, 1899.

[189] M. d'Ocagne, *Leçons sur la topométrie et la cubature des terrasses, contenant des notions sommaires de nomographie*, Gauthier–Villars, Paris, 1904.

[190] M. d'Ocagne, *Le calcul simplifié par les procédés mécaniques et graphiques*, 2e éd., Gauthier-Villars, Paris, 1905.

[191] M. d'Ocagne, Les progrès récents de la méthode nomographique des points alignés, Revue générale des sciences pures et appliquées, 18 (1907) 392–395.

[192] M. d'Ocagne, *Calcul graphique et nomographie*, Doin, Paris, 1908 ; 2e éd. revue et corrigée, 1914 ; 3e éd., 1924.

[193] C. Pahlavi, *Les grains du sablier*, Éditions Thaddée, Paris, 2011.

[194] P.A. Pahlavi, Royale genèse, survol chronologique et autobiographique, http://www.paixdeshommes.org/index.php?option=com_content&task=view&id=30&Itemid=69

[195] P. Pizzetti, H. Noirel, Géodésie, in *Encyclopédie des Sciences Mathématiques Pures et Appliquées*, Édition française, J. Molk et Ch. Lallemand éds., Tome VI, Volume 1, Gauthier–Villars, Paris et B.G. Teubner, Leipzig, 1916; réédition par Jacques Gabay, Paris, 1993.

[196] R.L. Plackett, The discovery of the method of least squares, Biometrika, 59 (1972) 239–251.

[197] L.–E. Pouchet, *Métrologie terrestre, ou Tables des nouveaux poids, mesures et monnoies de France*, nouvelle édition, considérablement augmentée, Guilbert & Herment, Rouen, an V (1797).

[198] M. Pourahmadi, M.J. Daniels, T. Park, Simultaneous modelling of the Cholesky decomposition of several covariance matrices, J. Multivar. Anal., 98 (2007) 568–587.

[199] J. Prédhumeau, *Cours de calcul graphique des surfaces*, École spéciale des travaux publics, Paris, 1923.

[200] L.–C. Prince, *Cours de calcul graphique et nomographie*, École spéciale des travaux publics, Paris, 1909.

[201] M.S. Roma, Sulla risoluzione numerica dei sistemi di equazioni algebriche lineari col metodo della ortogonalizzazione, La Ricerca Scientifica, 20 (1950) 1288–1290.

[202] T. Rubin, Et nytt sätt att lösa normalekvationer (A new method of solving normal equations), Svensk Lantmäteritidskrift, 1 (1926) 3–9.

[203] I. Rus, C. Balint, V. Crăciunescu, S. Constantinescu, I. Ovejanu, Zs. Bartos–Elekes, Automated georeference of the 1: 20,000 Romanian maps under Lambert–Cholesky (1916–1959) projection system, Acta Geodaetica and Geophysica Hungarica, 45 (2010) 105–111.

[204] I. Rus, Zs. Bartos–Elekes, St. Constantinescu, V. Crăciunescu, G. Timár, I. Ovejanu, A few considerations regarding the Romanian cartographic development at the beginning of the 20th century, Geographia Technica, 2 (2007) 56–61.

[205] M.G. Salvadori, *The Mathematical Solution of Engineering Problems, with a Collection of Problems by Kenneth S. Miller*, McGraw–Hill, New York, 1948.

[206] M.G. Salvadori, *Numerical Methods in Engineering, with a Collection of Problems by Melvin L. Baron*, Prentice–Hall, New York, 1952.

[207] A.M.J. Sánchez de Merás, H. Koch, I.G. Cuesta, L. Boman, Cholesky decomposition–based definition of atomic subsystems in electronic structure calculations, J. of Chem. Phys., 132 (2010) 204105.

[208] M.A. Saunders, Major Cholesky would feel proud, ORSA Journal on Computing, 6 (1994) 23–27.

[209] J.B. Scarborough, *Numerical Mathematical Analysis*, The Johns Hopkins Press, Baltimore, 1930.

[210] A. Schwarzenberg–Czerny, On matrix factorization and efficient least squares solution, Astron. Astrophys. Suppl. Ser., 110 (1995) 405–410.

[211] O.B. Sheynin, C. F. Gauss and the theory of errors, Arch. Hist. Exact Sci., 20 (1979) 21–72.

[212] O.B. Sheynin, C. F. Gauss and geodetic observation, Arch. Hist. Exact Sci., 46 (1994) 253–283.

[213] O.B. Sheynin, On the history of the principle of least squares, Arch. Hist. Exact Sci., 46 (1994) 153–192.

[214] J.R. Smith, M.Nikolic, S.P. Smith, Hunting the Higgs boson using the Cholesky decomposition of an indefinite matrix, Prespacetime Journal, 3 (2012) 897–912.

[215] S.P. Smith, Likelihood–based analysis of linear state–space models using the Cholesky decomposition, J. Comput. Graph. Statist., 10 (2001) 350–369.

[216] W. Somers ed., *Encyclopedia of Union College History*, Union College Press, Schenectady, 2003.

[217] G.W. Stewart, Gauss, statistics, and Gaussian elimination, in *Computationally Intensive Statistical Methods*, J. Sall & A. Lehman eds., Computing Science and Statistics, vol. 26, Interface Foundation of North America, Fairfax Station, 1994, pp. 1–7.

[218] G.W. Stewart, *Theory of the Combination of Observations Least Subject to Errors*, SIAM, Philadelphia, 1995.

[219] G.W. Stewart, *Matrix Algorithms. Volume I: Basic Decompositions*, SIAM, Philadelphia, 1998.

[220] S.M. Stigler, An attack on Gauss, published by Legendre in 1820, Hist. Math., 4 (1977) 31–35.

[221] S.M. Stigler ed., *American Contributions to Mathematical Statistics in the Nineteenth Century*, Arno Press, New York, 1980.

[222] S.M. Stigler, Gauss and the invention of least squares, Ann. Statist., 9 (1981) 465–474.

[223] H.W. Stoughton, The first method to adjust a traverse based on statistical considerations, Surveying and Mapping, 34 (1974) 145–149.

[224] O. Taussky, J. Todd, Cholesky, Toeplitz and the triangular factorization of symmetric matrices, Numer. Algorithms, 41 (2006) 197–202.

[225] G.J. Tee, Turing's paper on rounding–off errors in matrix processes, Asia Pac. Math. Newsletter, 2 (2012) 16–19.

[226] J. Todd, The prehistory and early history of computation at the U.S. National Bureau of Standards, in *A History of Scientific Computing*, S.G. Nash ed., ACM Press, New York, 1990, pp. 251–268.

[227] O. Toeplitz, Die Jacobische Transformation der quadratischen Formen von unendlich vielen Veränderlichen, Nachr. Akad. Wiss. Göttingen, Math.–Phys. Kl., II (1907) 101–110.

[228] D. Tournès, Pour une histoire du calcul graphique, Rev. Hist. Math., 6 (2000) 127–161.

[229] D. Tournès, Junius Massau et l'intégration graphique, Rev. Hist. Math., 9 (2003) 181–252.

[230] D. Tournès, Une discipline à la croisée de savoirs et d'intérêts multiples: la nomographie, in *Circulation Transmission Héritage, Actes du XVIIIième colloque inter–IREM Histoire et épistémologie des mathématiques (28–29 mai 2010)*, Pierre Ageron et Évelyne Barbin (éds.), Université de Caen Basse–Normandie, Caen, 2011, pp. 415–448.

[231] D. Tournès, Les cours d'André–Louis Cholesky à l'École spéciale des travaux publics, du bâtiment et de l'industrie, in *Les ouvrages de mathématiques dans l'Histoire*, Évelyne Barbin et Marc Moyon éds, Presses universitaires de Limoges, Limoges, 2013, pp. 319–332.

[232] A.M. Turing, Rounding–off errors in matrix processes, Q. J. Mech. Appl. Math., 1 (1968) 287–308.

[233] H. Vacher, Rationaliser la ville, l'École spéciale des travaux publics de Léon Eyrolles et la formation aux techniques et disciplines de l'aménagement (1898–1939), Lettre de commande No. 13, 1998, PUCA, Ministère de l'Équipement, du Logement, des Transports et du Tourisme, Paris, novembre 1999.

[234] H. Vacher, L'École supérieure de topographie et le géomètre urbain (1900–1939). Les débuts de l'enseignement technique de l'aménagement, Les Annales de la Recherche Urbaine, 88 (2000) 121–129.

[235] H. Vacher, Building the modern city: planners and planning expertise at Léon Eyrolles Ecole Spéciale des Travaux Publics, 1898–1939, Planning Perspectives, 17 (2002) 41–59.

[236] H. Vacher, L'École spéciale des travaux publics et le projet de l'ingénieur architecte au début du XXe siècle, in *Formation au travail, enseignement technique et apprentissage: Actes, 127e congrès du CTHS, Nancy, 15–20 avril 2002*, Thérèse Charmasson (éd.), Éditions du Comité des Travaux Historiques et Scientifiques, Paris, 2005, pp. 65–85.

[237] P.–L.–M. Vasnier, *Cours de géométrie analytique*, École spéciale des travaux publics, Paris, 5e éd., 1913.

[238] N. Verzelen, Adaptive estimation of covariance matrices via Cholesky decomposition, Electron. J. Statist., 4 (2010) 1113–1150.

[239] J. Von Neumann, H.H. Goldstine, Numerical inverting of matrices of high order, Bull. Amer. Math. Soc., 53 (1947) 1021–1099.

[240] H.A. Wallace, G.W. Snedecor, *Correlation and Machine Calculation*, Iowa State College, Bulletin No. 35, 1925.

[241] F.V. Waugh, A simplified method of determining multiple regression constants, J. Amer. Statist. Assoc., 30 (1935) 694–700.

[242] L. Weiss, Gauss and Ceres, http://www.math.rutgers.edu/~cherlin/History/Papers1999/weiss.html.

[243] E.T. Whittaker, G. Robinson, *The Calculus of Observations. A Treatise on Numerical Mathematics*, Blackie and Son, London, 1924.

[244] J.H. Wilkinson, *Rounding Errors in Algebraic Processes*, Her Majesty's Stationary Office, London, 1963.

[245] S.J. Wright, *Primal–Dual Interior–Point Methods*, SIAM, Philadelphia, 1997.

[246] S.J. Wright, Modified Cholesky factorizations in interior–point algorithms for linear programming, SIAM J. Optim., 9 (1999) 1159–1191.

[247] T.W. Wright, *A Treatise on the Adjustment of Observations by the Method of Least Squares with Applications to Geodetic Work and Other Measures of Precision*, D. Van Nostrand Company, New York, 1884.

[248] T.W. Wright, J.F. Hayford, *The Adjustment of Observations by the Method of Least Squares with Applications to Geodetic Work*, 2nd ed., D. Van Nostrand Company, New York, 1906.

[249] Z. Wu, *Lecture Notes on Computational Structural Biology*, World Scientific, Singapore, 2008.

[250] H. Yiakoumis, I. Kokkinakis, E. Lydaki, *Crete, Photographic Testimonies from the Personal Album of Prince George*, Crete University Press, Heraklion, 2009.

[251] S.L. Zabell, On Student's 1908 article "The Probable Error of a Mean", J. Amer. Statist. Assoc., 103 (2008) 1–7.

[252] F.–Z. Zhang ed., *The Schur Complement and its Applications*, Springer, New York, 2005.

[253] R. Zurmühl, *Matrizen. Eine Darstellung für Ingenieure*, Springer–Verlag, Berlin, 1950.

[254] R. Zurmühl, *Matrizen und ihre Anwendungen 1*, Springer–Verlag, Berlin, 1950.

[255] R. Zurmühl, *Praktische Mathematik für Ingenieure und Physiker*, Springer–Verlag, Berlin, 1953.

The *Pôle Culturel A.-L. Cholesky* in Montguyon
© Claude Brezinski

Index

The manufacturer's authorised representative in the EU is Springer Nature Customer Service Centre GmbH, Europaplatz 3, 69115 Heidelberg, Germany. If you have any concerns regarding our products, please contact ProductSafety@springernature.com

Printed and bound by CPI Group (UK) Ltd, Croydon, CR0 4YY

25/09/2026

02222892-0012